Springer Series on
ATOMIC, OPTICAL, AND PLASMA PHYSICS   31

Springer
Berlin
Heidelberg
New York
Barcelona
Hong Kong
London
Milan
Paris
Singapore
Tokyo

Physics and Astronomy ONLINE LIBRARY

http://www.springer.de/phys/

Springer Series on
ATOMIC, OPTICAL, AND PLASMA PHYSICS

The Springer Series on Atomic, Optical, and Plasma Physics covers in a comprehensive manner theory and experiment in the entire field of atoms and molecules and their interaction with electromagnetic radiation. Books in the series provide a rich source of new ideas and techniques with wide applications in fields such as chemistry, materials science, astrophysics, surface science, plasma technology, advanced optics, aeronomy, and engineering. Laser physics is a particular connecting theme that has provided much of the continuing impetus for new developments in the field. The purpose of the series is to cover the gap between standard undergraduate textbooks and the research literature with emphasis on the fundamental ideas, methods, techniques, and results in the field.

27 **Quantum Squeezing**
   By P.D. Drumond and Z. Spicek

28 **Atom, Molecule, and Cluster Beams I**
   Basic Theory, Production and Detection of Thermal Energy Beams
   By H. Pauly

29 **Polarization, Alignment and Orientation in Atomic Collisions**
   By N. Andersen and K. Bartschat

30 **Physics of Solid-State Laser Physics**
   By R.C. Powell
   (Published in the former Series on Atomic, Molecular, and Optical Physics)

31 **Plasma Kinetics in Atmospheric Gases**
   By M. Capitelli, C.M. Ferreira, B.F. Gordiets, A.I. Osipov

32 **Atom, Molecule, and Cluster Beams II**
   Cluster Beams, Fast and Slow Beams, Accessory Equipment and Applications
   By H. Pauly

Series homepage – http://www.springer.de/phys/books/ssaop/

Vols. 1–26 of the former Springer Series on Atoms and Plasmas are listed at the end of the book

M. Capitelli C.M. Ferreira
B.F. Gordiets A.I. Osipov

Plasma Kinetics in Atmospheric Gases

With 64 Figures and 60 Tables

Springer

Professor Mario Capitelli
University of Bari
Department of Chemistry and
CRN Research Center for Plasmachemistry
Via Orabona 4
70123 Bari, ITALY

Professor Boris F. Gordiets
Russian Academy of Sciences
Optical Department of
P.N. Lebedev Physical Institute
Leninsky prospect 53
117924 Moscow, RUSSIA

Professor Carlos M. Ferreira
Instituto Superior Tecnico
Department of Physics and
Centro de Fisica dos Plasmas
Av. Rovisco Pais, 1
1049-001 Lisboa, PORTUGAL

Professor Alexey I. Osipov
M.V. Lomonosov Moscow State University
Physical Department
Vorobyevy Gory
119899 Moscow, RUSSIA

ISSN 1615-5653

ISBN 3-540-67416-0 Springer-Verlag Berlin Heidelberg New York

Library of Congress Cataloging-in-Publication Data.

Plasma kinetics in atmospheric gases/M. Capitelli...[et al.]. p.cm.- (Springer series on atomic, optical, and plasma physics, ISSN 1615-5653; 31). Includes bibliographical references and index. ISBN 3540674160 (alk. paper). 1. Nonequilibrium plasmas. 2. Molecular gas lasers. 3. Kinetic theory of gases. 4. Nitrogen. 5. Oxygen. I. Capitelli, M. II. Series. QC718.5.E66 P53 2000   530.4'46–dc21   00-041961

This work is subject to copyright. All rights are reserved, whether the whole or part of the material is concerned, specifically the rights of translation, reprinting, reuse of illustrations, recitation, broadcasting, reproduction on microfilm or in any other way, and storage in data banks. Duplication of this publication or parts thereof is permitted only under the provisions of the German Copyright Law of September 9, 1965, in its current version, and permission for use must always be obtained from Springer-Verlag. Violations are liable for prosecution under the German Copyright Law.

Springer-Verlag Berlin Heidelberg New York
a member of BertelsmannSpringer Science+Business Media GmbH

© Springer-Verlag Berlin Heidelberg 2000
Printed in Germany

The use of general descriptive names, registered names, trademarks, etc. in this publication does not imply, even in the absence of a specific statement, that such names are exempt from the relevant protective laws and regulations and therefore free for general use.

Typesetting: Data conversion by Steingraeber Satztechnik GmbH, Heidelberg
Cover concept by eStudio Calmar Steinen
Cover design: *design & production* GmbH, Heidelberg

Printed on acid-free paper   SPIN: 10762824   57/3141/tr - 5 4 3 2 1 0

Preface

Investigation of the kinetic processes occurring in the atmospheric gases and low temperature plasmas of such gases is of great interest for numerous areas of modern physics and chemistry. These include discharge and laser physics, physics of the ionosphere, chemistry and optics of the atmosphere, laser and plasma chemistry, and nonequilibrium gas dynamics. Further, nonequilibrium gaseous media and low temperature plasmas formed from N_2, O_2, or N_2–O_2 mixtures are rich in active components such as O, N, and H atoms and excited metastable species, which fostered the utilization of such nonequilibrium systems in a variety of new technologies.

At present, several monographs devoted to the analysis of kinetic processes in molecular gases and plasmas are available in the literature. However, most of these works deal only with general physical-chemical kinetic aspects, and do not provide an analysis of basic kinetic theory and detailed investigation of kinetic processes in N_2, O_2 and their mixtures. A monograph devoted to such a thorough analysis for the atmospheric gases is still lacking in the literature. The present book was written to fill in this gap.

The book can be considered as a natural development of two former monographs, *Non-Equilibrium Vibrational Kinetics* (ed. by M. Capitelli, Springer–Verlag 1986) and *Kinetic Processes in Gases and Lasers* (B. Gordiets, A.I. Osipov, and L.A. Shelepin, Gordon and Breach 1988).

In these books, emphasis was given to the problem of nonequilibrium vibrational kinetics and its important influence on the properties of different media such as atmospheric plasmas and lasers. Since then, numerous papers have appeared in the literature stressing the need for the coupling between molecular and atomic kinetics, free electron kinetics and electrodynamics, for a complete description of an electrical discharge. Self-consistent solutions of the electron Boltzmann equation coupled to a system of vibrational and electronic state master equations, including dissociation and ionization reactions, have been obtained in conjunction with electrodynamics to describe the electrical discharge under steady and unsteady conditions. This numerical approach is now widely used by the scientific community, instead of the analytical approaches strongly developed in the past, mainly by Russian teams. However, the analytical approach is also widely used in the present book since it remains a powerful tool to understand the numerical results.

The book includes fundamental aspects of the physical-chemical kinetics, complete data on molecular parameters, cross sections and rate coefficients, and applications to conditions occurring in gas discharges and natural plasmas. We believe this work will be useful for scientists and graduate students interested in plasma physics, plasma chemistry, physics of ionosphere, chemistry and optics of the atmosphere, nonequilibrium aerodynamics, and physical-chemical kinetics.

Last but not the least, the book introduces the reader to the immense Russian literature on the subject. In this sense, it represents an attempt to unite Western and Eastern knowledge, for the enrichment of European Science.

Bari	*M. Capitelli*
Lisbon	*C.M. Ferreira*
Moscow	*B. Gordiets*
March 2000	*A. Osipov*

Contents

Introduction 1

1. **Translational Relaxation of Heavy Particles** 5
   References 11

2. **Rotational Relaxation of Molecules** 13
   References 21

3. **Vibrational Relaxation of Molecules** 23
   3.1 Estimation of Vibrational Relaxation Times 23
   3.2 Vibrational Kinetics of Harmonic Oscillators 26
       3.2.1 Vibrational Relaxation of Diatomic Molecules
             in an Inert Gas: V-T Exchange 26
       3.2.2 One-Component Systems: V-V Exchanges 29
       3.2.3 Binary Mixture of Diatomic Molecules:
             V-T, V-V and V-V' Exchanges 31
       3.2.4 Polyatomic Molecules 34
   3.3 Vibrational Kinetics of Anharmonic Oscillators 36
       3.3.1 V-T Relaxation in an Inert Gas 37
       3.3.2 V-V Exchanges. The Treanor Distribution 38
       3.3.3 V-V and V-T Exchanges.
             Moderate Deviation from Equilibrium 39
       3.3.4 Resonant V-V Relaxation
             Under High Exitation Conditions 41
       3.3.5 Time-Dependent Relaxation at High Excitation 45
   3.4 Vibrational Kinetics in the Presence of Chemical Reactions
       of Vibrationally Excited Molecules 47
   References 54

4. **Kinetics of Free Electrons** 59
   4.1 The Boltzmann Equation 59
   4.2 EEDF in Discharges Sustained by an Electric Field 66
   4.3 EEDF in Discharges Excited by an Electron Beam 74
   References 79

VIII Contents

5. Energetic and Spectroscopic Parameters of Atmospheric Species 85
References 93

6. Rates of Translational and Rotational Relaxation 95
6.1 Translational Relaxation and Diffusion Coefficients 95
6.2 Rotational Relaxation.................................... 99
References 103

7. Rate Coefficients for Vibrational Relaxation 105
7.1 Rate Coefficients of V-T Relaxation...................... 105
7.2 Rate Coefficients of V-V and V-V' Processes 112
References 116

8. Electron Rate Coefficients 119
8.1 Integral Scattering Cross-Sections 119
8.2 Electron Rate Coefficients and Power Balance in Plasmas Maintained by an Electric Field 123
8.3 Electron Loss Processes in the Plasma Bulk............... 139
8.4 Electron Rate Coefficients and Energy Balance in the Ionosphere Plasma................................ 142
8.5 Excitation, Dissociation and Ionization of N_2, O_2 and O Species by an Electron Beam 145
References 149

9. Electronic State Relaxation Rates 155
9.1 Radiative Lifetimes 155
9.2 Relaxation in Collisions with Atoms and Molecules 158
References 163

10. Rate Coefficients of Chemical Reactions 167
10.1 Reactions of Neutral Species 167
10.2 Ion–Molecule Reactions 178
10.3 Vibrational Excitation of Reaction Products 185
References 187

11. Interactions of Gas Phase Species with Surfaces 193
11.1 Balance Equations and Boundary Conditions at Walls....... 193
11.2 Wall Loss Probabilities γ 196
11.3 Kinetic Model for Surface Processes...................... 199
11.4 Molecular Dynamics Approach 206
References 207

12. Discharges in Pure N_2 and O_2 209
12.1 Discharge and Post-Discharge in N_2 209
12.2 Discharge in O_2 215
    12.2.1 The Positive Column in Electronegative Gases 216
    12.2.2 Chemical and Charge Kinetics in O_2 Discharges 221
References 224

13. Discharges in N_2-O_2 Mixtures 229
13.1 Modelling of Low Pressure N_2-O_2 Discharges.............. 229
    13.1.1 General Description of the Model................... 231
    13.1.2 Kinetics of Free Electrons 231
    13.1.3 Vibrational Kinetics 232
    13.1.4 Kinetics of Electronic States and Chemical Kinetics... 232
    13.1.5 Interaction with the Wall........................... 232
    13.1.6 Gas and Wall Temperatures 233
13.2 Modelling of High Pressure N_2-O_2 Discharges 241
    13.2.1 Cleaning of Polluted Atmospheric Gases............. 241
    13.2.2 N^{14} and N^{15} Isotope Separation 246
References 248

14. Kinetic Processes in the Ionosphere 251
14.1 Probabilities and Rates of Dissociation and Ionization....... 252
14.2 Chemical and Charge Kinetics 254
14.3 Vibrational Kinetics 257
References 264

15. Interaction of Space Vehicles with Atmospheric Gases 269
15.1 Free Molecular Regime 269
15.2 Hydrodynamic Regime. Vibrational
    and Chemical Kinetics During Space Shuttle Reentry 273
References 282

16. Acoustic and Shock Waves in Non-equilibrium Gases 287
16.1 Propagation of Small Perturbations in Non-equilibrium Gases 287
16.2 Nonlinear Hydrodynamic Waves in Non-equilibrium Gases ... 290
16.3 Propagation of Shock Waves in Non-equilibrium Gases 292
References 295

Index 299

Introduction

The understanding of kinetic processes in low temperature plasmas of atmospheric gases is of great interest in various branches of modern physics and chemistry, such as discharge physics, plasma chemistry, chemistry and optics of the atmosphere, nonequilibrium gas dynamics, laser physics and laser chemistry, and physics of the ionosphere. In particular, nonequilibrium low temperature plasmas in pure N_2 and O_2 or in N_2–O_2 mixtures, in which atomic and molecular active components are produced in large amounts, are now widely used in a variety of new technologies such as microelectronics, film deposition, surface processing, material treatment, depollution devices, and space technologies. For many years, the investigation of kinetic processes in nonequilibrium plasmas did not receive much attention from the plasma physics community, traditionally more interested in fully ionized plasmas where molecules are absent and the electrons are Maxwellian. Present plasma-based technologies fostered however increasing interest in the field of nonequilibrium plasma kinetics, and significant progress in this area has been achieved in recent years as a result of detailed investigations of state-to-state molecular kinetics coupled to the electron Boltzmann equation in order to obtain the nonequilibrium distributions of atomic and molecular internal states and electron energy in such systems.

The present technological markets, including the space-related one, call for fundamental investigations of nonequilibrium plasmas with ionization degrees ranging from 10^{-7} to 10^{-3}, and mean electron energies from 0.1 to 10 eV. To describe accurately such situations, one needs to construct complex numerical models that couple in a self-consistent way the equations describing the energy distributions of free electrons and heavy particle species. Models of such a kind for low temperature plasmas of atmospheric gases will be described in this book. However, to understand these complicated numerical models, the basic principles on which they are founded need to be understood. For this reason, this book also discusses some analytical approaches based on rigorous kinetic theory which provide a physically sound description of the energy distributions for the various species. Such analytical formulations are useful to interpret the numerical results of the complex numerical models.

The book seeks to balance theory and numerical calculations, and provides data for numerous elementary processes. Its contents can be divided into four main parts. The first part (Chaps. 1–4) deals with the basic theory of kinetic processes in molecular gases and low temperature plasmas. The processes of translational, rotational and vibrational relaxation, and the free electron kinetics are analysed here. To a first approximation, the relaxation times τ of these processes can usually be obtained from a linear equation of the type

$$\frac{dE}{dt} = -\frac{E - E_0}{\tau}, \tag{1}$$

where E is the energy of the relevant degree of freedom and E_0 is the corresponding equilibrium value.

Particular attention is paid in these first chapters to examining the hierarchy of relaxation times for the different processes. Usually, these times obey the following hierarchy:

$$\tau_{TT} \leq \tau_{RT} \ll \tau_\epsilon \leq \tau_{VT}, \tau_{ET}, \tau_{Ch}, \tag{2}$$

where τ_{TT}, τ_{RT}, τ_{VT}, τ_{ET} are the typical times for attaining equilibrium between, respectively, translational degrees of freedom, translational and rotational degrees of freedom, and vibrational and electronic degrees of freedom of atoms, molecules and ions; τ_ϵ is the typical energy relaxation time for plasma free electrons; and τ_{Ch} is the typical time for chemical conversion.

When this hierarchy of relaxation times is satisfied, the analysis of nonequilibrium processes is somewhat simplified. For example, to determine the electron energy distribution function or analyse the vibrational, electronic and chemical kinetics in a low temperature nonequilibrium plasma, the translational and rotational relaxation of atoms and molecules need not be considered in detail; Maxwell–Boltzmann distributions with the same local temperature can be assumed for translational and rotational degrees of freedom of the heavy particles. However, under many circumstances the time scales of vibrational relaxation, relaxation of electronic metastable states, and chemical conversion can be of the same order of magnitude so that a simultaneous analysis of vibrational, electronic and chemical kinetics is compulsory. On the other hand, short τ_ϵ times allows in many cases a stationary solution of the electron Boltzmann equation to be used for describing the time evolution of chemical species and excited states.

The basic elementary processes and the kinetic parameters of the atmospheric gases are reviewed in the second part (Chaps. 5–11). Data for molecular parameters and cross-sections and rate coefficients of numerous energy exchange processes induced by electrons, photons and heavy particles are provided. These data are necessary in order to model and interpret nonequilibrium kinetic processes in plasmas of N_2, O_2 and their mixtures. Analytical approximations for some rate coefficients are also presented, which can be useful for practical modelling work. Interaction of gas phase species with surfaces is an important problem since surface kinetics contributes to the creation and destruction of a number of species, and thus can significantly affect the whole discharge kinetics. Chapter 11 is dedicated to gas-surface interactions in plasma columns of atmospheric gases, particularly to heterogeneous recombination of nitrogen and hydrogen atoms on discharge tube walls. A phenomenological surface kinetic model of such processes is described in detail. A molecular dynamic approach which was recently developed to investigate this problem is also briefly presented.

The third part (Chaps. 12–14) is devoted to the description of recent developments in the modelling of kinetic and hydrodynamic processes in low temperature plasmas of the atmospheric gases. The chemical and the vibrational kinetics, and the kinetics of ions and electronic levels in discharge and post-discharge plasma columns in N_2, O_2 and N_2–O_2 are analysed in detail. The models presented here, which couple the bulk and the surface kinetics in a consistent way, are shown to provide a very complete physical interpretation of recent experimental data and a consistent description of the basic discharge workings. Complementary, Chap. 14 is dedicated to the analysis of kinetic processes in the ionospheric plasma.

The fourth part of the book (Chaps. 15 and 16) deals with the interaction processes of space vehicles and atmospheric gases at altitudes above 50 km, and the propagation of sound and shock waves in a nonequilibrium gas. In particular, it is shown that the reentry of a space vehicle in the atmosphere induces coupled nonequilibrium vibrational kinetic and dissociation and recombination processes which in many aspects are similar to those occurring in gas discharges. The main difference is that in the latter case the vibrational quanta are primarily pumped by electrons while during reentry they are pumped by recombination processes that populate the top of the vibrational ladder.

In conclusion, this book focus on the basic elementary processes occurring in gas discharge columns of the atmospheric gases and in the natural ionospheric plasma, in order to provide a unifying, basic theoretical background for all these situations. The emphasis is on the bulk kinetics and the coupling between the kinetics of electrons, ions, and vibrational and electronic states, but the analysis of heterogeneous surface reactions which is also provided is useful for both plasma-tube wall and Shuttle-atmosphere interactions. A number of problems of gas discharge physics which can be important for technical applications of some types of discharges (for example, capacitively coupled RF discharges), as sheath and pre-sheath phenomena or nonlocal effects on the electron kinetics to mention just a few, are out of the scope of this book, but references on the subject are presented which interested readers can refer to for further information.

1. Translational Relaxation of Heavy Particles

Translational relaxation, that is, the process of evolution towards equilibrium of the translational degrees of freedom of heavy particles (atoms, molecules, ions), is one of the fastest collision relaxation processes. To estimate the characteristic time of this process one can use, for example, the general relaxation equation (0.1) discussed in the Introduction. Let $E(t)$, in this equation, denote the translational energy per unit volume and assume relaxing particles initially at rest, that is, $E(0) = 0$. Then, the energy relaxation time, τ_{TT}, due to translational-translational energy exchanges is

$$\tau_{TT} = \frac{E^0}{(\mathrm{d}E/\mathrm{d}t)_{E=0}}, \tag{1.1}$$

where E^0 is the equilibrium value of E. The quantity $(\mathrm{d}E/\mathrm{d}t)_{E=0}$ can be expressed as

$$(\mathrm{d}E/\mathrm{d}t)_{E=0} = ZN\overline{\Delta E_T}, \tag{1.2}$$

where N is the density of relaxing particles, Z is the number of collisions per unit time of a particle at rest with the other moving gas particles, and $\overline{\Delta E_T}$ is the average translational energy transferred in such collisions.

Considering (1.2), and taking into account that $1/Z = \tau_0$, where τ_0 is the mean free time, the translational relaxation time (the relaxation time of translational energy) can be rewritten as

$$\tau_{TT} = \tau_0 \frac{E^0}{N\overline{\Delta E_T}}. \tag{1.3}$$

Equation (1.3) means that the number of collisions necessary to reach equilibrium is given by $E^0/N\overline{\Delta E_T}$.

To estimate τ_{TT}, let us assume [1] that a small number of atomic particles of type B, with mass m_B and initial velocity $c_B = 0$, are injected into an atomic gas of type A (atoms of mass m_A), which is in equilibrium at temperature T_A. Let N_B and N_A stand for the particle number densities of theses gases, respectively. To calculate τ_{TT} from (1.1), one has to find $(\mathrm{d}E/\mathrm{d}t)$ at $E = 0$, i.e., to find the rate of increase of the translational energy per unit volume of the admixed particles. The increase in kinetic energy for particles B due to collisions with particles A is

1. Translational Relaxation of Heavy Particles

$$\frac{dE}{dt} = N_B \int \Delta \epsilon_T(c) dZ(c) , \tag{1.4}$$

where $dZ(c)$ is the number of collisions per unit time of a particle B at rest with particles A with absolute velocity lying in the interval c, $c + dc$, and $\Delta\epsilon_T$ is the mean energy transferred to a particle B in such a collision.

Using the classical hard-sphere model to calculate the collision dynamics of two particles with diameters d_A and d_B, respectively, one has

$$\frac{\Delta\epsilon}{\epsilon} = \frac{4m_A m_B}{(m_A + m_B)^2}; \quad \epsilon = \frac{m_A c^2}{2} \tag{1.5}$$

for a head-on collision. Taking into account all possible impact parameters, the average energy transferred per collision, $\overline{\Delta\epsilon}$, is given by

$$\frac{\overline{\Delta\epsilon}}{\epsilon} = \frac{2m_A m_B}{(m_A + m_B)^2} . \tag{1.6}$$

For a Maxwellian velocity distribution function of the particles A, it can easily be derived that

$$dZ(c) = N_A d_{AB}^2 (2\pi)^{1/2} \left(\frac{m_A}{kT_A}\right)^{3/2} \exp\left(-\frac{m_A c^2}{2kT_A}\right) c^3 dc , \tag{1.7}$$

where k is the Boltzmann's constant and $d_{AB} = 1/2(d_A + d_B)$. Therefore, taking into account (1.4), (1.6) and (1.7), we obtain [2]

$$\frac{dE}{dt} = N_B \frac{4m_A m_B}{(m_A + m_B)^2} kT_A Z , \tag{1.8}$$

where

$$Z = 2d_{AB}^2 N_A \left(\frac{2\pi kT_A}{m_A}\right)^{1/2} \tag{1.9}$$

is the number of collisions of a particle B at rest with the particles A per unit time.

Taking into account that $E^0 = (3/2)kT_A N_B$, one readily obtains from (1.1) and (1.8)

$$\tau_{TT} = \frac{3}{8} \frac{(m_A + m_B)^2}{m_A m_B} Z^{-1} . \tag{1.10}$$

Since Z^{-1} is the mean free time τ_0 of a B particle in the gas A, we can approximately write, for $m_A \sim m_B$,

$$\tau_{TT} \sim \tau_0 , \tag{1.11}$$

that is, the translational energy relaxation time is of the order of magnitude of the mean free time.

When $m_A \gg m_B$ (Lorentz gas) or $m_A \ll m_B$ (Rayleigh gas) the energy exchange is much less efficient, and the relaxation time increases by a factor m_A/m_B or m_B/m_A, respectively.

To confirm this estimation it is necessary to analyze the Boltzmann gas-kinetic equation which governs the evolution of the one-particle velocity distribution function $f(\mathbf{c}, \mathbf{r}, t)$ at point $\mathbf{r}$ in space, in a non-equilibrium rarefied gas. For a one-component system of particles with mass m this equation reads [2,3]

$$\frac{\partial f}{\partial t} + \mathbf{c}\frac{\partial f}{\partial \mathbf{r}} + \frac{\mathbf{F}}{m}\frac{\partial f}{\partial \mathbf{c}} = \int\int\int (f'f_1' - ff_1)gb\,db\,d\varphi\,d\mathbf{c_1}, \qquad (1.12)$$

where $f' = f(\mathbf{c'}, \mathbf{r}, t)$, $f_1 = f(\mathbf{c_1}, \mathbf{r}, t)$, $f_1' = f(\mathbf{c_1'}, \mathbf{r}, t)$; $\mathbf{c}, \mathbf{c_1}$ are the velocities of the colliding particles and $\mathbf{c'}$, $\mathbf{c_1'}$ are the velocities of the product particles; $g = |\mathbf{c} - \mathbf{c_1}|$ is the absolute value of the relative velocity, which is invariant for elastic collisions; b is the impact parameter; φ is the azimuthal scattering angle; and $\mathbf{F}$ denotes the external force acting on a particle.

To evaluate the relaxation time of translational energy, let us consider a spatially homogeneous case in the absence of external forces ($\mathbf{F} = 0$) and take $\partial f/\partial t \sim f/\tau_{\mathrm{TT}}$. The term on the right-hand side of (1.12), the so-called collision integral, J, is of the same order of magnitude as each of the two subtracting terms, hence we can assume that

$$|J| \sim \int\int\int ff_1 gb\,db\,d\varphi\,d\mathbf{c_1}. \qquad (1.13)$$

The latter integral is of the order of $fN\sigma\bar{g}$, where $\bar{g}$ is the mean relative velocity and $\sigma \sim \pi d^2$ is the hard-sphere collision cross-section, d denoting the molecular diameter. This yields

$$\tau_{\mathrm{TT}} \sim \frac{1}{N\sigma\bar{g}} \sim \tau_0, \qquad (1.14)$$

which is the same as (1.11).

The application of the Boltzmann equation to investigate the translational relaxation of heavy particles will not be carried out in detail here. We will examine only certain peculiarities, which will be used in the subsequent discussion. Originally, (1.12) was obtained by Boltzmann upon balance considerations on the number of particles in an element of volume of phase space. In this case, the Boltzmann equation has a clear physical sense. It describes the rate of change in the number of particles in an element of volume due to the particle motion in phase space and to the effects of binary collisions between the particles. The integral term involving the product $f'f_1'$ is the rate of increase of f due to the scattering of particles into the examined element of volume owing to collisions. The other integral term involving the product ff_1 is, conversely, the rate of decrease of f due the particles that are scattered out of that element of volume by collisions. These considerations of balance are rather general and allow the kinetic equation for the populations of rotational and vibrational molecular levels to be straightforwardly written (see Chaps. 2 and 3).

Mathematically, the Boltzmann equation is rather difficult to solve. Therefore, some simplifications are widely used depending on the physical situation under examination. Essentially, there are two types of physical situations of interest. The first type concerns the relaxation problem, that is, the evolution in time of a non-equilibrium distribution towards equilibrium (the Cauchy problem). Exact solutions for this case were found only in 1975 [4], that is, almost a century after the Boltzmann equation was derived. The second type of situation concerns forced non-equilibrium states, in which the factors causing a deviation from equilibrium (e.g., external forces, spatial gradients, chemical reactions and other sources or sinks of particles) are constantly present in the system.

Let us first examine the relaxation problem. To deal with this problem, the ordinary relaxation approximation (also called the "strong collision" approximation) and the diffusion approximation are of particular interest.

In the strong collision approximation, assuming a spatially homogeneous case the Boltzmann equation is written as [5]:

$$\frac{\partial f}{\partial t} = -\frac{1}{\tau_{\mathrm{TT}}}(f - f_{\mathrm{eq}}) + q - \frac{1}{\tau_{\mathrm{s}}} f , \qquad (1.15)$$

where $f_{\mathrm{eq}}(c) = N \left(m/2\pi kT\right)^{3/2} \exp\left(-mc^2/2kT\right)$ is the equilibrium (Maxwellian) distribution function, and q and f/τ_{s} account for the particle sources and sinks, respectively.

For $\tau_{\mathrm{TT}} = \mathrm{const.}$, $q = 0$, and $1/\tau_{\mathrm{s}} = 0$ the solution of (1.15) is simply

$$f(c,t) = f(c,0) e^{-t/\tau_{\mathrm{TT}}} + \left(1 - e^{-t/\tau_{\mathrm{TT}}}\right) f_{\mathrm{eq}} . \qquad (1.16)$$

This solution just describes a transition of exponential type from any initial distribution $f(c,0)$ towards the equilibrium Maxwellian distribution f_{eq}. The deviation from the equilibrium distribution vanishes exponentially in time with a time constant given by τ_{TT}.

In the general case, equation (1.15) is non-linear, because the final particle density, N, and temperature, T, of the equilibrium Maxwellian distribution are to be determined consistently with the time-dependent equations

$$N = \int f \mathrm{d}c, \qquad \frac{3}{2} NkT = \int \frac{mc^2}{2} f \mathrm{d}c , \qquad (1.17)$$

which involve the unknown distribution $f(t)$ itself. However, in an isolated system, where the total energy and the total number of particles are conserved, N and T are constants, and their magnitudes are determined by the initial conditions. In this particular situation, equation (1.15) therefore becomes linear. In open systems, where exchanges of energy and particles with the environment are possible, it is necessary in general to keep in equation (1.15) the terms describing such exchanges (that is, the sources q and sinks f/τ_{s}).

When sources and sinks have to be accounted for, a simple steady-state solution of (1.15) can readily be derived for the case $\tau_{\mathrm{TT}}/\tau_{\mathrm{s}} \ll 1$. This solution is

$$f \simeq f_{eq} + q\tau_{TT}. \tag{1.18}$$

As seen from (1.18), a source of particles can strongly influence the distribution function f in a region of particle velocities (or energies ϵ) where $f_{eq} \sim \exp(-\epsilon/kT) \leq q\tau_{TT}$. The "strong collision" approximation under stationary conditions has been used, for example, in [6] to estimate the influence of hot N atoms on the rate of the reaction $N+O_2 \to NO+O$. Such hot N atoms are produced through the exothermic reaction $O+N_2(v \geq 13) \to NO+N$ in N_2-O_2 discharges.

Another useful simplification of (1.12) is the diffusion approximation. Sometimes it is possible to replace the integro-differential equation (1.12) with a differential equation of the Fokker-Planck type. For example, the Boltzmann equation for heavy particles B admixed in small amounts into a light gas A (Rayleigh gas, $m_B \gg m_A$), can be written as [7]:

$$\frac{\partial f(\epsilon, t)}{\partial t} = \frac{\partial}{\partial \epsilon}\left[\left(A\epsilon + \frac{3}{2}B\right) f(\epsilon, t)\right] + \frac{\partial^2}{\partial \epsilon^2}[B\epsilon f(\epsilon, t)], \tag{1.19}$$

where the coefficients A and B are determined by the distribution function of the light particles, F, according to the expressions

$$A = \frac{4}{3}\frac{m_A}{m_B} J_1; \quad B = \frac{4}{3} m_B \left(\frac{m_A}{m_B}\right)^2 J_2;$$

$$J_1 = \int\int\int \frac{\partial F(g)}{\partial g} 4\pi g^4 \sin^2\left(\frac{\zeta}{2}\right) b\, db\, d\varphi\, dg;$$

$$J_2 = \int\int\int F(g) 4\pi g^5 \sin^2\left(\frac{\zeta}{2}\right) b\, db\, d\varphi\, dg. \tag{1.20}$$

Herein, $F(g) \equiv F(\mathbf{g})$ and $\zeta = \zeta(b)$ is the scattering angle of the relative velocity due to a collision; φ is angle parameter of collision.

Two interesting physical results follow from equation (1.19):

(a) The Maxwellian distribution $f_{eq} \sim \sqrt{\epsilon}e^{-\frac{\epsilon}{kT^*}}$, where $kT^* = -B/A$, $A < 0$, is the steady-state solution of (1.20). In this case, the distribution of the heavy particles evolves in time towards a Maxwellian and this fact is independent of the distribution $F(\mathbf{c})$ of the light particles. However, the temperature T^* depends both on this distribution $F(\mathbf{c})$ and on the type of interaction of the heavy particles with the light ones. In particular, for the hard-sphere model one has $kT^* = m_A \overline{c^3}/(4\bar{c})$, where $\overline{c^3}$ and $\bar{c}$ are velocity moments of light particles obtained from the distribution $F(c)$ (for example, $\bar{c} = \int_0^\infty cF(c)4\pi c^2 dc$).
(b) If the distribution f of the heavy particles is initially Maxwellian, the relaxation described by (1.19) proceeds through a succession of Maxwellian stages ("canonical invariance"), with a time evolution of the temperature given by

$$T(t) = T^* + (T(0) - T^*)e^{-t/\tau_{\text{TT}}}, \tag{1.21}$$

where $\tau_{\text{TT}} = -1/A$.

It should be stressed that the peculiarities of the above relaxation process apply only to the Rayleigh gas.

The diffusion approximation is also widely used for the analysis of the behaviour of plasma electrons, and in the theory of rotational relaxation (see Chaps. 2 and 4).

Let us now examine forced non-equilibrium situations. If the characteristic time τ_{TT} of "maxwellization" is considerably less than the characteristic time τ_s of the perturbing factors, a quasi-stationary solution of the Boltzmann equation can be found as a power series of the small parameter τ_{TT}/τ_s. The first (dominant) term of this series corresponds to a local Maxwellian distribution. The second and all subsequent terms take into account the deviations from the equilibrium distribution owing to the above-mentioned factors (see also (1.18)). Detailed mathematical formulations were developed to analyze small perturbations of the equilibrium distribution function caused by, e.g., finite spatial gradients of temperature, or hydrodynamic velocities and concentrations. Such formulations are globally referred to as the theory of transport phenomena [2,3]. The problem of deviations from the Maxwellian distribution caused by chemical reactions have also been thoroughly studied (see, for example, [8,9]). In this context, it should be emphasized that chemical reactions can lead to significant perturbations of the distribution function at relatively high energies of the order of the reaction activation energy. Very recently Bruno et al. [10] have found a depletion in the tail of the translational energy distribution function for the model system Xe+H_2 due to the dissociation process. This result was obtained by DSMC (Direct Simulation Monte Carlo) [11], a numerical method able to follow any kind of relaxation in rarefied gases including translational relaxation [10–13]. This method is based on a particle simulation of neutral flow fields in the phase space, where interparticle collision processes are included by random selection of couples of particles as potential collision partners and subsequent evaluation of the collision frequency.

The problem of the effects of particle sources or sinks on the distribution function has received little attention up to now. In this field the only fundamental works are by Stupochenko [14,15]. It should be noted that sources and sinks of hot particles generally lead to distinct perturbations of the distribution function. To understand this, consider sources or sinks of particles in the vicinity of the energy ϵ_0 in the energy space, with ϵ_0 such that $e^{-\epsilon_0/kT} \ll 1$. In the case of a sink (due, for example, to a chemical reaction with activation energy ϵ_0), the distortion of the distribution function is concentrated essentially in the vicinity of energy ϵ_0. At lower energies the distribution remains nearly Maxwellian, due to the efficient energy exchanges between the particles by elastic collisions. By contrast, in the energy range $\epsilon > \epsilon_0$ the relative particle density is small (because $e^{-\epsilon_0/kT} \ll 1$), and the distribution function may deviate considerably from a Maxwellian one.

A different situation occurs in presence of a source of particles. Particle sources with energy ϵ_0 perturb the distribution function both in the regions $\epsilon \leq \epsilon_0$ and $\epsilon \geq \epsilon_0$. In the former region, the presence of the source leads to the formation of a gently sloping plateau. This behaviour of the distribution function is understandable if one takes into account that most of the collisions undergone by the created particles lead to a decrease in their energy. Consequently, the majority of the energetic particles created will move down along the energy axis, increasing the particle density in the region $\epsilon \leq \epsilon_0$. However, for $\epsilon \geq \epsilon_0$ the distribution function is nearly Maxwellian, but there will be an increased number of particles as compared to the case where the source is absent.

The behaviour of the distribution function over time will also be different in the presence of particle sources or sinks. For example, in the case of a δ-type sink in energy space, the distribution may essentially remain Maxwellian in shape, but with a depletion at the energy of the sink. Only the total number of particles will change in time. In the presence of a source the situation is different. Here the number of particles will increase in time, and the distribution will approach a Maxwellian one only after sufficient time has elapsed. This is easy to understand if all particles are sorted according to their ages. After a time of the order of τ_{TT}, the "old" particles, which have been subjected to many collisions, will have a nearly Maxwellian distribution. But the "young" particles will not. The difference between the actual distribution and the Maxwellian distribution (the perturbation function [14,15]) is invariant during this quasi-stationary stage. Subsequently, the relative value of the perturbation function describing the distribution of the non-Maxwellian particles will progressively vanish (that is, during this evolution the relative number of non-Maxwellian particles decreases). In the course of this final evolution stage the total distribution therefore approaches the Maxwellian form. This explanation can be understood from (1.17), (1.18). The equilibrium distribution f_{eq} increases in magnitude due to the rise of the total density N ($dN/dt = q > 0$).

Most of the above-discussed physical and mathematical aspects of translational relaxation apply to problems of rotational and vibrational relaxation as well. However, in these processes all phenomena proceed on other time scales (especially for vibrational relaxation) as discussed in the following chapters.

References

1. Gordiets B.F., Osipov A.I. and Shelepin L.A. (1988) *Kinetic Processes in Gases and Molecular Lasers*, Gordon and Breach, New York London Paris (translated from Russian)
2. Chapman S. and Cowling T.G. (1970) *The Mathematical Theory of Non-Uniform Gases*, 3$^{\rm rd}$ Edition, Cambridge University Press, Cambridge

3. Ferziger J. H. and Kaper H. G. (1972) *The Mathematical Theory of Transport Processes*, North–Holland, Amsterdam
4. Bobylev A.B. (1975) Rep. Acad. Sci. USSR **225**, 1041 (*in Russian*); Bobylev A.B. (1975) ibid **225**, 1296 (*in Russian*)
5. Gross E., Bhatnagar D. and Krook M. (1954) Phys. Rev. **94**, 511
6. Nahorny J., Ferreira C.M., Gordiets B., Pagnon D., Touzeau M. and Vialle M. (1994) J. Phys. D: Appl. Phys. **28**, 738
7. Osipov A.I. (1982) Chemical Physics **N 1**, 59 (*in Russian*)
8. Osipov A.J. (1980) Engineering-Physical Journal **38**, 351 (*in Russian*)
9. Shizgal B. and Karplus M. (1970) J. Chem. Phys. **52**, 4262; (1971) ibid **54**, 4345
10. Bruno D., Capitelli M. and Longo S. (1998) Chem. Phys. Letters **289**, 141
11. Bird G.A. (1995) *Molecular Gas Dynamics and the Direct Simulation of Gas Flows*, Clarendon Press, Oxford
12. Stark J.P.W. (1999) in *Rarefied Gas Dynamics*, eds. R. Brun, R. Campargue, R. Gatignol, and J. C. Lengrand, Cépadues Editions, Toulouse, Vol. 1, p. 199.
13. Koura K. (1999) Phys. Fluids **11**, 3174
14. Stupochenko E.V. (1947) Rep. Acad. Sci. USSR **67**, 447 (*in Russian*); (1947) ibid **67**, 635 (*in Russian*)
15. Stupochenko E.V. (1949) J. Exper. Theor. Phys. **19**, 493 (*in Russian*); (1953) Bull. Moscow State Univ., Ser. Phys. Astron. **8**, 57 (*in Russian*)

2. Rotational Relaxation of Molecules

Rotational relaxation, the process of evolution towards equilibrium between rotational and translational degrees of freedom (R-T processes), takes place, for the majority of molecular gases, with a characteristic time of the same order as the time of translational relaxation. This is evident from (1.1), if the rotational relaxation time τ_{RT} is rewritten as:

$$\tau_{RT} = \frac{E_{rot}^0}{(dE_{rot}/dt)_{E_{rot}=0}} . \tag{2.1}$$

Here, E_{rot} is the rotational energy of molecules per unit volume and E_{rot}^0 is the equilibrium value of E_{rot}. To estimate τ_{RT}, let us assume, as in Chap. 1, that relaxing molecules BC (rotators with mass $2m_B$) are admixed as a small impurity of density N_{BC} in an atomic gas A (atoms with mass m_A) assumed in equilibrium at temperature T_A. As $E_{rot}^0 = N_{BC}kT_A$, the order of magnitude of the ratio τ_{RT}/τ_{TT} will be determined by the ratio $\Delta\epsilon_{rot}(c)/\Delta\epsilon(c)$, where $\Delta\epsilon_{rot}(c)$ and $\Delta\epsilon(c)$ are the rotational and translational energy, respectively, transferred to a molecule BC, at rest and non-rotating, in a single collision with an atom A of velocity c in absolute value.

To estimate $\Delta\epsilon_{rot}(c)/\Delta\epsilon(c)$, let us consider the most favorable configuration for the transfer of rotational energy. This configuration corresponds to the case in which the atom A moves perpendicularly to the axis of the rotator, directly towards one of its atoms, as shown in Fig. 2.1.

Assume that the angular velocity ω and the center of mass velocity c_2 of the rotator are equal to zero before the collision, and the velocity of the incident atom is c_1. Let c_1', c_2', ω' denote the corresponding values after the collision. We have:

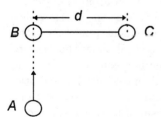

Fig. 2.1. Collision between atom A and rotator BC

$$\frac{\Delta\epsilon_{\text{rot}}(c)}{\Delta\epsilon(c)} = \frac{J(\omega')^2}{2} \bigg/ \frac{2m_B(c'_2)^2}{2} = \frac{d^2(\omega')^2}{4(c'_2)^2} \,, \qquad (2.2)$$

where $J = 2m_B(d/2)^2$ is the moment of inertia of the rotator relative to its center of mass. From conservation of linear and angular momentum, one has

$$m_A c_1 = m_A c'_1 + 2m_B c'_2$$

$$m_A c_1 \frac{d}{2} = m_A c'_1 \frac{d}{2} + J\omega' \,, \qquad (2.3)$$

so it follows that $c'_2 = d\omega'/2$. Thus, for the chosen collision configuration, we have $\Delta\epsilon_{\text{rot}}(c)/\Delta\epsilon(c) = 1$, and so

$$\frac{\tau_{\text{RT}}}{\tau_{\text{TT}}} \sim 1 \,. \qquad (2.4)$$

Since not all collision configurations are that most favorable for the excitation of rotational degrees of freedom, it follows that

$$\tau_{\text{RT}} \geq \tau_{\text{TT}} \,. \qquad (2.5)$$

These estimations are correct for impulsive collisions, when only two particles A and B are in fact colliding. For light molecules (H_2, D_2, HD) or for molecules with a large rotational constant, B_e, the rotation frequency is high (consequently, the Massey parameter is larger than unity), which implies that particle A is actually interacting with a fast-rotating molecule BC. In this situation, $\Delta\epsilon_{\text{rot}}$ may be much smaller than $\Delta\epsilon$ and $\tau_{\text{RT}} \gg \tau_{\text{TT}}$. For example, the rotational relaxation time corresponds to approximately 350 collisions for H_2 and 200 collisions for D_2, at room temperature [1].

We have to distinguish between two cases for the kinetic description of rotational relaxation [2]. For the majority of molecules, the rotational and the translational relaxation times are of a similar order of magnitude as already noted. This is why these two processes should not be investigated separately. Problems related to translational–rotational non-equilibrium may arise in experiments on anomalous dispersion of sound waves and ultrasonic absorption, as well as in the study of fast-expanding flows [3] and shock waves [4]. However, few theoretical works have as yet delt with combined translational–rotational relaxation problems. One example of such works is [5] where translational–rotational relaxation in a gas mixture of heavy rotators and light atoms was investigated using the diffusion approximation. In this situation, the translational and the rotational relaxation processes have the same characteristic time, although being independent (which is rather an exception than the rule). Numerical calculations [5] show that the rotational and translational relaxation time, and the non-equilibrium translational and rotational distribution functions, depend on the initial distribution of translational–rotational energy. In spite of this simultaneity of translational and rotational relaxation processes, the synchronism of their rates is broken for high rotational levels. Indeed, rotational excitation and deactivation of such levels is slower than translational relaxation.

The fact that rotational relaxation of light molecules and high rotational levels is slower than translational relaxation implies that the former relaxation can be considered separately once the latter is complete (that is, the distribution function of the translational degrees of freedom becomes Maxwellian).

The most general way to describe rotational relaxation is based on the gas kinetic balance equations (the master equations) for the populations of rotational levels. For diatomic molecules admixed as a small impurity in an atomic gas, the master equations for the number N_j of molecules in each rotational level j can be written as

$$\frac{dN_j}{dt} = Z \sum_i (P_{ij} N_i - P_{ji} N_j) , \tag{2.6}$$

where P_{ij} is the probability of the $i \to j$ rotational transition per collision, averaged over a Maxwellian velocity distribution (it is assumed that translational relaxation is complete), and Z is the number of atom-molecule collisions per unit time. The system of equations (2.6) takes into account rotational-translational (R-T) energy exchanges in collisions of molecules with structureless atoms only. Mutual collisions between molecules have been neglected, since low molecular concentration was assumed.

At present, there are no exact and general theoretical expressions for the mean probabilities P_{ij}. This is why various semi-empirical formulas have been proposed in the literature, as, for example, [6]:

$$P_{ij} = G_0 \exp\left(-\frac{a_1 |\Delta\epsilon_{ij}|}{kT}\right) \quad \text{or} \quad P_{ij} = G_1 \left(\frac{kT}{|\Delta\epsilon_{ij}|}\right)^{a_2} . \tag{2.7}$$

Here, a_1 and a_2 are empirical constants, $\Delta\epsilon_{ij} = \epsilon_i - \epsilon_j$, ϵ_j is the rotational energy of level j, and the factors G_0 and G_1 account for the degeneracy of the rotational levels. For rotational transitions of CO molecules induced by collisions with H_2 molecules, the following expression has been proposed [7]

$$\frac{Z}{[H_2]} P_{ij} = a_3 \frac{g_j}{g_i} \left(1 + \frac{|\Delta\epsilon_{ij}|}{kT}\right) \exp\left[-a_4 \left(\frac{|\Delta\epsilon_{ij}|}{kT}\right)^{1/2}\right] , \tag{2.8}$$

where g_j is the statistcal weight of the rotational level j, and a_3 and a_4 are numerical coefficients depending on $|i - j|$, namely, $a_3 = 1.66; 2.8; 1.19; 1.0;$ 1.3 (in 10^{-10} cm^{3}s^{-1}) and $a_4 = 1.67; 1.47; 1.85; 1.55; 2.24$, for $|i - j|=1; 2;$ 3, 4; 5, respectively. Expressions (2.7) and (2.8) illustrate that the rotational transition probabilities depend heavily on the transition energy $|\Delta\epsilon_{ij}|$.

Analytical solutions of system (2.6) are unknown when multi-quantum rotational transitions are taken into account. A method to solve this system of equations approximately was first proposed in [8,9] (see also [6]). It is based on a transformation of the unknowns (the rotational level populations) into a set of more convenient adiabatic variables allowing for diagonalization of the system of equations.

Investigation of rotational relaxation can be carried out analytically if some approximations and simplifications of (2.6) are made, such as the so-called "strong collision" approach. In this approach, in the master equation for a given level j the populations of all other levels $i \neq j$ are replaced with the equilibrium ones, that is,

$$N_i^0 = q_i N_{\text{rot}}; \qquad q_i = \frac{1}{Q_{\text{rot}}} g_i \exp\left(-\frac{\epsilon_i}{kT}\right), \tag{2.9}$$

where N_{rot} is the total density of rotators, g_i is the statistical weight of rotational level i, and Q_{rot} is the rotational partition function. For diatomic molecules, $\epsilon_i = B_e i(i+1)$, $q_i = 2i+1$, $Q_{\text{rot}} = kT/B_e$, where B_e is the rotational constant. By inserting (2.9) into (2.6) and noting that, according to the principle of detailed balance,

$$g_i P_{ij} \exp\left(-\frac{\epsilon_i}{kT}\right) = g_j P_{ji} \exp\left(-\frac{\epsilon_j}{kT}\right), \tag{2.10}$$

one readily obtains

$$\frac{dN_j}{dt} = \frac{1}{\tau_{\text{RT}}}(q_j N_{\text{rot}} - N_j) \equiv \frac{1}{\tau_{\text{RT}}}(N_j^0 - N_j), \tag{2.11}$$

where

$$\frac{1}{\tau_{\text{RT}}} = \frac{1}{Z}\sum_i P_{ji}. \tag{2.12}$$

The "strong collision" approach and equations similar to (2.11), with additional terms accounting for other sources and sinks of rotationally excited molecules, have been used to analyze the effect of rotational relaxation on the absorption of laser radiation in molecular gases by vibrlational–rotational transitions [10,11] and to study gas cooling through such absorption [12] (see also [13,14]).

Other analytical solutions of system (2.6) can also be obtained using the diffusion approximation. To use this approximation, it is required that $\Delta\epsilon_{j+1,j}/kT < 1$, but due to the smallness of the rotational quanta this condition is usually met over a wide range of physical situations. For diatomic molecules, the system (2.6) in the diffusion approximation can be rewritten in the form of the following Fokker–Planck-type equation [6]:

$$\frac{\partial N(\epsilon_{\text{rot}})}{\partial t} = \frac{\partial}{\partial \epsilon_{\text{rot}}}\left[D(\epsilon_{\text{rot}})\left(\frac{\partial N(\epsilon_{\text{rot}})}{\partial \epsilon_{\text{rot}}} + \frac{1}{kT}N(\epsilon_{\text{rot}})\right)\right], \tag{2.13}$$

where $N(\epsilon_{\text{rot}})$ is a continuous rotational distribution function connected with the level populations N_j by the relationship $N_j = N(\epsilon_{\text{rot}})d\epsilon_{\text{rot}}/dj$, with $\epsilon_{\text{rot}} = B_e j(j+1)$, and $D(\epsilon_{\text{rot}})$ is a diffusion coefficient in the space of rotational energies ϵ_{rot}.

Expressions for this diffusion coefficient are known in a few cases only. Considering, for example, impulsive collisions of rotators of mass M with atoms of mass m ($m \ll M$), one has [15]:

$$D(\epsilon_{\rm rot}) = b\epsilon_{\rm rot}; \qquad b = \frac{32}{3}\frac{m}{M} N_{\rm b} kT \Omega_{12}^{(11)}, \qquad (2.14)$$

where $N_{\rm b}$ is the particle number density of the buffer gas, and $\Omega_{12}^{(11)}$ is the gas kinetic collision integral. For the hard-sphere model,

$$b = \frac{8}{3}\frac{m}{M}\frac{kT}{\tau_0}; \qquad \tau_0 = \frac{1}{2N_{\rm b} d^2 \sqrt{2\pi kT/m}}, \qquad (2.15)$$

where $d = (d_1 + d_2)/2$ is the collision diameter, and d_1 and d_2 are the diameters of the rotators and the atoms, respectively.

Another explicit expression for $D(\epsilon_{\rm rot})$ can be obtained from the general relationship between $D(\epsilon_{\rm rot})$ and the probabilities P_{ij}, namely,

$$D(\epsilon_{\rm rot}) = \frac{1}{2}\langle \Delta^2 \rangle Z$$

$$\langle \Delta^2 \rangle = \frac{1}{Z} \int_0^\infty (\epsilon_{\rm rot} - \epsilon'_{\rm rot})^2 P(\epsilon_{\rm rot}, \epsilon'_{\rm rot}) d\epsilon'_{\rm rot}, \qquad (2.16)$$

where

$$P(\epsilon_{\rm rot}, \epsilon'_{\rm rot}) = \frac{P_{j,i}}{B_e(2i+1)} \qquad (2.17)$$

is the probability density of a transition from energy $\epsilon_{\rm rot}$ into energy $\epsilon'_{\rm rot}$ in rotational energy space.

Assuming that $P(\epsilon_{\rm rot}, \epsilon'_{\rm rot})$ depends only on the transferred energy $|\epsilon_{\rm rot} - \epsilon'_{\rm rot}|$ (see (2.7)) and decreases sharply with increasing $|\epsilon_{\rm rot} - \epsilon'_{\rm rot}|$, one readily obtains [16]

$$D(\epsilon_{\rm rot}) = {\rm const}. \qquad (2.18)$$

If the probabilities (2.8) are used in (2.16),(2.17) to calculate the diffusion coefficient $D(\epsilon_{\rm rot})$, this coefficient becomes non-linear on $\epsilon_{\rm rot}$ [17,18].

Taking into account (2.14) and (2.15), some peculiarities of rotational relaxation arise from (2.13).

1. An equation for the total rotational energy $E_{\rm rot} = \int_0^\infty \epsilon_{\rm rot} N(\epsilon_{\rm rot}) d\epsilon_{\rm rot}$ can be easily derived from equation (2.13) with the diffusion coefficient (2.14). This equation is a simple linear relaxation equation of the form

$$\frac{dE_{\rm rot}}{dt} = -\frac{E_{\rm rot} - E_{\rm rot}^0}{\tau_{\rm RT}}, \qquad (2.19)$$

where $E_{\rm rot}^0 = N_{\rm rot} kT$, and $\tau_{\rm RT} = kT/b$.

Thus, in the context of the diffusion approximation, one obtains the simple relaxation equation (3) discussed in the Introduction, which was used (see

(2.1)) to estimate the rotational relaxation time. It should be stressed that, in the framework of the above model, quations (2.1),(2.19) are correct for finite deviations from equilibrium, $|E_{\rm rot} - E_{\rm rot}^0|/E_{\rm rot}^0 \sim 1$. For any other molecular model, such equations will in general be correct only for small deviations from equilibrium, $|E_{\rm rot} - E_{\rm rot}^0|/E_{\rm rot}^0 \ll 1$.

2. Equation (2.13), with the diffusion coefficient (2.14), has the canonical invariance property. Consequently, for initial conditions

$$N(\epsilon_{\rm rot}, t = 0) = \frac{N_{\rm rot}}{kT_{\rm rot}(0)} \exp\left(-\frac{\epsilon_{\rm rot}}{kT_{\rm rot}(0)}\right),$$

$$N_{\rm rot} = \int_0^\infty N(\epsilon_{\rm rot}) d\epsilon_{\rm rot}, \tag{2.20}$$

with $T_{\rm rot}(0) \neq T$, a Boltzmann distribution function with temperature $T_{\rm rot}(t)$, satisfying (2.19), is a solution of (2.13), for which $E_{\rm rot} = N_{\rm rot} k T_{\rm rot}(t)$.

For arbitrary initial conditions the solution is

$$N(\epsilon_{\rm rot}, t) = \sum_n C_n \exp\left(-\frac{nt}{\tau_{\rm RT}}\right) L_n\left(\frac{\epsilon_{\rm rot}}{kT}\right) \exp\left(-\frac{\epsilon_{\rm rot}}{kT}\right), \tag{2.21}$$

where $L_n(x)$ are the Laguerre polinomials, and

$$C_n = \frac{1}{kT} \int_0^\infty N(\epsilon_{\rm rot}, t = 0) L_n\left(\frac{\epsilon_{\rm rot}}{kT}\right) d\epsilon_{\rm rot}.$$

For a constant diffusion coefficient (2.18), the solution of equation (2.13) takes the form

$$N(\epsilon_{\rm rot}, t) = \int_0^\infty G(\epsilon_{\rm rot}, \epsilon', t) N(\epsilon', t = 0) d\epsilon', \tag{2.22}$$

where the Green function G is given by

$$G(\epsilon_{\rm rot}, \epsilon', t) = \frac{1}{2kT} \left(\frac{\pi t}{\tau_{\rm RT}}\right)^{-1/2} \exp\left[-\left(\frac{\epsilon_{\rm rot} - \epsilon'}{kT} + \frac{t}{\tau_{\rm RT}}\right)^2 \frac{\tau_{\rm RT}}{4t}\right]$$

$$+ \frac{1}{2kT} \left(\frac{\pi t}{\tau_{\rm RT}}\right)^{-1/2} \exp\left[-\left(\frac{\epsilon_{\rm rot} + \epsilon'}{kT} + \frac{t}{\tau_{\rm RT}}\right)^2 \frac{\tau_{\rm RT}}{4t}\right] \exp\left(\frac{\epsilon'}{kT}\right)$$

$$+ \frac{1}{2kT} \left\{1 - \Phi\left[\left(\frac{\epsilon_{\rm rot} + \epsilon'}{kT} - \frac{t}{\tau_{\rm RT}}\right) \frac{1}{2}\sqrt{\frac{\tau_{\rm RT}}{t}}\right]\right\} \exp\left(-\frac{\epsilon_{\rm rot}}{kT}\right). \tag{2.23}$$

Here, $\Phi(z) = \frac{2}{\sqrt{\pi}} \int_0^z \exp(-y^2) dy$ and

$$\tau_{\rm RT} = \frac{(kT)^2}{D}. \tag{2.24}$$

The solution (2.22),(2.23) has a clear physical meaning. For example, if an initial δ- function distribution ($N(\epsilon_{\rm rot}, 0) = N_{\rm rot} \delta(\epsilon_{\rm rot} - \epsilon_{\rm rot}^0)$) is chosen, the solution $N(\epsilon_{\rm rot}, t) = N_{\rm rot} G(\epsilon_{\rm rot}, \epsilon_{\rm rot}^0, t)$ describes a continuous spreading of the particles over the entire rotational energy space. The narrow initial distribution centered around energy $\epsilon_{\rm rot}^0$ (terms inside the first curved brackets in (2.23)) decays, while a Boltzmann distribution builds up for the bulk of particles (terms inside the second curved brackets). If $\epsilon_{\rm rot}^0 > kT$, the relaxation of the δ -distribution corresponds to a relatively slow spreading of the particles, which is however accompanied by rapid thermalization of particles with low rotational energies. This explains interesting experimental results obtained with expanding jets, which show that the effective rotational temperature for the higher rotational levels is larger than for the lower ones. In the case of expansion in nozzles, the temperature of the upper levels at the output of the nozzle is close to the temperature in the upstream chamber, while that of the lower levels "follows" the translational temperature. In these conditions and if the expansion is fast enough, a population inversion between upper and lower rotational levels can be established [16,19].

Rotational relaxation in the diffusion approximation has also been investigated in order to analyze the influence of radiative transitions between rotational levels in polar molecules [17,18,20]. This problem is useful in understanding some processes in interstellar clouds. In particular, analytical expressions have been obtained for the radiative cooling rate in the rotational band of CO molecules [18], an important mechanism for the energy balance of interstellar clouds.

The previous discussion was restricted to the relaxation of rotators considered as a small impurity in an atomic gas. To analyze rotational relaxation in one-component systems or in the presence of high concentrations of rotators, it is required to take into account both R-R (exchanges of rotational energy in collisions between rotators) and R-T energy exchanges in molecular collisions. The probabilities of R-R exchanges have been calculated in [6]. The probabilities and diffusion coefficients thus obtained are not convenient in solving the kinetic equation. Nonetheless, the influence of R-R exchanges on rotational relaxation can be understood on the basis of the following theoretical considerations. The probabilities of R-R and R-T exchanges are comparable for the lower rotational levels. Since the rate of thermalization due to R-T exchanges is high for the lower levels, R-R exchanges have practically no influence on the kinetics. The probabilities of R-R exchanges for upper levels are higher than those of R-T processes. However, R-R exchanges are not effective as a whole, because the number of particles in the upper levels is small, and collisions between them are rare. R-R exchanges can however play a considerable role at intermediate energies (thermal energy and somewhat above). In this case, non-resonant R-R exchanges can accelerate relaxation and lead to the formation of a quasi-stationary distribution of rotational energy, in a similar way as occurs with the distribution of vibrational energy for anharmonic oscillators (see Chap. 3).

There have been few numerical computations of the distribution function of rotational energy for real systems. The most complete data calculated to date concern hydrogen, nitrogen, carbon oxide and halogen-hydrogen molecules. The results of these calculations are discussed in detail in [6]. Here, we will briefly discuss only some particular aspects of the rotational kinetics in a few systems.

Calculations of rotational relaxation in pure para-hydrogen have been carried out for three situations [21]: a gas at rest, an expanding jet, and beyond a shock wave front. In the first case, the system of relaxation equations for the rotational level populations together with the energy conservation equation has been numerically solved by the Gear method. Only four lower rotational levels, with $j = 0, 2, 4, 6$, were taken into account. Using the calculated populations, the application of (2.19) to describe the evolution of the rotational energy $E_{\rm rot} = \sum_j \epsilon_j N_j(t)$ has been analyzed. The results of these calculations, presented in Fig. 2.2, show that the relaxation time $\tau_{\rm RT}$ is not constant during the relaxation process for $\Delta T = T(0) - T_{\rm rot}(0) < 0$. The slow relaxation of the populations of upper rotational levels is the reason for the increase in $\tau_{\rm RT}$. For $\Delta T > 0$, the value of $\tau_{\rm RT}$ remains practically unchanged. These calculations further show that the rotational relaxation of H_2 does not have the canonical invariance property. This can be clearly illustrated by introducing a "pair rotational temperature"

$$kT_{\rm rot}^{ij}(t) = \frac{\epsilon_j - \epsilon_i}{\ln[(2j+1)N_i(t)/(2i+1)N_j(t)]}$$

for each pair of levels. The evolution of several of these pair rotational temperatures during relaxation is illustrated in Fig. 2.3.

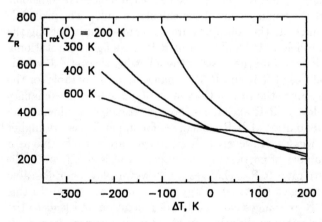

Fig. 2.2. Average number of collisions $Z_R = 1/\tau_{\rm RT}$ required to establish rotational equilibrium in H_2 versus initial deviation $\Delta T = T(0) - T_{\rm rot}(0)$, where $T(0)$ and $T_{\rm rot}(0)$ are the initial translational and rotational temperatures [21]

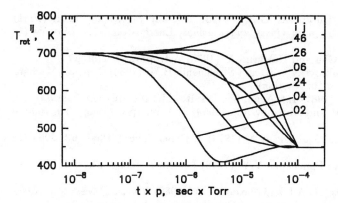

Fig. 2.3. Evolution of pair rotational temperatures T_{rot}^{ij} during relaxation in H_2. The initial distribution is a Boltzmann distribution with $T_{\text{rot}}(0) = 700$ K. The gas temperature $T(0) = 300$ K [21]

It can be seen from Fig. 2.3 that all T_{rot}^{ij} are different during relaxation. Therefore, the concept of rotational temperature is meaningful only both at the beginning of the relaxation process, when $N_i(0) = N_i(T_{\text{rot}}(0))$, and at its end, close to equilibrium.

Rotational relaxation in an N_2-Ar mixture in a freely expanding jet has been calculated in [22]. A system of relaxation equations of the type (2.6) was solved numerically. It was found that the rotational "temperatures" T_{rot}^{ij} differ from each other and from the gas temperature. Moreover, the rotational temperatures of the upper levels can be frozen in the case of strong gas expansion and cooling.

Numerical solutions of (2.6) (with additional terms describing radiative transitions between rotational levels) were also obtained in [7,23] to investigate energy loss rates of microwave radiation due to excitation of rotational transitions of CO molecules in collapsing interstellar clouds. In this case, the rotational distribution function is far from equilibrium due to the very low gas density. Consequently, radiative transitions between rotational levels have a strong influence upon level populations.

References

1. Bamford C.H. and Tipper C.F.H, eds. (1969) *Comprehensive Chemical Kinetics. The Formation and Decay of Excited Species*, Elsevier, Amsterdam
2. Osipov A.I. (1985) Engineering-Physical Journal **49**, 154 (*in Russian*)
3. Zarvin A.E. and Sharafudrinov R.G. (1981) J. Appl. Mechan. Tech. Phys. **N5**, 9 (*in Russian*)
4. Robben F. and Talbot L. (1966) Phys. Fluids **9**, 653
5. Yoshikava K.K. (1978) in *Proceed. Int. Symp. Rarefied Gas Dynamics*, Paris, Vol. 1, p. 389

6. Bogdanov A.B., Dubrovsky G.B., Osipov A.I. and Strelchenia B.M. (1991) *Rotational Relaxation in Gases and Plasmas*, Energoatomizdat, Moscow (*in Russian*)
7. De Jong, Shin-I Chu and Dalgarno A. (1975) Astrophys. J. **199**, 69
8. Vasiliev A.P., Dubrovsky G.B. and Strelchenia B.M. (1984) J. Appl. Mechan. Tech. Phys. **N5**, 6 (*in Russian*)
9. Dubrovsky G.B. (1987) Inf. Acad. Sci. USSR, MJG **N6**, 167 (*in Russian*)
10. Kabashnikov V.P. and Rubanov A.S. (1969) J. Appl. Spectroscopy **10**, 760 (*in Russian*)
11. Letokhov V.S. and Makarov A.A. (1972) J. Exper. Theor. Phys. **63**, 2064 (*in Russian*)
12. Gordiets B.F. and Panchenko V.Y. (1978) Letters J. Techn. Phys. **4**, 1396 (*in Russian*)
13. Gordiets B.F., Osipov A.I. and Shelepin L.A. (1986) *Kinetic Processes in Gases and Molecular Lasers*, Gordon and Breach, New York London Paris
14. Gordiets B.F. and Zhdanok S. (1986) in *Non-equilibrium Vibrational Kinetics*, ed. M.Capitelli, Springer, Berlin Heidelberg, p. 47
15. Safarian M.N. and Stupochenko E.V. (1964) J. Appl. Mechan. Techn. Phys. **N4**, 29 (*in Russian*)
16. Goroshkov A.A. and Osipov A.I. (1980) Chem. Phys. Lett. **74**, 345
17. Gordiets B.F. and Stepanovich A.N. (1987) in *Mathematical Problems of Applied Aeronomy*, ed. M. Y. Marov, Inst. Appl. Math. 111, Moscow (*in Russian*)
18. Gordiets B.F. and Stepanovich A.N. (1989) Astronom. J. **66**, 30 (*in Russian*)
19. Osipov A.I. and Chaikina Y.A. (1985) Chemical Physics **4**, 167 (*in Russian*)
20. Gordiets B.F., Stepanovich A.N., Chaikina Y.A. and Osipov A.I. (1983) Chem. Phys. Lett. **102**, 189
21. Rabitz H. and Lam S.H. (1975) J. Chem. Phys. **62**, 1425
22. Koura K. (1962) J. Chem. Phys. **77**, 5141
23. Goldreich P. and Kwan J. (1974) Astrophys. J. **189**, 441

3. Vibrational Relaxation of Molecules

Vibrational kinetics investigates the relaxation processes of molecular vibrational levels and their influence on various properties of gases. It is a fundamental part of physical-chemical kinetics for the study of different phenomena in non-equilibrium molecular gases. Vibrational relaxation has a special role among relaxation processes. First of all, this is because under ordinary conditions, i.e., for gas temperatures below a few thousand degrees, vibrational relaxation is a slow process compared with translational and rotational relaxation. Secondly, much of the energy can be stored in vibrational levels (recall that a nonlinear molecule with N atoms has $3N - 6$ independent vibrational modes). Non-equilibrium vibrationally excited molecules can be easily created and have a wide range of practical applications.

Detailed studies on vibrational kinetics and its applications in gas dynamics and laser physics which were carried out before 1985, were presented in various works [1–8]. The role of vibrational kinetics in plasma chemistry has also been discussed in [6,7,9].

3.1 Estimation of Vibrational Relaxation Times

The physical reason for the slow energy exchange between vibrational and translational degrees of freedom (V-T processes) in collisions between molecules is connected with the adiabacity of collisions. In fact, as was mentioned in the Introduction, the Massey parameter for V-T exchanges is, as a rule, much higher than unity. To prove this, one can use the following simple model. Consider a simple system consisting of inert gas atoms A and a small impurity of relaxing diatomic molecules BC. Let us assume that the center of mass velocities of all the particles have a Maxwellian distribution. To estimate the vibrational relaxation time, τ_{VT}, we can use (1.1), rewritten for V-T exchanges:

$$\tau_{VT} = \frac{E_{vib}^0}{(dE_{vib}/dt)_{E_{vib}=0}}. \tag{3.1}$$

3. Vibrational Relaxation of Molecules

Using the harmonic oscillator model, one has

$$E^0_{\text{vib}} = \hbar N_{\text{BC}}\omega \left[\exp\left(\frac{\hbar\omega}{kT}\right) - 1\right]^{-1}, \tag{3.2}$$

where N_{BC} is the density of relaxing molecules. The rate of change of vibrational energy is defined by an expression similar to (1.5):

$$\frac{dE_{\text{vib}}}{dt} = N_{\text{BC}} \int \Delta\epsilon_{\text{vib}}(c) dZ(c), \tag{3.3}$$

where $dZ(c)$ is determined by (1.7), with m_A replaced with the reduced mass $\mu = m_A m_{\text{BC}}/(m_A + m_{\text{BC}})$.

To determine dE_{vib}/dt, it is necessary to solve the dynamical problem of the oscillator-atom collision in order to find $\Delta\epsilon_{\text{vib}}(c)$. For simplicity, let us consider the most favorable collision configuration for the transfer of vibrational energy (Fig. 3.1). As a first approximation, an oscillator-atom collision can be modelled by considering that the atom exerts a perturbing force $F(t)$ on the oscillator. The equation of motion of the oscillator is therefore written

$$\ddot{y} + \omega^2 y = \frac{1}{m}F(t), \tag{3.4}$$

where y is the elongation (deviation from the equilibrium position), ω is the oscillator frequency, and $m = m_B m_C/(m_B + m_C)$ is the reduced mass of the oscillator. Consistently with (3.1), we will assume that the oscillator is initially not excited, that is, $\epsilon_{\text{vib}} = 0$, $y(-\infty) = 0$, $\dot{y}(-\infty) = 0$. The transferred energy $\Delta\epsilon_{\text{vib}}$ is determined by

$$\Delta\epsilon_{\text{vib}} = \frac{m}{2}(\dot{y}^2 + \omega^2 y^2) \quad \text{for} \quad t = \infty. \tag{3.5}$$

Introducing the variable $\xi = \dot{y} + i\omega y$, equation (3.4) can be rewritten as

$$\frac{d\xi}{dt} - i\omega\xi = \frac{1}{m}F(t), \tag{3.6}$$

whose solution is

$$\xi(t) = e^{i\omega t}\int_{-\infty}^{t} \frac{1}{m}F(t')e^{-i\omega t'} dt'. \tag{3.7}$$

From (3.5) and (3.7), we have

$$\Delta\epsilon_{\text{vib}} = \frac{m}{2}|\xi|^2_{t=\infty} = \frac{1}{2m}\left|\int_{-\infty}^{\infty} F(t)e^{-i\omega t} dt\right|^2. \tag{3.8}$$

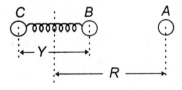

Fig. 3.1. Collision between an atom A and an oscillator BC

Therefore, the transferred energy is determined by the squared modulus of the Fourier-component of the exerted force at the proper frequency of the oscillator. In other words, only this Fourier-component of the force is effective in exciting vibrational degrees of freedom.

To calculate $\Delta\epsilon_{\text{vib}}$ explicitly, the actual interaction potential which determines $F(t)$ is needed. If one takes into account only the short-range interaction between the neighboring particles A and B, assumes an interaction potential of the form $V = Ce^{-r_{AB}/L}$, and neglects the influence of vibrational excitation on the trajectory of the colliding particles, one obtains [4,10]:

$$F(t) = -\frac{1}{L}\frac{\mu c^2}{2}\frac{m_C}{m_B + m_C}\cosh^{-2}\left(\frac{ct}{2L}\right) ; \qquad (3.9)$$

$$\Delta\epsilon_{\text{vib}}(c) = \frac{1}{L^2}\frac{\mu^2 c^4 m_C^2}{(m_B + m_C)^2 8m}\left[\frac{4\pi\omega L^2}{c^2}\bigg/\sinh\left(\frac{\pi\omega L}{c}\right)\right]^2 . \qquad (3.10)$$

Since the Massey parameter for vibrational excitation is $\pi\omega L/c \gg 1$, equation (3.10) can be approximately written as

$$\Delta\epsilon_{\text{vib}}(c) = \frac{8\pi^2\omega^2\mu^2 m_C^2 L^2}{(m_B + m_C)^2 m}\exp\left(-\frac{2\pi\omega L}{c}\right) . \qquad (3.11)$$

Therefore, $\Delta\epsilon_{\text{vib}} \sim e^{-2\omega\tau_{\text{col}}}$, where $\tau_{\text{col}} = \pi L/c$ is the duration of the collision. To calculate dE_{vib}/dt, we must now average (3.11) over a Maxwellian velocity distribution according to (3.3). The integrant in (3.3) has a sharp maximum at the point where the argument of the exponential function passes through a minimum. This point can be found from the equation

$$\frac{d}{dc}\left(\frac{2\pi\omega L}{c} + \frac{\mu c^2}{2kT}\right) = 0 . \qquad (3.12)$$

It follows that the velocity c^* corresponding to the maximum of the integrant is

$$c^* = \left(\frac{2\pi kT\omega L}{\mu}\right)^{1/3} . \qquad (3.13)$$

The major contribution to the value of the integral in (3.3) arises from the neighborhood of point c^*. Thus,

$$\overline{\Delta\epsilon_{\text{vib}}} \sim \exp\left(-\frac{2\pi\omega L}{c^*} - \frac{\mu c^{*2}}{2kT}\right) \equiv \exp(-3\chi) , \qquad (3.14)$$

where

$$\chi = \frac{\mu c^{*2}}{2kT} = \frac{\pi\omega L}{c^*} = \left(\frac{\pi^2\mu\omega^2 L}{2kT}\right)^{1/3} . \qquad (3.15)$$

Note that $\chi \simeq 5-10$ for most molecules in the temperature range of practical interest.

For a quantum harmonic oscillator $\overline{\Delta\epsilon_{\text{vib}}} = \sum_{n=0} n\hbar\omega P_{0n}(T)$, where P_{0n} is the probability of a vibrational transition $0 - n$ of the molecule BC in a collision with an atom A. Since $P_{01} \gg P_{0n}$ for $n \geq 2$, one can write

$$\overline{\Delta\epsilon_{\text{vib}}} \cong \hbar\omega P_{01} . \tag{3.16}$$

By equating (3.16) to the classical result (3.14), an expression for the probability P_{01} can be readily derived.

From (3.1)–(3.3), (3.10), (3.14)–(3.16) the relaxation time τ_{VT} can be now expressed as

$$\tau_{\text{VT}} = \frac{1}{ZP_{10}(1 - e^{-\hbar\omega/kT})} , \tag{3.17}$$

where, according to the principle of detailed balance, the probability P_{10} for the $1-0$ transition is related to P_{01} by the expression

$$P_{01} = P_{10} \exp\left(-\frac{\hbar\omega}{kT}\right) . \tag{3.18}$$

Expression (3.17), with P_{10} determined as explained above, is known as the Landau–Teller formula [11]. An important physical result that follows from (3.14)–(3.18) is

$$\ln[\tau_{\text{VT}} Z(1 - e^{-\hbar\omega/kT})] \sim T^{-1/3} . \tag{3.19}$$

In many cases, the dependence on temperature predicted by (3.19) correlates very well with experimental data.

If (3.17) is rewritten as

$$\tau_{\text{VT}} \sim \frac{1}{P_{10}} \tau_0 , \tag{3.20}$$

where $\tau_0 = 1/Z$, it becomes clear that

$$\tau_{\text{VT}} \gg \tau_0 \sim \tau_{\text{TT}} \sim \tau_{\text{RT}} . \tag{3.21}$$

Owing to the above scaling of relaxation times, one can usually treat the problem of vibrational relaxation by assuming that both the rotational and the translational degrees of freedom have already attained equilibrium.

The kinetic description of vibrational relaxation is based on the balance equations for the populations of the various vibrational levels.

3.2 Vibrational Kinetics of Harmonic Oscillators

3.2.1 Vibrational Relaxation of Diatomic Molecules in an Inert Gas: V-T Exchange

Let us consider the vibrational relaxation of molecules through vibrational-translational energy exchanges (V-T processes) in binary molecule-atom collisions. Let N_n stand for the population of the n-th vibrational level and P_{nm}

for the probability of the vibrational transition $n \to m$ in one such collision. The system of master kinetic equations for the vibrational level populations in a spatially homogeneous gas can therefore be written as

$$\frac{dN_n}{dt} = Z \sum_m (P_{mn} N_m - P_{nm} N_n) , \qquad (3.22)$$

where Z is the molecule-atom collision frequency.

Assuming a Maxwellian velocity distribution function with temperature T, the probabilities of direct and reverse transitions, from the principle of detailed balance, are related through the expression

$$P_{nm} \exp\left(-\frac{E_n}{kT}\right) = P_{mn} \exp\left(-\frac{E_m}{kT}\right) , \qquad (3.23)$$

where E_n and E_m are the energies of the two vibrational levels.

Typically, only single-quantum transitions with probabilities

$$P_{n+1,n} = (n+1) P_{10} \qquad (3.24)$$

are important for a harmonic oscillator [1–4,7,10]. Taking into account (3.23) and (3.24), the kinetic equations (3.22) can be rewritten as

$$\frac{dN_n}{dn}$$
$$= Z P_{10} \{(n+1)[N_{n+1} - N_n \exp(-E_{10}/kT)]$$
$$- n[N_n - N_{n-1} \exp(-E_{10}/kT)]\} , \qquad (3.25)$$

where $E_{10} = E_1 - E_0$ is the energy of the vibrational quanta. This system has two important peculiarities.

1. Equations (3.25) have the property of canonical invariance with respect to the Boltzmann distribution function [12]. In fact, if the vibrational populations $N_n(0)$ are initially distributed according to the Boltzmann law, that is,

$$N_n(0) = N(1 - e^{-\vartheta_0}) e^{-n\vartheta_0} , \qquad (3.26)$$

where $\vartheta_0 = E_{10}/kT_V(0)$, $T_V(0)$ denoting the initial vibrational temperature of the oscillators, the solution of (3.25) at any time t will also be a Boltzmann distribution given by

$$N_n(t) = N(1 - e^{-\vartheta(t)}) e^{-n\vartheta(t)} , \qquad (3.27)$$

where $\vartheta(t) = E_{10}/kT_V(t)$ and $T_V(t)$ is the vibrational temperature at time t. One has

$$\frac{d\vartheta(t)}{dt} = Z P_{10} (1 - e^{-\vartheta})(e^{-\vartheta} - e^{-E_{10}/kT}) e^{\vartheta} , \qquad (3.28)$$

whose solution is

$$\vartheta(t) = \ln \frac{e^{-t/\tau_{VT}}(1 - e^{E_{10}/kT - \vartheta_0}) - e^{E_{10}/kT}(1 - e^{-\vartheta_0})}{e^{-t/\tau_{VT}}(1 - e^{E_{10}/kT - \vartheta_0}) - (1 - e^{-\vartheta_0})} \qquad (3.29)$$

with

$$\tau_{VT} = [ZP_{10}(1 - e^{-E_{10}/kT})]^{-1} . \qquad (3.30)$$

For an arbitrary initial distribution, the solution of (3.25) is [12]

$$N_n(t) = \sum_m c_m l_n(m) e^{-mt/\tau_{VT}} , \qquad (3.31)$$

where

$$l_n(m) = e^{-nE_{10}/kT} \sum_{i=0}^{\infty} (1 - e^{-E_{10}/kT})^i \binom{n}{i} \binom{m}{i} \qquad (3.32)$$

with

$$\binom{n}{i} = \begin{cases} \frac{n!}{i!(n-i)!} & \text{for } n \geq i \\ 0 & \text{for } n < i \text{ or } i < 0 \\ 1 & \text{for } i = 0 \end{cases}$$

are the Gotlieb polynomials, and the parameters c_m are determined by the initial conditions as follows:

$$c_m = (1 - e^{-E_{10}/kT}) \sum_{n=0}^{\infty} N_n(0) l_m(n) .$$

Using the hypergeometric functions, the Gotlieb polynomials can be written as $l_n(m) = F(-n, m+1, 1; 1 - e^{-E_{10}/kT})$.
The relations of orthogonality for the Gotlieb polynomials

$$\sum_{i=0}^{\infty} e^{-iE_{10}/kT} l_n(i) l_m(i) = \begin{cases} 0 & n \neq m \\ e^{-nE_{10}/kT}(1 - e^{-E_{10}/kT})^{-1} & n = m; \end{cases}$$

$$\sum_{n=0}^{\infty} e^{nE_{10}/kT} l_n(i) l_n(m) = \begin{cases} 0 & m \neq i \\ e^{iE_{10}/kT}(1 - e^{-E_{10}/kT})^{-1} & m = i \end{cases} \qquad (3.33)$$

are useful for practical calculations.

2. An equation for the relaxation of the total vibrational energy can be easily derived from the system (3.25). The vibrational energy per unit volume is

$$E_{\text{vib}} = \sum_n E_n N_n , \qquad (3.34)$$

where $E_n = nE_{10}$ for harmonic oscillators. Multiplying (3.25) by nE_{10} and summing the resulting equations over n yields the energy relaxation equation

$$\frac{dE_{\text{vib}}}{dt} = -\frac{E_{\text{vib}} - E_{\text{vib}}^0}{\tau_{VT}} , \qquad (3.35)$$

where $E_{\text{vib}}^0 = NE_{10}(e^{E_{10}/kT} - 1)^{-1}$ is the equilibrium value of E_{vib} at gas temperature T. Equation (3.35) was first derived by Landau and Teller [11].

For constant T, the solution of (3.35) has the exponential form

$$E_{\text{vib}}(t) = E_{\text{vib}}^0 + (E_{\text{vib}}(0) - E_{\text{vib}}^0)e^{-t/\tau_{\text{VT}}} \qquad (3.36)$$

and depends only on the initial value of the vibrational energy. Equation (3.35) is frequently written in a different form, in terms of the average number of vibrational quanta α (if $N = \text{const.}$):

$$\frac{d\alpha}{dt} = -\frac{\alpha - \alpha^0}{\tau_{\text{VT}}}, \qquad (3.37)$$

where

$$\alpha = \frac{1}{N}\sum_n nN_n \equiv \frac{E_{\text{vib}}}{NE_{10}}. \qquad (3.38)$$

Equations (3.35) and (3.37) are equivalent for harmonic oscillators. For anharmonic oscillators, (3.35) and (3.37) are approximately correct only, and they are also not equivalent, because

$$E_{\text{vib}} = \sum[nE_{10} - \Delta E n(n-1)]N_n \neq NE_{10}\alpha, \qquad (3.39)$$

where ΔE is the anharmonicity of the molecule.

3.2.2 One-Component Systems: V-V Exchanges

In collisions of diatomic molecules, energy exchanges between vibrational degrees of freedom (V-V exchanges) take place as well.

If the probabilities of V-V and V-T exchanges are small, in a first approximation they can be considered independent from each other. In this case, the master equations for the vibrational level populations, accounting for both V-V and V-T exchanges, can be written as

$$\frac{dN_n}{dt}$$
$$= Z\sum_m (P_{mn}N_m - P_{nm}N_n) + \frac{Z}{N}\sum_{m,s,l}(Q_{mn}^{sl}N_sN_m - Q_{nm}^{ls}N_lN_n). \qquad (3.40)$$

Here, $N = \sum_{m=0} N_m$; Q_{mn}^{sl} is the probability of V-V exchange in a collision after which two molecules in the states m and s transfer into the n and l states, respectively.

From the principle of detailed balance, the probabilities of V-V exchange are related through the expression

$$Q_{mn}^{sl}\exp\left(-\frac{E_s + E_m}{kT}\right) = Q_{nm}^{ls}\exp\left(-\frac{E_l + E_n}{kT}\right). \qquad (3.41)$$

3. Vibrational Relaxation of Molecules

For a harmonic oscillator, this yields [4,7,10]:

$$Q_{mn}^{sl} = \begin{cases} (m+1)sQ_{10} & n = m+1,\ l = s-1, \\ m(s+1)Q_{10} & n = m-1,\ l = s+1, \\ 0 & n \neq m \pm 1,\ l \neq s \mp 1. \end{cases} \quad (3.42)$$

Taking into account (3.23), (3.24), (3.41) and (3.42), equations (3.40) can be rewritten as

$$\frac{dN_n}{dn} = ZP_{10}\{(n+1)[N_{n+1} - N_n e^{(-E_{10}/kT)}] - n[N_n - N_{n-1}e^{(-E_{10}/kT)}]\}$$
$$+ ZQ_{10}\{(n+1)[(1+\alpha)N_{n+1} - \alpha N_n] - n[(1+\alpha)]N_n - \alpha N_{n-1}]\}, \quad (3.43)$$

where

$$\alpha = \frac{1}{N}\sum_{n=0} nN_n \quad (3.44)$$

and $Q_{10} \equiv Q_{10}^{01} \equiv Q_{01}^{10}$. System (3.43) is nonlinear, since α depends on N_n. Moreover, the translational temperature, and consequently P_{10} and Q_{10}, are time-dependent, since

$$\left(\frac{5}{2}kT + E_{10}\alpha\right)N = \text{const.}$$

in accordance with conservation of total (transllational–rotational and vibrational) energy. Vibrational relaxation of harmonic oscillators, described by system (3.43), has two important peculiarities.

1. There exist two different relaxation time scales, because usually

$$\tau_{VV} \approx \frac{1}{ZQ_{10}} \ll \frac{1}{ZP_{10}} \approx \tau_{VT}. \quad (3.45)$$

Condition (3.45) allows fast and slow stages in the relaxation process to be distinguished.

During the fast stage ($t \leq \tau_{VV}$), V-V exchanges play the main role, and V-T exchanges can be neglected. At this stage the relaxation can be described by the simplified system

$$\frac{dN_n}{dn} \approx ZQ_{10}\{(n+1)$$
$$\times [(1+\alpha)N_{n+1} - \alpha N_n] - n[(1+\alpha)]N_n - \alpha N_{n-1}]\}. \quad (3.46)$$

For harmonic oscillators, single-quantum V-V exchanges are resonant (resonance energy defect $(E_s + E_m) - (E_l + E_n) = 0$). Therefore, the total number of vibrational quanta is conserved in such processes. This can be readily proved by multiplying (3.46) by n and summing over all n. Therefore, at the V-V exchange stage ($t \leq \tau_{VV}$) equations (3.46) are linear (because $\alpha = \text{const.}$).

Noting that, according to detailed balancing, one has

$$(1+\alpha)N_{n+1} - \alpha N_n = 0 , \qquad (3.47)$$

the steady state solution of (3.46) can be easily derived. If we let

$$\alpha/(1+\alpha) = e^{-\vartheta} \equiv \exp\left(-\frac{E_{10}}{kT_V}\right) , \qquad (3.48)$$

it follows that a Boltzmann distribution with vibrational temperature T_V will be the solution of (3.47) [13–15]. Thus, the establishment of a Boltzmann distribution is the result of the fast V-V exchange stage. This fact illustrates the fundamental role of V-V exchanges in vibrational kinetics. The second, slow stage of vibrational relaxation is described by the complete system of equations (3.43). For times $t \gg \tau_{VV}$, the fast V-V exchanges need not be considered, and only the ultimate result of these processes - the formation of a Boltzmann distribution with a temperature determined by the total number of vibrational quanta - is essential for the subsequent evolution of the system. Therefore, the solution of (3.43) on the time scale $t \gg \tau_{VV}$ is a Boltzmann distribution function with a vibrational temperature $T_V(t)$ which evolves in time according to (3.29) [15].

2. The relaxation of total vibrational energy E_{vib} (see (3.34)) in a one-component system of harmonic oscillators can be described by the relaxation equation (3.35). This can be easily proved if (3.43) is multiplied by nE_{10} and summed over all n. This is because single-quantum V-V exchanges do not change the total vibrational energy of the system of harmonic oscillators. Therefore, only the V-T processes contribute to the relaxation of vibrational energy.

The harmonic oscillator model describes vibrational relaxation very well for low vibrationally excited levels, for which anharmonicity effects are negligible. However, such effects grow with increasing average vibrational energy stored per molecule. anharmonicity effects will be considered later in this chapter.

3.2.3 Binary Mixture of Diatomic Molecules: V-T , V-V and V-V' Exchanges

Vibrational energy exchanges in collisions between molecules of different types (V-V' exchanges) usually need to be considered in mixtures of diatomic gases. If the probabilities of V-T, V-V and V-V' processes are small, they can be considered, in a first approximation, as independent. In this case, the master equations for the vibrational level populations include separate terms accounting for V-T, V-V and V-V' processes, respectively. The terms describing V-T and V-V processes are the same as discussed above. The collision operator for V-V' processes is similar to the operator of V-V exchange (see (3.40), with a small modification accounting for the fact that the two colliding molecules are of different types. If N_n^A and N_n^B denote the number densities of molecules of type A and B, respectively, excited in the n-th vibrational level the change in N_n^A due to V-V' processes will be given by

$$\left(\frac{dN_n^A}{dt}\right)_{VV'} = \frac{Z_{AB}}{N_B} \sum_{m,s,l} (Q_{mn}^{'sl} N_s^B N_m^A - Q_{nm}^{'ls} N_l^B N_n^A), \qquad (3.49)$$

where: $N_A = \sum_n N_n^A$; $N_B = \sum_n N_n^B$; $Z_{AB} = N_B Z_{AB}^0$; Z_{AB}^0 is the mean number of collisions of a molecule A with molecules B at unit B-gas density; $Q_{mn}^{'sl}$ is the probability of V-V' exchange in a collision whereupon molecule A changes from vibrational state m to state n, and molecule B from s to l.

From the principle of detailed balance, the probabilities of V-V' exchanges, for Maxwellian velocity distribution functions with temperature T, are related through the expression

$$Q_{mn}^{'sl} \exp\left(-\frac{E_s^B + E_m^A}{kT}\right) = Q_{nm}^{'ls} \exp\left(-\frac{E_l^B + E_n^A}{kT}\right), \qquad (3.50)$$

where E_i^M is the energy of the i-th vibrational level of molecule M (M=A;B).

Typically, the probabilities of single-quantum V-V' exchanges between two harmonic oscillators with quantum energies such that

$$E_{10}^A > E_{10}^B; \quad E_{10}^A < 2E_{10}^B \qquad (3.51)$$

are given by (3.42). In this case, considering (3.50), equation (3.49) can be written as

$$\left(\frac{dN_n^A}{dt}\right)_{VV'} = Z_{AB}^0 N_B Q_{10}' \left\{ (n+1)\left[(1+\alpha_B)e^{\frac{E_{10}^A}{kT}} N_{n+1}^A - \alpha_B e^{\frac{E_{10}^B}{kT}} N_n^A\right] \right.$$
$$\left. - n\left[(1+\alpha_B)e^{\frac{E_{10}^A}{kT}} N_n^A - \alpha_B e^{\frac{E_{10}^B}{kT}} N_{n-1}^A\right] \right\}, \qquad (3.52)$$

where $\alpha_A = 1/N_A \sum_n n N_n^A$, $\alpha_B = 1/N_B \sum_n n N_n^B$. The corresponding equation for N_n^B is simply obtained from (3.52), by exchanging the indexes A and B.

Vibrational relaxation through V-V, V-V' and V-T processes has two important peculiarities.

1. Equations (3.43) with the additional V-V' term on the right-hand-side of (3.52), either for A and or for B molecules, have three relaxation times:

$$\tau_{VV}^A \sim \frac{1}{Z_{AA}^0 N_A Q_{10}}, \quad \tau_{VV'}^A \sim \frac{1}{Z_{AB}^0 N_B Q_{10}'}, \quad \tau_{VT}^A \sim \frac{1}{Z_A P_{10}^A}; \qquad (3.53)$$

$$\tau_{VV}^B \sim \frac{1}{Z_{BB}^0 N_B Q_{10}}, \quad \tau_{VV'}^B \sim \frac{1}{Z_{BA}^0 N_A Q_{10}'}, \quad \tau_{VT}^B \sim \frac{1}{Z_B P_{10}^B}. \qquad (3.54)$$

Different solutions can be obtained, depending on the scaling of the relaxation times τ_{VV}, $\tau_{VV'}$ and τ_{VT}. When the concentrations N_A and N_B are not significantly different and the following inequalities hold

3.2 Vibrational Kinetics of Harmonic Oscillators

$$\tau_{VV}^A, \tau_{VV}^B, \ll \tau_{VV'}^A, \tau_{VV'}^B, \ll \tau_{VT}^A, \tau_{VT}^B, \tag{3.55}$$

the relaxation proceeds over three different time scales. On the time scale $t \leq \tau_{VV}^A, \tau_{VV}^B$, V-V exchanges lead to the establishment, for each component, of a Boltzmann distribution with a vibrational temperature determined by the total vibrational energy stored in that component. On the time scale $\tau_{VV}^A, \tau_{VV}^B \leq t \leq \tau_{VV'}^A, \tau_{VV'}^B$, V-V' exchanges are effective and establish a relationship between the vibrational temperatures T_V^A and T_V^B. The two vibrational temperatures become synchronized. In fact, by the action of V-V' exchanges quasi-stationary vibrational distributions are formed. For single-quantum transitions, considering detailed equilibrium, these distributions must satisfy the relationship

$$Q_{m,m+1}^{'s+1,s} N_{s+1}^B N_m^A = Q_{m+1,m}^{'s,s+1} N_s^B N_{m+1}^A. \tag{3.56}$$

Taking into account (3.50), equation (3.56) can be rewritten as

$$e^{E_{10}^B/kT} N_{s+1}^B N_m^A = e^{E_{10}^A/kT} N_s^B N_{m+1}^A.$$

When the Boltzmann distributions

$$N_m^A = N_0^A \exp\left(-\frac{E_m^A}{kT_V^A}\right); \quad N_s^B = N_0^B \exp\left(-\frac{E_s^B}{kT_V^B}\right), \tag{3.57}$$

are substituted in (3.56), one readily obtains the relationship [16]

$$\frac{E_{10}^A}{kT_V^A} - \frac{E_{10}^B}{kT_V^B} = \frac{E_{10}^A - E_{10}^B}{kT}, \tag{3.58}$$

which connects T_V^A, T_V^B, and the translational temperature T.

Finally, on a time scale $t \sim \tau_{VT}$ full statistical equilibrium with equal temperatures for all degrees of freedom will be reached.

2. If equation (3.52) is multiplied by n and summed over all n, one gets

$$\left(\frac{d\alpha_A}{dt}\right)_{VV'} = Z_{AB}^0 N_B Q_{10}'$$
$$\times \left[\alpha_B(1+\alpha_A)e^{\frac{E_{10}^B}{kT}} - \alpha_A(1+\alpha_B)e^{\frac{E_{10}^A}{kT}}\right]. \tag{3.59}$$

A similar equation holds also for α_B, namely,

$$\left(\frac{d\alpha_B}{dt}\right)_{VV'} = Z_{AB}^0 N_A Q_{10}'$$
$$\times \left[\alpha_A(1+\alpha_B)e^{\frac{E_{10}^A}{kT}} - \alpha_B(1+\alpha_A)e^{\frac{E_{10}^B}{kT}}\right]. \tag{3.60}$$

Equations (3.59) and (3.60) are valid to describe the system in the time range $t \leq \tau_{VV'} \ll \tau_{VT}$. If we now multiply (3.59) by N_A, (3.60) by N_B and add both equations, we obtain:

$$N_A \alpha_A + N_B \alpha_B = \text{const}. \tag{3.61}$$

This equation means that single-quantum V-V' exchanges do not change the total number of vibrational quanta in the system, which is obvious. The system (3.59), (3.60) describes the kinetics of synchronization of the vibrational temperatures mentioned above. The stationary solution of (3.59), (3.60) yields the relationship [16]

$$\frac{\alpha^A}{1+\alpha^A} = \frac{\alpha^B}{1+\alpha^B} \exp\left(\frac{E_{10}^B - E_{10}^A}{kT}\right). \tag{3.62}$$

Since each of the vibrational systems has already reached a Boltzmann distribution owing to fast V-V exchanges, one has

$$\frac{\alpha^A}{1+\alpha^A} = \exp\left(-\frac{E_{10}^A}{kT_V^A}\right); \quad \frac{\alpha^B}{1+\alpha^B} = \exp\left(-\frac{E_{10}^B}{kT_V^B}\right). \tag{3.63}$$

Substituting (3.63) into (3.62) again yields (3.58).

Apart from the time scaling case (3.55), there exist other situations worthy of interest. This is the case, for example, when for one of the components (say, the B component) the following inequalities hold:

$$\tau_{VV}^B \ll \tau_{VT}^B \ll \tau_{VV'}^B. \tag{3.64}$$

In this case, after a time of the order of τ_{VT} the vibrational distribution of molecules B has already relaxed to a near equilibrium with the translational mode, that is, $T_V^B \simeq T$. Then, for component A, V-V' exchanges with molecules B in fact play the role of an additional V-T process with the relaxation time $[Z_{AB}^0 N_B Q_{10}' e^{E_{10}^A/kT}(1+\alpha_B)]^{-1}$. This fact can be easily proved from (3.59). Such a regime of vibrational relaxation is typical for N_2 molecules in low-temperature discharge plasmas in N_2-O_2 mixtures. In fact, the O_2 vibrational level populations are often close to equilibrium with the translations because of the fast V-T processes of O_2 molecules with O atoms formed by electron impact dissociation of O_2 in the discharge (see Chap. 7).

3.2.4 Polyatomic Molecules

Polyatomic molecules are different from diatomic ones due to the presence of several vibrational degrees of freedom. For relatively low vibrational excitation, the vibrational motion of the nuclei can be represented as a superposition of normal vibration modes, each of which corresponds to a harmonic oscillator. Therefore, in this approximation, vibrational relaxation in one-component systems (or in mixtures of polyatomic molecules) can be considered as vibrational relaxation in a multi-component mixture of harmonic oscillators. It can be assumed that V-V energy exchanges within every mode occur with considerably higher probabilities than V-V' exchanges between different modes or V-T processes, as in the case of binary mixtures of harmonic oscillators (see above). Under these conditions, the establishment of

3.2 Vibrational Kinetics of Harmonic Oscillators

a Boltzmann distribution for every mode, with a particular vibrational temperature, due to V-V exchanges, is the first stage of vibrational relaxation processes in mixtures of polyatomic molecules. For times longer than the V-V relaxation times, V-V relaxation processes need no longer be considered; the vibrational temperatures of the different modes first become synchronized due to V-V' exchanges, then total statistical equilibrium will be reached at a later stage via V-T exchanges. To describe this process, it is convenient to analyze the evolution of the mean number of vibrational quanta α_s of vibrational mode s. An equation describing this evolution was derived in detail in [17]. It reads

$$\frac{d\alpha_s}{dt} = Z^0_{AB} N_B Q_{AB} \begin{Bmatrix} l_i \to 0 \\ 0 \to l_j \end{Bmatrix} l_s \prod_{i=1}^{k}(r_i + \alpha_i^0)^{-l_i} \prod_{j=k+1}^{L}(\alpha_j^0)^{-l_j}$$

$$\times \left\{ \prod_{i=1}^{k}[\alpha_i^0(r_i + \alpha_i)]^{l_i} \prod_{j=k+1}^{L}[\alpha_j(r_j + \alpha_j^0)]^{l_j} \right.$$

$$\left. - \prod_{i=1}^{k}[\alpha_i(r_i + \alpha_i^0)]^{l_i} \prod_{j=k+1}^{L}[\alpha_j^0(r_j + \alpha_j)]^{l_j} \right\}, \qquad (3.65)$$

where $\alpha_s = r_s/[\exp(E_{10}^s/kT_V^s) - 1]$, r_s is the degeneracy and T_V^s is the vibrational temperature of mode s, and α_s^0 is the equilibrium value of α_s, i.e., the value of α_s for $T_V^s = T$, T standing for the translational temperature. When deriving the above equation, only collisions between molecules A and B and transitions of the system A+B (with L normal vibration modes) from vibrational state $(v_1, v_2, ..., v_L)$ to state $(v_1 \pm l_1, ..., v_k \pm l_k, v_{k+1} \mp l_{k+1}, ..., v_L \mp l_L)$ were taken into account.

In (3.65), $Q_{AB} \begin{Bmatrix} l_i \to 0 \\ 0 \to l_j \end{Bmatrix}$ is the transition probability, calculated without consideration of degeneracy according to the ordinary method by Schwartz, Slawsky, and Herzfeld (SSH theory) [1,10,18,19].

Equation (3.65) describes the relaxation of the average number of vibrational quanta of mode s through a single channel specified by the numbers $l_1, l_2, ..., l_L$. In the presence of several relaxation channels, the right-hand side of (3.65) must be summed over all possible values of $l_1, l_2, ..., l_L$. In a general case, vibrational relaxation in a mixture of polyatomic molecules will be described by a system of equations like (3.65), including an equation for each mode.

Equation (3.65) is the most general relaxation equation of vibrational energy accounting for V-V' and V-T exchanges in collisions of different molecules. This equation transforms into the Landau-Teller relaxation equation (3.35), (3.37) under V-T exchanges, for $L = 1$, $l_i = l_j = 0$ (with $i \neq s$), and $l_s = 1$. Furthermore, (3.65) is identical to the relaxation equation (3.59) under single-quantum V-V' exchanges in a binary mixture, for $L = 2$, $l_i = 1$, $l_j = 1$, $l_s = 1$, $r_s = 1$, and $r_j = 1$.

3.3 Vibrational Kinetics of Anharmonic Oscillators

At the present time, particular attention is given to the analysis of vibrational kinetics at low temperatures under conditions of large non-equilibrium stocks of vibrational energy. The harmonic oscillator model was found to be unsuitable for describing the vibrational kinetics under such conditions, and anharmonicity must be taken into consideration. Anharmonicity is important for high vibrational levels and influences the populations of these levels, the vibrational energy relaxation rates, and the rates of chemical reactions involving vibrationally excited molecules.

In general, the analysis of vibrational relaxation taking anharmonicity effects into account requires the use of numerical methods. Nevertheless, let us consider simplifications that allow analytical solutions to be obtained. To solve, for example, a nonlinear system of equations like (3.40), one can take into consideration the large differences between the probabilities of the different processes and assume simple analytical dependences of these probabilities on the vibrational quantum numbers. Over a wide range of gas temperatures, the probabilities of single-quantum processes P_{mn}, Q^{ij}_{mn} for many vibrational levels are much greater than those of multi-quantum processes. This allows system (3.40) to be simplified as follows:

$$\frac{1}{N}\frac{dN}{dt}f_n + \frac{df_n}{dt}$$
$$= Z[(P_{n+1,n}f_{n+1} - P_{n,n-1}f_n) - (P_{n,n+1}f_n - P_{n-1,n}f_{n-1})]$$
$$+ Z\sum_i [(Q^{i,i+1}_{n+1,n}f_i f_{n+1} - Q^{i,i+1}_{n,n-1}f_i f_n)]$$
$$- Z\sum_i [(Q^{i+1,i}_{n,n+1}f_{i+1} f_n - Q^{i+1,i}_{n-1,n}f_{i+1} f_{n-1})] \,. \qquad (3.66)$$

Here, f_n is the normalized vibrational distribution function defined by the relation $N_n = Nf_n$, where N_n is the number density of molecules in level n and N is the total concentration of molecules. Further simplifications can be made using the Morse anharmonic oscillator model. According to this model, the energy E_n of the n-th level in a diatomic molecule is given by

$$E_n = nE_{10}\left[1 - \frac{\Delta E}{E_{10}}(n-1)\right] \,, \qquad (3.67)$$

where ΔE is the anharmonicity, and the parameters E_{10} and ΔE are related to the spectroscopic molecular constants ω_e and x_e through the expressions $E_{10} = \hbar(\omega_e - 2\omega_e x_e)$, $\Delta E = \hbar\omega_e x_e$. Using the SSH theory to calculate the probabilities, as is customary in most analytical studies [1,6,7,10,18,19], and taking into consideration the influence of anharmonicity on the value of the adiabatic factor only (through the changes in the transition energies), one obtains (compare with (3.24), (3.42) for a harmonic oscillator; see also (7.5)–(7.9), (7.34)–(7.39)):

$$P_{n+1,n} \approx (n+1)P_{10}\exp(\delta_{VT}n), \tag{3.68}$$

$$Q_{n+1,n}^{i,i+1} \approx (n+1)(i+1)Q_{10} \\ \times \exp(-\delta_{VV}|n-i|)[3/2 - 1/2\exp(-\delta_{VV}|n-i|)], \tag{3.69}$$

$$Q_{n+1,n-1}^{01} \approx (n+1)nQ_{10}\frac{\Delta E}{2E_{10}}\exp\left(-2\delta_{VV}\left|\frac{E_{10}}{4\Delta E}-n\right|\right), \tag{3.70}$$

where $\delta_{VV} = 0.427 \times L(\frac{\mu}{T})^{\frac{1}{2}}\Delta E$; μ is the reduced mass of the colliding particles in a.e.; L is the constant in the exponential repulsive potential of the molecular interaction (in Å); ΔE and T are measured in K. A similar expression is used for δ_{VT}; however, μ and L may be different if the V-T processes involve collisions with an impurity gas.

Expression (3.70), unlike (3.68)(3.69), applies to two–quantum exchanges. It is given here since, in some cases, this process may have an influence on the vibrational level populations.

In some circumstances, for more accurate calculations, it is necessary to take into account the influence of anharmonicity on the pre-exponential factors in the expressions of the probabilities. In this case, equations (3.68)–(3.70) should be multiplied, respectively, by:

$$(1-x_e n)^{-1};\ (1-x_e n)^{-1}(1-x_e i)^{-1};\ (1-x_e n)^{-1}(1-2x_e n)^3. \tag{3.71}$$

3.3.1 V-T Relaxation in an Inert Gas

Consider the evolution of the vibrational distribution function and of the stock of vibrational energy of diatomic molecules admixed as a small impurity in an inert gas. Sources and sinks of vibrationally excited molecules are assumed to be negligible. In this case, the rate of change in vibrational level populations is described by (3.66), retaining on the right-hand-side the terms within the first square brackets only. An analytical solution of this system can then be obtained for anharmonic oscillators at low gas temperatures $kT < E_{10}$. If the initial (at $t = 0$) vibrational distribution of molecules is a Boltzmann distribution with $T_0 \neq T$ ($kT_0 < E_{10}$), for times $t > 1/ZP_{10}$ one obtains [20]

$$f_n \simeq f_0 \exp\left[-\frac{E_n}{kT} + \frac{E_1}{r_n}\left(\frac{1}{kT} - \frac{1}{kT_1(t)}\right)\right], \tag{3.72}$$

where

$$r_n = \prod_{i=2}^{n}\left(1 - \frac{P_{10}}{P_{i,i-1}}\right). \tag{3.73}$$

For a harmonic oscillator, $E_n = nE_{10}$, $\delta_{VT} = 0$, $r_n = 1/n$, and thus $f_n(t)$ is a Boltzmann distribution with a time-dependent temperature $T_1(t)$. For an anharmonic oscillator with probabilities (3.68), $r_n > 1/n$. In this case, as follows from (3.72), the populations of levels $n \geq 2$ are different from the Boltzmann ones; they are smaller if $T_1(t) > T$ (deactivation regime), and they are larger if $T_1(t) < T$ (thermal regime of vibrational excitation).

The time evolution of the distribution function f_n also provides information on the relaxation rate of vibrational energy $E_{vib} = N \sum_n E_n f_n$. Multiplying (3.66) (truncated as above) by E_n and using a Boltzmann distribution $f_n^B = f_0 \exp(-nE_{10}/kT_1)$ (this is a good approximation for low levels) we obtain the Losev formula [21] for the rate of change of E_{vib}:

$$\frac{dE_{vib}}{dt} = -\frac{E_{vib} - E_{vib}^0}{\tau_{VT}^{anh}}, \qquad (3.74)$$

$$\tau_{VT}^{anh} = \tau_{VT}^{h} \left[\frac{1 - \exp(-E_{10}/kT_1 + \delta_{VT})}{1 - \exp(-E_{10}/kT_1)} \right]^2, \qquad (3.75)$$

where E_{vib}^0 is the equilibrium value of E_{vib}, and τ_{VT}^{anh} and τ_{VT}^{h} are the V-T relaxation times for the anharmonic and harmonic oscillator models, respectively. Expression (3.75) was improved in [20] by accounting for the deviation of the actual distribution from the Boltzmann one for $n > 2$, under conditions close to equilibrium (see also [65]).

At high gas temperatures, multi-quantum V-T transitions may dominate the relaxation processes, and the single-quantum approximation may then fail to predict the evolution of the system accurately. In this case, vibrational relaxation can be better described by the Fokker–Planck diffusion equation. Vibrational kinetics of anharmonic oscillators in an inert gas was investigated in detail in [22,23] using the diffusion approximation.

3.3.2 V-V Exchanges. The Treanor Distribution

In a one-component system of anharmonic oscillators (or in a system of such oscillators partially diluted in an inert gas), V-V exchanges dominate other processes for the lower vibrational levels. In this case, the terms within the second square brackets on the right-hand-side of (3.66) must be kept to find the vibrational distribution function. The steady-state non-equilibrium vibrational distribution function satisfying this system of equations was first derived by Treanor, Rich and Rehm [24]. Later, this solution became known as the Treanor distribution. Its simplest derivation follows from the detailed balancing for any pair of direct and reverse V-V transitions. This means that the term in the second square brackets on the right-hand-side of (3.66) is equal to zero for any i, n. This yields

$$f_n^{\text{Tr}} = f_0 \exp\left(-\frac{nE_{10}}{kT_1} + \frac{nE_{10} - E_n}{kT}\right)$$
$$= f_0 \exp\left\{-n\left[\frac{E_{10}}{kT_1} - (n-1)\frac{\Delta E}{kT}\right]\right\}. \tag{3.76}$$

Note that the Treanor distribution (3.76) is independent of the transition probabilities, and therefore has a general thermodynamic nature. It can be also derived from basic principles of statistical physics, using the Gibbs canonical distribution [24,25], the expressions for the entropy of a vibrational subsystem interacting with a thermostat [26], and the Boltzmann H-theorem [27].

For harmonic oscillators $\Delta E = 0$, and (3.76) reduces to a Boltzmann distribution with vibrational temperature T_1. For anharmonic oscillators and $T_1 \neq T$, the distribution differs from a Boltzmann distribution, the deviation growing with increasing n. For $T_1 < T$, anharmonicity causes the depletion of upper vibrational levels, while for $T_1 > T$, it causes overpopulation of such levels. For the purposes of plasma chemistry and laser physics, the regime $T_1 > T$ is of particular interest. In this case, at low gas temperatures T, strong overpopulation of the upper levels may be observed. Moreover, for $T_1 > T$ the Treanor distribution has a minimum for some level n_0, above which (i.e., for $n > n_0$) the level populations are inverted. The value of n_0 may be easily derived from the condition $df_n^{\text{Tr}}/dn = 0$, which yields

$$n_0 = \frac{E_{10}}{2\Delta E}\frac{T}{T_1} + \frac{1}{2}. \tag{3.77}$$

In equilibrium ($T_1 = T$), the level n_0 coincides with the upper bound vibrational level of a Morse oscillator. With increasing T_1/T the level n_0 shifts downwards. Note, however, that it may be rather difficult to observe inversion of populations for $n > n_0$ in real systems. This is because dissipation of vibrational quanta due to V-T and other processes increases for the upper levels (see below).

3.3.3 V-V and V-T Exchanges. Moderate Deviation from Equilibrium

V-V processes are predominated only for relaxation of the lower levels of an anharmonic oscillator. In fact, for the upper levels the V-T probabilities become much greater than those of V-V exchanges, according to (3.68) and (3.69). As a result, the relative populations of these levels are characterized by a vibrational temperature equal to the gas temperature.

To find the complete vibrational distribution function for a one-component system of anharmonic oscillators, equations (3.66) has to be solved taking into account all the terms on the right-hand-side. Approximate analytical solution of such a system can be found only for particular conditions concerning the degree of vibrational excitation and the gas temperature.

First, let us consider regimes of moderate deviations from equilibrium [25,28–30], in which the excited level populations are so small that V-V exchanges involve only the lower states. Expressed mathematically, this means that

$$Q^{i,i+1}_{n+1,n} f_i f_{n+1} \gg Q^{n,n+1}_{n+1,n} f_n f_{n+1}, \qquad \text{for} \quad i \ll n \qquad (3.78)$$

on the right-hand-side of (3.66). When these inequalities hold, the main V-V processes which can affect the populations of higher levels ($n \gg 1$) are non-resonant exchanges with low vibrational levels. System (3.66) can then be simplified on account of the weak dependence of $\sum_i Q^{i,i+1}_{n+1,n} f_i$ on the distribution type. Since V-V exchanges with the lower levels dominate other processes, the terms with small i are the most significant. They correspond to a distribution that is still close to the Boltzmann one. Summation of the right-hand-side of system (3.66) over i with the Boltzmann distribution f_i yields a system of linear equations for the relative populations f_n. Under quasi-stationary conditions the solution is the following [28–30]:

$$f_n = f_n^{\text{Tr}} \prod_{i=1}^{n-1} \varphi_{i+1} \qquad n \geq 2, \qquad (3.79)$$

$$\varphi_{i+1} = \frac{(3/2)\beta Q^{01}_{i+1,i} + P_{i+1,i} \exp[E_{10}/kT_1 - E_{10}/kT]}{(3/2)\beta Q^{01}_{i+1,i} + P_{i+1,i}}, \qquad (3.80)$$

$$\beta = \frac{1 - \exp(-E_{10}/kT)}{[1 - \exp(-E_{10}/kT + \delta_{\text{VV}})]^2}. \qquad (3.81)$$

Expressions (3.79) and (3.80) can be simplified by dividing the system of vibrational levels into groups according to the dominant mechanisms. For lower levels i such that $i < n^*$, $\varphi_{i+1} \simeq 1$, therefore (3.79) reduces to a Treanor distribution. For $i > n^{**}$, $\varphi_{i+1} \simeq \exp(-\Theta)$, where

$$\Theta = \frac{E_{10}}{kT} - \frac{E_{10}}{kT_1}; \qquad (3.82)$$

therefore (3.79) reduces to the following Boltzmann distribution at the translational temperature T and $n > n^{**}$:

$$f_n \simeq f_0 \left[\frac{3}{2}\beta \frac{Q_{10}}{P_{10}} \right]^{\frac{\Theta}{\delta_{\text{VV}} + \delta_{\text{VT}}}} \exp\left\{ \frac{\Theta^2}{2(\delta_{\text{VV}} + \delta_{\text{VT}})} \right\} \exp\left(-\frac{E_n}{kT} \right). \qquad (3.83)$$

From (3.80), the quantum numbers n^* and n^{**} determining the boundaries between the different groups of levels are given by the expressions

$$n^* = \frac{1}{\delta_{\text{VV}} + \delta_{\text{VT}}} \ln\left(\frac{3}{2}\beta \frac{Q_{10}}{P_{10}} \right), \qquad (3.84)$$

$$n^{**} = n^* + \frac{\Theta}{\delta_{VV} + \delta_{VT}} + \frac{1}{2} . \tag{3.85}$$

The above analysis was concerned only with single-quantum transitions. However, for an anharmonic oscillator, the energy gap between two successive levels decreases with increasing quantum level number. Therefore, for quantum numbers $n_r \simeq ((k-1)/k)(E_{10}/2\Delta E)$, the exchange of k upper vibrational quanta with one lower quantum of energy E_{10} will be resonant. Under such conditions, the probabilities of two- and multi-quantum V-V exchange processes may increase greatly (formula (3.70) illustrates this fact for two-quantum exchanges), and such processes should therefore be considered. Analytical calculations of the vibrational distribution function taking into account both single- and two–quantum exchanges were performed in [29] under moderate excitation conditions.

3.3.4 Resonant V-V Relaxation Under High Exitation Conditions

Situations characterized by strong deviations from equilibrium, at gas temperatures $kT \ll E_{10}$, in which considerable amounts of energy are stocked in vibrational levels, are of great practical interest for plasma chemistry, laser chemistry, laser physics and physics of the upper atmosphere. Several analytical works have investigated vibrational kinetics under such conditions by considering one-component systems of anharmonic oscillators in steady-state [31–37].

To describe such a regime, it is convenient to use the diffusion approximation, i.e., assuming that f_n changes smoothly as a function of the level quantum number n. This is mathematically expressed as

$$\frac{f_{n+1} - f_n}{f_n} = \frac{\Delta f_n}{f_n \Delta n} = \frac{d(\ln f_n)}{dn} \ll 1 . \tag{3.86}$$

The flux Π_n of vibrational quanta at an arbitrary level n caused by single-quantum V-V exchanges can be expressed as [34]

$$\Pi_n = \sum_{i=1}^{n+i-1} \sum_{m=n} Q_{m,m+1}^{m-i+1,m-i}$$

$$\times \left[f_m f_{m-i+1} - f_{m+1} f_{m-i} \exp\left(-\frac{2\Delta E i}{kT}\right) \right] . \tag{3.87}$$

First, let us consider the steady-state case, when the gas temperature is such that

$$\frac{2\Delta E}{kT} \ll 1 . \tag{3.88}$$

For $T \geq 200$ K, condition (3.88) is satisfied for all molecules except H_2 and D_2. Inserting the probabilities (3.69) in (3.87) and expanding the functions

$\exp[-2\Delta E(i-n)/kT]$ and f_i in a power series of $(i-n)$, after replacing the sums with integrals and integration one obtains

$$\Pi_n = \frac{3Q_{10}}{\delta_{VV}^3}(n+1)^2 f_n^2 \left[\frac{2\Delta E}{kT} - \frac{d^2 \ln f_n}{dn^2}\right]. \tag{3.89}$$

The derivation of the above equation assumes that the major contribution to the integral is given by the neighborhood of n (i.e., resonant and near-resonant V-V exchanges). Therefore, the situation considered is called the resonant V-V relaxation regime.

It follows from (3.89) that f_n is a Treanor distribution when $\Pi_n = 0$. However, under non-equilibrium conditions Π_n always differs from zero due to the dissipation of vibrational quanta (vibrational energy) at upper vibrational levels (e.g., through V-T processes or dissociation). In many cases, it may be assumed to a good accuracy that there is a group of levels for which $\Pi_n = $ const. Analysis of solutions of (3.89) in the presence of this flux shows that the Treanor distribution (3.76) is valid only for levels $n \leq n_0$ (with n_0 given by (3.77)). The ascending branch of the Treanor distribution is not stable and the actual distribution instead has the form of a gently sloping plateau for $n > n_0$ [31–34]. In this region

$$\frac{2\Delta E}{kT} \gg \frac{d^2 \ln f_n}{dn^2} \tag{3.90}$$

and it follows from (3.89) that

$$f_n \simeq \frac{1}{n+1}\sqrt{\frac{kT}{6\Delta E}\frac{\Pi}{Q_{10}}\delta_{VV}^3} \equiv \frac{\Gamma}{n+1}. \tag{3.91}$$

This distribution is valid up to some level n_1 which is to be determined from the analysis of vibrational quantum dissipation by V-T processes at the upper levels, when

$$E_{10} - 2\Delta n \gg kT. \tag{3.92}$$

Such an analysis was performed [32,33] for steady-state conditions assuming single-quantum V-V and V-T processes only. In this case, taking into account (3.50), (3.86), and (3.92), the system (3.66) has the form

$$\frac{df_n}{dt} = 0 = J_n^{VV} - J_{n+1}^{VV} + P_{n+1,n}f_{n+1} - P_{n,n-1}f_n, \tag{3.93}$$

where J_n^{VV} is the molecular flux due to V-V processes in the space of vibrational quantum numbers n:

$$J_{n+1}^{VV} = \sum_i Q_{n+1,n}^{i,i+1}\left\{f_n f_{i+1}\exp\left[\frac{2\Delta E(n-i)}{kT}\right] - f_{n+1}f_i\right\}. \tag{3.94}$$

Since the probabilities $Q_{n+1,n}^{i,i+1}$ have a sharp maximum for $i = n$ and f_n changes smoothly, the main contribution to the sum in (3.94) arises from the vicinity of n, i.e., from resonant or near-resonant V-V processes. By performing the same mathematical operations as in obtaining (3.89), the flux J_{n+1}^{VV} can be rewritten as

$$J_{n+1}^{\text{VV}} = -\frac{3Q_{10}}{\delta_{\text{VV}}^3}\frac{\text{d}}{\text{d}n}\left[(n+1)^2 f_n^2\left(\frac{2\Delta E}{kT} - \frac{\text{d}^2 \ln f_n}{\text{d}n^2}\right)\right]. \qquad (3.95)$$

Inserting (3.68) and (3.95) into (3.93), and summing the equations so obtained from 0 to n yields for f_n [32,33]:

$$\frac{3Q_{10}}{\delta_{\text{VV}}^3}\frac{\text{d}}{\text{d}n}\left[(n+1)^2 f_n^2\left(\frac{2\Delta E}{kT} - \frac{\text{d}^2 \ln f_n}{\text{d}n^2}\right)\right]$$
$$+ P_{10} e^{\delta_{\text{VT}} n}(n+1)f_n = 0. \qquad (3.96)$$

Equation (3.96) has the following approximate analytical solution:

$$f_n \approx \begin{cases} f_n^{\text{Tr}} \exp[-\frac{1}{2}(\frac{n}{n_0})^2] & \text{for } n \leq n_0 \qquad (3.97\text{a}) \\ \frac{\Gamma}{n+1} - \frac{P_{10}}{Q_{10}}\frac{kT}{12\Delta E}\frac{\delta_{\text{VV}}^3}{\delta_{\text{VT}}}\frac{\exp(\delta_{\text{VT}} n)}{(n+1)} & \text{for } n_0 < n < n_1 \qquad (3.97\text{b}) \end{cases}$$

where f_n^{Tr} and n_0 are given by (3.76), (3.77) and the constant Γ is determined by merging (3.97a) and (3.97b) at point n_0.

The level n_1 specifies the plateau length of the distribution function. For $n > n_1$, the distribution should tend towards a Boltzmann distribution at the gas temperature T. Therefore, n_1 can be estimated from condition $\frac{1}{f_n}\text{d}f_n/\text{d}n\big|_{n_1} = \frac{1}{f_n^B}\text{d}f_n^B/\text{d}n\big|_{n_1}$, where f_n^B is the Boltzmann distribution at the gas temperature T. This gives the following equation for n_1:

$$P[1 + B(n_1)] \exp(\delta_{\text{VT}} n_1)$$
$$= (n_1 + 1)\exp\left(-\frac{\Delta E}{kT}n_0^2 - \frac{1}{2}\right) + P\exp(\delta_{\text{VT}} n_0), \qquad (3.98)$$

where

$$P = \frac{kT}{12\Delta E}\frac{\delta_{\text{VV}}^3}{\delta_{\text{VT}}}\frac{P_{10}}{Q_{10}}, \qquad (3.99)$$

$$B(n_1) = \delta_{\text{VT}}\left\{\frac{n_1[E_{10} - (n_1 - 1)\Delta E]}{kT} - \frac{1}{n_1 + 1}\right\}^{-1} \qquad (3.100)$$

and n_0 is given by (3.77).

Equation (3.98) has a simple solution for the typical case $B(n_1) \ll 1$, namely

$$n_1 \approx \frac{1}{\delta_{\text{VT}}} \ln\left[\frac{n_0 + 1}{P}\exp\left(-\frac{\Delta E}{kT}n_0^2 - \frac{1}{2}\right) + \exp(\delta_{\text{VT}} n_0)\right]. \qquad (3.101)$$

Note that the regime of resonant V-V relaxation with the distribution function (3.97) can be realized in practice only if

$$\frac{T_1}{T} > \frac{E_{10}}{2\Delta E(n_{\min} - 0.5)}, \qquad (3.102)$$

where

$$n_{\min} = \min(n_2, n^*), \qquad (3.103)$$

with n^* determined by (3.84) and n_2 standing for the solution of the following equation:

$$\frac{8}{\delta_{VV}^3} \frac{\Delta E}{kT}(n_2 + 1)\left[1 - \exp\left(-\frac{2\Delta E}{kT}n_2 + \frac{\Delta E}{kT} + \delta_{VV}\right)\right]^2$$

$$\approx \exp\left[\frac{\Delta E}{kT}n_2^2 - \left(\delta_{VV} + \frac{2\Delta E}{kT}\right)n_2 + \frac{\Delta E}{kT} + \frac{1}{2}\right]. \qquad (3.104)$$

To estimate n_2, an approximate solution of (3.104) can be used, which yields

$$n_2 \approx 1 + \frac{\delta_{VV}}{2}\frac{kT}{\Delta E}$$

$$+ \left\{\left(1 + \frac{\delta_{VV}}{2}\frac{kT}{\Delta E}\right)^2 + \left[\ln\left(\frac{8}{\delta_{VV}^2}\right) - \frac{1}{2}\right]\frac{kT}{\Delta E}\right\}^{1/2}. \qquad (3.105)$$

The influence on the distribution function of two-quantum V-V exchanges under highly non-equilibrium conditions was investigated in [36]. It was found that two-quantum exchanges reduce the plateau length of the distribution from $n_1 - n_0$ to $n_r - n_0$, where $n_r = E_{10}/4\Delta E$. In the region $n < n_r$ the plateau retains the same form, but its level (i.e., the constant Γ in (3.97)) increases by a factor of $\sqrt{2}$.

Let us turn our attention to important macroscopic parameters such as the average number of non-equilibrium vibrational quanta, α, and the "vibrational" temperature T_1. Under strongly non-equilibrium conditions the relationship between α and T_1 is obtained from the trivial equation $\alpha = \sum_n n f_n$ upon substitution of the distribution (3.97). This yields:

$$\alpha \approx [\exp(E_{10}/kT_1) - 1]^{-1} + f_{n_0}^{\text{Tr}} e^{-1/2}(n_0 + 1)(n_1 - n_0). \qquad (3.106)$$

Herein, the second term on the right-hand-side accounts for the quanta stored in the plateau. Expression (3.106) can also be employed in practice under moderate excitation conditions. In this case, the second term is always less than the first, and the stock of vibrational quanta is close to that of an harmonic oscillator.

In steady-state, α can be found assuming that the excitation rate of vibrations due to external sources, q, and the dissipation rate of vibrational quanta, $(d\alpha/dt)^{\text{anh}}$, balance each other, that is

$$q + \left(\frac{d\alpha}{dt}\right)^{\text{anh}} = 0 \qquad (3.107)$$

For strongly non-equilibrium conditions, equations (3.68), (3.69), (3.96) and (3.97) yield [33]:

$$q = -\left(\frac{d\alpha}{dt}\right)^{\text{anh}} \approx \frac{\alpha - \alpha_0}{\tau_{\text{VT}}^{\text{anh}}} + \frac{2.2 Q_{10}}{\delta_{\text{VV}}^3} \frac{\Delta E}{kT}(n_0+1)^2 (f_{n_0}^{\text{Tr}})^2 \,, \qquad (3.108)$$

where $\tau_{\text{VT}}^{\text{anh}}$ is determined from (3.75) and α_0 is the equilibrium value of α. The first term on the right-hand-side of (3.108) describes relaxation under small deviations of α from equilibrium. Far from equilibrium, however, the second term which accounts for the flux caused by V-V processes at level n_0 dominates. The nature of the relaxation changes since the dissipation of quanta is now determined by the V-V exchange probability, rather than by the V-T one. This has a simple physical explanation. In fact, V-T dissipation actually occurs at the upper levels $n > n_0$, but the flux of vibrational quanta towards these levels is determined by the resonant V-V process. This flux is limited, as in a bottle-neck, by the "narrow region" of minimum level populations, i.e., the region around the minimum of the Treanor distribution at $n = n_0$.

Though the second term on the right-hand-side of (3.108) describes strongly non-equilibrium conditions, it can be retained for "moderate" excitation conditions since in this case it is always much less than the first term. This means that (3.108) can be used over a wide range of parameters T_1 and T. It constitutes an extension to anharmonic oscillators of the well-known Landau–Teller formula describing the relaxation of α for a harmonic oscillator (see (3.35) and (3.37)).

The above analysis of relaxation for strongly non-equilibrium conditions was based on the probabilities (3.68), (3.69). It may be necessary in some circumstances to take into account the correction factors (3.71) for the probabilities to describe the processes more accurately. In this case, appropriate corrections should be made in (3.89), (3.91), (3.95)–(3.98), (3.100), (3.101) and (3.104).

3.3.5 Time-Dependent Relaxation at High Excitation

The steady-state regime of vibrational relaxation was examined in Sects. 3.3.2–3.3.4. Here, the unsteady regime will be investigated, following [38,39], under strongly non-equilibrium conditions. Our interest will be limited to the study of the distribution function f_n in the plateau region, where V-T relaxation can be neglected. It will be assumed that there are no chemical conversions of molecules, i.e., the term $\frac{1}{N}\frac{dN}{dt} f_n$ in (3.66) is equal to zero. Then, the evolution of f_n is described by the system of equations

$$\frac{\partial f_n}{\partial t} = -[J_{n+1}^{\text{VV}} - J_n^{\text{VV}}] \,, \qquad (3.109)$$

where the molecular flux J_{n+1}^{VV} due to V-V exchange in the space of vibrational numbers is given by (3.95).

First, let us replace the variable t with τ defined as

$$\tau = \int_{t_0}^{t} \frac{6\Delta E}{\delta_{VV}^3 kT} Q_{10} dt', \qquad (3.110)$$

where t_0 is an instant of time to be specified according to the particular conditions under examination.

If the gas temperature and density are constant, then $(6\Delta E/\delta_{VV}^3 kT)Q_{10}$ is time-independent and

$$\tau = \frac{6\Delta E}{\delta_{VV}^3 kT} Q_{10}(t - t_0). \qquad (3.111)$$

By introducing the function $F_n = (n+1)f_n$, one easily obtains from (3.95), (3.109), and (3.110) the following equation for F_n:

$$\frac{\partial F_n}{\partial \tau} = (n+1)\frac{\partial^2 F_n^2}{\partial n^2}. \qquad (3.112)$$

Solutions of (3.112) for various vibrational excitation regimes were presented in [38]. We discuss below some typical cases.

(a) Consider a constant excitation which is initiated at t_0. In this case,

$$f_n = \frac{\Gamma}{n+1} - \frac{1}{2\tau}. \qquad (3.113)$$

It can be seen that (3.113) goes over to the steady-state solution (3.91) as $t \to \infty$ (i.e., $\tau \to \infty$).

(b) Consider that the excitation which maintains a steady-state distribution (3.91) is switched off at time t_0. Then, for t such that $\tau \gg (n+1)/\Gamma$, we have

$$f_n \simeq \left[\frac{\Gamma}{(n+1)\tau}\right]^{1/2}. \qquad (3.114)$$

(c) Consider pulsed excitation of the lower vibrational levels over a short time interval. This results in a non-equilibrium stock of vibrational quanta α in these levels. When the excitation is switched off, a redistribution of vibrational level populations take place with $\alpha = $ const through the spreading of vibrational excitation over all vibrational levels by V-V exchanges. The populations of the upper vibrational states therefore begin to increase. We have for $f_n(t)$:

$$f_n \simeq \frac{1}{2\tau}\left[\frac{(12\alpha\tau)^{1/4}}{(n+1)^{1/2}} - 1\right]. \qquad (3.115)$$

Expression (3.115) describes the propagation of an excitation wave towards large n and its damping due to redistribution of the initially available quanta α among many vibrational levels. The maximum of $f_n(t)$ for the level n occurs at the time

$$\tau_{\max} = \left(\frac{4}{3}\right)^4 \frac{(n+1)^2}{12\alpha}. \qquad (3.116)$$

Note that equations (3.113) and (3.115) for, respectively, continuous and pulsed excitation are valid for times greater than t_0 (see (3.111)). This time t_0 however is not determined within the framework of the theory described in [38]. This theory also does not determine the number of quanta α_1 remaining on low levels and determining the last stage of evolution of the distribution when (3.115) no longer holds. The values t_0 and α_1 were obtained in [39] by comparing exact numerical calculations of f_n with expressions (3.111), (3.113) and (3.115). Under the continuous excitation conditions (**a**) it was found that

$$t_0 \simeq \frac{1}{W} \left[\exp\left(\frac{E_{10}}{kT_1}\right) - 1 \right]^{-1}, \qquad (3.117)$$

where W is the excitation probability (in s^{-1}). Under the pulsed excitation conditions (**c**), the remainder of quanta α_1 at $n < n_0$ were found to be

$$\alpha_1 \simeq \left[\exp\left(4.6\sqrt{\frac{\Delta E}{kT}}\right) - 1 \right]^{-1}, \qquad (3.118)$$

which corresponds to a vibrational "temperature" and a vibrational number for the minimum of the Treanor distribution, respectively, given by

$$\frac{kT_1}{E_{10}} \simeq \left(4.6\sqrt{\frac{\Delta E}{kT}}\right)^{-1}, \qquad (3.119)$$

$$n_0 \simeq 2.3\sqrt{\frac{kT}{\Delta E}}. \qquad (3.120)$$

The time t_0 in the case (**c**) can be evaluated from

$$t_0 \simeq \frac{kT\delta_{VV}^3 n_0}{12\Delta E Q_{10}\alpha_1}, \qquad (3.121)$$

where α_1 and n_0 are determined by (3.118) and (3.120).

Thus, expression (3.115) is applicable when $t > t_0$, $n > n_0$, and a number of quanta $\alpha > \alpha_1$ are produced by an excitation pulse.

3.4 Vibrational Kinetics in the Presence of Chemical Reactions of Vibrationally Excited Molecules

As has been mentioned, the vibrational kinetics of anharmonic oscillators is of great importance for plasma and laser chemistry. Selective "heating-up" of molecular vibrations by electric currents or radiation can appreciably increase the rate of chemical reactions involving vibrationally excited molecules as reactants. Such an increase can be obtained if molecules accumulate a vibrational energy exceeding the reaction activation energy. This involves in general excitation up to high vibrational levels, where vibrational anharmonicity must be taken into consideration. Thus, to calculate reaction rate coefficients under selective vibrational excitation, the kinetic analysis must consider anharmonic oscillators.

A chemical reaction involving vibrationally excited molecules constitutes a sink for such molecules. Such a sink can be taken into account in (3.40) or (3.66) by adding a term $-ZL_n N_n$ or $-ZL_n f_n$, respectively, where L_n is the probability (per collision) of the reaction for a molecule in level n. The effective probability L_{dis} of the reaction is defined as

$$L_{\text{dis}} = \sum_n L_n f_n \ . \tag{3.122}$$

The aim of the present section is to find L_{dis} explicitly for certain reactions and various regimes of vibrational non-equilibrium.

If the rates $L_n f_n$ are smaller than those for V-V or V-T processes, the chemical reaction has no influence on the vibrational distribution function f_n. However, in many cases, the probabilities L_n are so large that the chemical reaction distorts the distribution f_n. These cases are most interesting since they provide us with maximum non-equilibrium values of the effective reaction probability L_{dis}. A typical example of such a reaction is molecular dissociation from upper vibrational levels.

Below we will analyze a model accounting for dissociation from a single level, the upper bound vibrational level of the molecule, n_{b}. The dissociation model involves level-to-level single-quantum transitions to the bound level, from which transition to a continuous vibrational energy spectrum occurs, i.e., the molecule dissociates.

The effective reaction probability L_{dis} of such a process under non-equilibrium conditions can be calculated if the populations N_n of all vibrational states $n \leq n_{\text{b}}$ are determined. However, this is in general difficult to achieve. The analysis can be simplified considering a quasi-steady regime. In this case, the vibrational distribution f_n depends only on the mean number of non-equilibrium quanta, α, and the gas temperature T. The level populations N_n will vary with time as

$$N_n(t) = N(t) f_n; \qquad \sum_{n=0}^{n_{\text{b}}} f_n = 1 \ . \tag{3.123}$$

This means that we can write $\frac{1}{N} \frac{dN}{dt} f_n \equiv -Z L_{\text{dis}} f_n$ on the left-hand-side of (3.66). This system of equations can be solved analytically, for both moderate and strong deviations from equilibrium. The basic problem is to find the vibrational distribution function f_n corrected for the dissociation effect as follows: $f_n = f_n^0 (1 - \chi_n)$. For moderate deviations from equilibrium, the distribution f_n^0 is determined by expressions (3.79)–(3.81). Using (3.79), (3.80) one obtains [29,30]:

$$\chi_n = L_{\text{dis}} \psi_n \ , \tag{3.124}$$

3.4 Vibrational Kinetics in the Presence of Chemical Reactions

$$\psi_n = \sum_{m=1}^{n} \frac{1}{f_m^0(\beta Q_{m,m-1}^{01} + P_{m,m-1})} \,, \qquad (3.125)$$

where β is determined by (3.81). Equations (3.124), (3.125) can be used to calculate the relative population of the level n_b and the effective probability L_{dis} for dissociation. This yields:

$$L_{\text{dis}}(T, T_1) = \frac{L_{n_b} f_{n_b}^0}{1 + L_{n_b} f_{n_b}^0 \psi_{n_b}} \,, \qquad (3.126)$$

where L_{n_b} is the probability for dissociation from level n_b. Note that the non-equilibrium effective probability L_{dis} depends on T and T_1, since both $f_{n_b}^0$ and ψ_{n_b} depend on these temperatures. It is useful to express $L_{\text{dis}}(T, T_1)$ as

$$L_{\text{dis}}(T, T_1) = L_{\text{dis}}^0(T) \Phi(T, T_1) \,, \qquad (3.127)$$

where L_{dis}^0 is the effective probability for dissociation in equilibrium, that is, when $T_1 = T$, and $\Phi(T, T_1)$ is the non-equilibrium factor. For moderate deviations from equilibrium, one readily obtains from (3.79), (3.80), (3.124)–(3.127):

$$\Phi(T, T_1) \simeq \frac{1 - \exp(-E_{10}/kT_1)}{1 - \exp(-E_{10}/kT)} \exp\left[n^{**} E_{10} \left(\frac{1}{kT} - \frac{1}{kT_1}\right)\right] \,, \qquad (3.128)$$

where n^{**} is given by (3.85).

The kinetics of dissociation of diatomic molecules for moderate non-equilibrium conditions (for $T > T_1$) was also investigated in [4,8,40–47]. The non-equilibrium factors $\Phi(T, T_1)$ calculated for N_2 using different models of dissociation are presented in Fig. 3.2.

The parameter $\Phi(T, T_1)$ can be also estimated for the regime of resonant V-V relaxation with strong deviation from equilibrium. In this case [47]

$$\Phi(T, T_1) \simeq \frac{1 - \exp(-E_{10}/kT_1)}{1 - \exp(-E_{10}/kT)} \frac{B(n_1)}{n_1 + 1} P \exp\left\{\frac{E_{n_1}}{kT} + \delta_{\text{VT}} n_1\right\} \,, (3.129)$$

where $B(n_1)$, P and n_1 are determined by (3.98)–(3.101).

Expressions (3.128) and (3.129) provide the dependence of the non-equilibrium dissociation coefficient on the vibrational temperature T_1, i.e., on the number of non-equilibrium vibrational quanta, α, related to this temperature as in (3.106). The balance equation for α must now be analyzed to relate these quantities to the rate of vibrational excitation q due to the external source. When a molecule dissociates at level n_b, n_b vibrational quanta are lost from the vibrational reservoir. Therefore, this balance equation is simply obtained by adding the term $(n_b - \alpha) L_{\text{dis}}(T, T_1)$, which accounts for dissipation of α due to dissociation, to the right-hand-side of (3.108).

In the case of diatomic molecules, which according to the Morse oscillator model dissociate at the level $n_b = E_{10}/2\Delta E$, the term $(n_b - \alpha) L_{\text{dis}}(T, T_1)$ in the equation for α is often negligible if $T_1 > T$, and $T \sim 300$ K. Although this

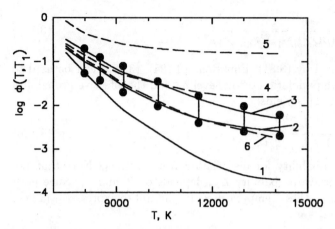

Fig. 3.2. Non–equilibrium factors $\Phi(T, T_1)$ for dissociation of N_2 molecules as a function of gas temperature T for vibrational temperature $T_1 = 3800$ K. Curves are for different kinetic models: 1 – from [45]; 2 – from [42]; 3 – from [43]; 4 – from [41]; 5 – from [40]; 6 – from [47]. Vertical lines with points are from experimental data [48]

seems to be the most favorable regime to achieve fast dissociation, it is not in fact the case. The reason is that the V-T relaxation rate of anharmonic oscillators is very large and their populations for the upper levels $n \geq n^*$ (which are governed by the gas temperature) therefore are very small.

However, for other reactions involving lower vibrational states, high reaction rates can be achieved at low (room) gas temperatures through selective "heating-up" of vibrations. Let us find the rate coefficient for a reaction involving a level $n_b < n^*$, n_1, where n^* and n_1 are given by (3.84) and (3.101). The probability of the reaction, L_{dis}, is actually determined in this case by the vibrational kinetics in a subsystem of levels (truncated anharmonic oscillator) for which the effects of V-T processes on the distribution function f_n can be ignored. First, let us consider a moderate deviation from equilibrium. In this case, in expression (3.126) for L_{dis} the quantity ψ_n is obtained from (3.125) with $P_{m,m-1}$ set equal to zero, and f_m^0 is the Treanor distribution (3.76). If the vibrational temperature T_1 of this distribution is such that $n^* > n_0 \equiv \frac{T}{T_1}\frac{E_1}{2\Delta E} + \frac{1}{2} > n_b$, then for $L_{n_b} \gg Q^{01}_{n_b, n_b-1}$ we have:

$$L_{\mathrm{dis}} \simeq \beta Q^{01}_{n_b, n_b-1} f^0_{n_b}$$

$$\simeq \frac{3}{2}\beta Q_{10} \exp\left[-\left(\delta_{\mathrm{VV}} + \frac{E_{10}}{kT_1}\right)n_b + \frac{\Delta E}{kT}n_b(n_b-1)\right]. \quad (3.130)$$

If the difference between T_1 and T is such that $n_0 < n_b$, then

$$L_{\mathrm{dis}} \simeq \frac{3}{2}\beta Q_{10} e^{-\delta_{\mathrm{VV}} n_0} f^{\mathrm{Tr}}_{n_0} \quad (n_0 < n_b < n^*). \quad (3.131)$$

3.4 Vibrational Kinetics in the Presence of Chemical Reactions

Expressions (3.130) and (3.131) show that for $n_b < n_0$ the reaction rate coefficient is specified by the level n_b, and for $n_b > n_0$ by level n_0. This is due to the existence of a minimum in the Treanor population distribution at level n_0. Note that, for $n_b > n_0$, the vibrational distribution function is strongly distorted in the range $n_0 < n < n_b$ by the molecular flux to the level n_b where the molecules react. Actually, for $n > n_0$ the function f_n takes the form of a gently sloping plateau.

The formation of a plateau in a distribution f_n is a general phenomenon, when there is a molecular flux towards large n in number space, and the quasi-equilibrium distribution (the Treanor distribution in our case) has a minimum. For example, it is known from the theory of condensation [49] that when there is a condensation cluster flux in the space of the number of molecules per cluster n, the cluster distribution function vs. n exhibits a gently sloping plateau for $n > n_0$, where n_0 is the number of molecules in the critical-size cluster.

The mutual influence of vibrational kinetics and chemical reactions of vibrationally excited molecules for strong deviations from equilibrium was analyzed in [50,51] using "truncated anharmonic oscillators". The analysis considered the case when a chemical reaction takes place for levels $n > n_0$ only, and the reaction rate coefficient increases either linearly or quadratically with n. The vibrational distribution functions for theses cases were found, and the effective rate coefficient and energetic efficiency of the reaction were also calculated.

It is clear that reaction efficiency is maximum when the reaction rate coefficient for level n_b is large and significantly exceeds the rate coefficient of V-T processes. In the limiting case of an "infinitely large" reaction rate coefficient for level n_b (the population of which, in this case, vanishes) the vibrational distribution function f_n in the range $n_0 < n < n_b$ is [51]

$$f_n \simeq \frac{\Gamma}{n+1}\sqrt{1 - \frac{n}{n_b}}. \tag{3.132}$$

The effective reaction probability L_{dis} can be estimated from the balance of vibrational quanta. The flux of these quanta, determined by expression (3.89) or by the second term on the right-hand-side of (3.108), compensates in this case for the loss of quanta by chemical reactions at levels $n \geq n_0$, and not by V-T processes as was the case before (when considering complete anharmonic oscillators without chemical reactions). Bearing this in mind, we have for L_{dis}:

$$L_{\text{dis}} \simeq \frac{1}{n_b}\frac{2.2 Q_{1,0}}{\delta_{\text{VV}}^3}\frac{\Delta E}{kT}(n_0+1)^2 (f_{n_0}^{\text{Tr}})^2 \quad (n_0 < n_b). \tag{3.133}$$

Comparison of (3.131) and (3.133) shows that, when $n_b > n_0$, for both moderate and strong deviations from equilibrium, the reaction rate is determined by the minimum population in the system at level n_0. However, due to the resonant nature of V-V processes, for a strongly non-equilibrium regime the reaction rate is proportional to $(f_{n_0}^{\mathrm{Tr}})^2$ (see (3.133)), while it is proportional to $f_{n_0}^{\mathrm{Tr}}$ for moderate deviations from equilibrium (see (3.131)).

Expressions (3.130), (3.131), (3.133) for L_{dis} have an important peculiarity. This is that L_{dis} increases with decreasing gas temperature T, unlike the ordinary Arrhenius law. This is due to the fact that under the non-equilibrium conditions considered, provided the Treanor distribution is valid, the overpopulation of the upper levels increases with decreasing T. This behaviour allows L_{dis} to increase either by "heating" the vibrations or by gas cooling.

It is interesting to note that under the dominant dissipation of vibrational energy by a chemical reaction, the problem of determining L_{dis} as a function of the excitation rate of vibrations by the external source, q, is considerably simplified. The magnitude of L_{dis} in this case is practically independent of the adopted molecular model and can be found to within an accuracy of a factor ~ 3 from the vibrational energy balance:

$$ZL_{\mathrm{dis}} \simeq \frac{q}{(n_{\mathrm{b}} - \alpha)} \,. \tag{3.134}$$

Numerical calculations of vibrational distribution functions in atmospheric gases were performed in many works (for example, [52–72] for N_2 and [73–75] for O_2). Both unsteady and steady regimes were examined. Numerous problems were investigated: comparisons with analytical models, with measured vibrational populations, and with measured gas heating rates; dissociation rate calculations; influence of vibrational excitation on the electron energy distribution function in plasmas; mechanisms of associative ionization involving vibrationally excited molecules; non-equilibrium vibrational distributions in boundary layers close to the surface of a supersonic body moving through a gas; etc. Examples of numerical calculations of vibrational distribution functions of N_2 in nitrogen discharges and post-discharges are shown in Figs. 3.3 and 3.4. These figures show strongly non-equilibrium distributions characterized by the formation of a gently sloping plateau at intermediate vibrational levels. Similar results have been reported for the CO system pumped by electrons [76] or infrared photons [77,78]. For this system, good agreement between theory and experiments has generally been found.

The above analysis of the vibrational kinetics of anharmonic oscillators was concerned with a one-component system, or with a diatomic gas partially diluted in an inert gas. However, in practice selective "heating" of vibrations is often carried out in molecular gas mixtures including polyatomic molecules. An analytical investigation of the vibrational distribution function in a binary mixture of diatomic gases, under highly non-equilibrium conditions, was performed in [79,80]. An investigation of the vibrational relaxation of polyatomic anharmonic molecules was made in [81]. These problems were also discussed in [7] and reviewed in [81].

Note that the analysis of vibrational kinetics in binary mixtures of isotopic molecular gases is of particular importance for isotope separation by non-equilibrium chemical reactions involving vibrationally excited molecules

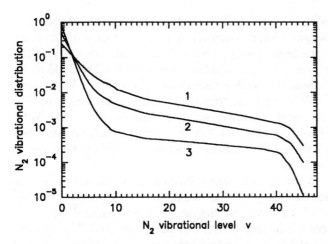

Fig. 3.3. $N_2(v)$ stationary vibrational distribution function for $T = 400$ K and the following values of T_1 (in K): 6000 (1), 4000 (2) and 3000 (3) [65]

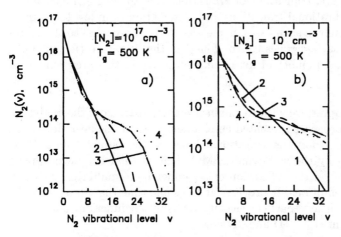

Fig. 3.4. $N_2(v)$ vibrational distribution function in nitrogen discharge with electron density $N_e = 10^{11}$ cm^{-3} (a); 10^{12} cm^{-3} (b) for a residence time $t = 10^{-3}$ s (curves 1) and several times Δt in post-discharge: $\Delta t = 10^{-3}$ s (curves 2); 3.10^{-3} s (curves 3); 5.10^{-3} s (curves 4) [55]

(see Chap. 13). The effect of anharmonicity on the isotope separation coefficient by these processes has been studied in [79,82–88]. The use of polyatomic molecules is attractive for carrying out effective collision non-equilibrium dissociation. The reason is that molecular bonds can be often broken in this case by predissociation at relatively low levels of a certain vibrational mode. Dissociation therefore occurs within the subsystem of "truncated anharmonic oscillators", whose populations are unaffected by V-T processes, which enables effective dissociation at low gas temperatures.

It is difficult to make rigorous calculations of non-equilibrium dissociation of polyatomic molecules since interactions of different vibrational modes must be considered due to the high density of vibrational levels over part of the energy range. In order to simplify the problem, only a single vibrational mode is usually considered, assuming weak interactions with others. The analysis reduces, in this case, to the study of "truncated anharmonic oscillators" of diatomic molecules, and the dissociation rate coefficient can be calculated by expressions (3.130), (3.131), (3.133). The kinetics of polyatomic vibrationally excited molecules in plasmas was discussed in detail in reviews [89,90] and in a book [9].

In conclusion, analytical approximate solutions of the master equations for vibrational level populations assuming single-quantum vibrational transitions have been discussed in the present chapter. The problem of dissociation from a single upper bound vibrational level was also considered. The approximations used are realistic for many cases but may be unsuitable for very high gas temperature regimes (for example, behind strong shock wave fronts) and for non-adiabatic transitions or transitions due to chemical exchanges. Multi-quantum vibrational transitions and dissociation from low levels must be taken into account in such cases. Some analytical approaches to the vibrational kinetics in such regimes can be found in the literature (for example, [22,23,45,46]), but a number of interesting problems in this area remain to be solved.

To end this chapter, we wish to emphasize that, apart from the analytical and numerical approaches presented here, a new method has been developed in recent years to study the vibrational kinetics: the *Direct Monte Carlo Simulation* known by the acronyme DSMC. This method has been applied to study, for example, the relaxation of N_2 under non–equilibrium [91] and equilibrium [92] conditions. It provides at the same time the vibrational distribution and the heavy–particle velocity distribution function, and it can easily be implemented for 1D and 2D flows.

References

1. Herzfeld K.F. and Litovitz T.A. (1959) *Absorption and Dispersion of Ultrasonic Waves*, Academic Press, New York
2. Cottrell T.L. and McCoubrey J.C. (1961) *Molecular Energy Transfer in Gases*, Butterworths, London
3. Clark J.F. and McChesney M. (1964) *The Dynamics of Real Gases*, Butterworths, London
4. Stupochenko E.V., Losev S.A. and Osipov A.I. (1967) *Relaxation Processes in Shock Waves*, Springer, Berlin Heidelberg
5. Zeldovich Ya.B. and Raizer Yu.P. (1967) *Physics of Shock Waves and High-Temperature Hydrodynamic Phenomena*, Academic Press, New York
6. Gordiets B.F. and Zhdanok S. (1986) in *Non-equilibrium Vibrational Kinetics*, ed. M.Capitelli, Springer, Berlin Heidelberg, p. 47

7. Gordiets B.F., Osipov A.I. and Shelepin L.A. (1988) *Kinetic Processes in Gases and Molecular Lasers*, Gordon and Breach, New York
8. Park C. (1990) *Non-equilibrium Hypersonic Aerodynamics*, John Wiley and Sons, New York
9. Rusanov V.D. and Fridman A.A. (1984) *Physics of Chemically Active Plasmas*, Nauka, Moscow (*in Russian*)
10. Nikitin E.E. (1974) *Theory of Elementary Atomic and Molecular Processes in Gases*, Clarendon Press, Oxford
11. Landau L. and Teller E. (1936) Phys. Z. Sow. **10**, 34
12. Montroll E.W. and Shuler K.E. (1957) J. Chem. Phys. **26**, 454
13. Osipov A.I. (1960) Rep. Acad. Sci. USSR **130**, 523 (*in Russian*)
14. Shuler K.E. (1960) J. Chem. Phys. **32**, 1692
15. Rankin C.C. and Light J.C. (1967) J. Chem. Phys. **46**, 1305
16. Osipov A.I. (1964) J. Appl. Mechan. Techn. Phys. N**1**, 41 (*in Russian*)
17. Biryukov A.S. and Gordiets B.F. (1972) J. Appl. Mechan. Techn. Phys. N**6**, 29 (*in Russian*)
18. Schwartz R.N., Slawsky I. and Herzfeld K.F. (1952) J. Chem. Phys. **20**, 1591
19. Keck J. and Carrier G. (1965) J. Chem. Phys. **43**, 2284
20. Naidis G.N. (1976) J. Appl. Mechan. Techn. Phys. **2**, 38 (*in Russian*)
21. Losev S.A., Shatalov O.P. and Yalovik M.S. (1970) Rep. Acad. Sci. USSR, **195**, 585 (*in Russian*)
22. Safarian M.N. and Pruchkina N.M. (1970) Theor. Exper. Chem. **6**, 306 (*in Russian*)
23. Safaryan M.N. and Skrebkov O.V. (1975) Phys. Combust. Explos. **4**, 614 (*in Russian*)
24. Treanor C.E., Rich I.W. and Rehm R.G. (1968) J. Chem. Phys. **48**, 1798
25. Kuznetsov N.M. (1971) Theor. Exper. Chem. **7**, 22 (*in Russian*)
26. Savva B.A. (1971) J. Appl. Spectroscopy **17**, 992 (*in Russian*)
27. Likalter A.A. (1975) J. Appl. Mechan. Techn. Phys. N**3**, 8 (*in Russian*)
28. Gordiets B.F., Osipov A.I. and Shelepin L.A. (1970) J. Exp. Theor. Phys. **59**, 615 (*in Russian*)
29. Gordiets B.F., Osipov A.I. and Shelepin L.A. (1971) J. Exp. Theor. Phys. **60**, 102 (*in Russian*)
30. Gordiets B.F., Osipov A.I. and Shelepin L.A. (1971) J. Exp. Theor. Phys. **61**, 562 (*in Russian*)
31. Brau S.A. (1972) Physics **58**, 533
32. Gordiets B.F. and Mamedov Sh.S. (1974) J. Appl. Mechan. Techn. Phys. N **3**, 13 (*in Russian*)
33. Gordiets B.F., Mamedov Sh.S. and Shelepin L.A. (1974) J. Exp. Theor. Phys. **67**, 1287 (*in Russian*)
34. Zhelezniak M.B., Likalter A.A. and Naidis G.B. (1976) J. Appl. Mechan. Techn. Phys. N **6**, 11 (*in Russian*)
35. Zhelezniak M.B. and Naidis G.B. (1976) Theor. Exper. Chem. **12**, 71 (*in Russian*)
36. Demianov A.B., Kochetov I.B., Napartovich A.P., Pevgov B.G. and Starostin A.N. (1980) Thermophys. High Temper. **18**, 918 (*in Russian*)
37. Eletsky A.B. and Zaretsky N.P. (1981) Rep. Acad. Sci. USSR, **260**, 591 (*in Russian*)
38. Zhdanok S.A., Napartovich A.P. and Starostin A.N. (1979) J. Exp. Theor. Phys. **76**, 130 (*in Russian*)
39. Demianov A.B., Zhdanok S.A., Kochetov I.B., Naportovich L.G., Pevgov V.G. and Starostin A.N. (1981) J. Appl. Mechan. Techn. Phys. N **3**, 5 (*in Russian*)
40. Hammerling P., Teare J.D. and Kivel B. (1959) Phys. Fluids **2**, 422

41. Marrone P.V. and Treanor C.E. (1963) Phys. Fluids **6**, 1215
42. Losev C.A. and Generalov N.A. (1961) Rep. Acad. Sci. USSR **141**, 69 (*in Russian*)
43. Kuznetsov N.M. (1982) *Kinetics of Monomolecular Reactions*, Nauka, Moscow (*in Russian*)
44. Kuznetsov N.M. and Sergievskaya A.L. (1994) Chemical Physics **13**, 15 (*in Russian*)
45. Macheret S.O., Fridman A.A., Adamovich I.V., Rich I.W. and Treanor C.E. (1994) AIAA Paper **94-1984**
46. Adamovich I.V., Macheret S.O., Rich J.W., Treanor C.E. and Fridman A.A. (1995) in *Molecular Physics and Hypersonic Flows*, ed. M.Capitelli, Kluwer, Dordrecht, NATO ASI Ser., Vol. 482, p. 85
47. Gordiets B.F. and Sergievskaia A.L. (1997) Chemical Physics **16**, 11 (*in Russian*)
48. Iyalovik M.S. and Losev S.A. (1972) Proceed. Inst. Mechan. Moscow State Univ. **18**, 4 (*in Russian*)
49. Gordiets B.F., Shlepin L.A. and Shmotkin Yu.S. (1982) Chemical Physics **N 2**, 1391 (*in Russian*)
50. Krasheninnikov S.I. and Nikiforov V.A. (1982) in *Plasma Chemistry*, ed. B.M. Smirnov, Energoizdat, Moscow, Vol. 9, p. 179 (*in Russian*)
51. Garrido H.D. and Zhdanok S.A. (1984) Chem. High Energy **18**, 539 (*in Russian*)
52. Bray K.N.C. (1968) J. Phys. B: At. Mol.Phys. **1**, 705
53. Polak L.S., Sergeev P.A. and Slovetsky D.I. (1973) Chem. High Energy, **7**, 38 (*in Russian*)
54. Capitelli M. and Dilonardo M. (1977) Chem. Phys. **24**, 427
55. Capitelli M. and Dilonardo M. (1978) Revue Phys. Appl. **13**, 115
56. Capitelli M., Dilonardo M. and Gorse C. (1981) Chem. Phys. **56**, 29
57. Capitelli M. and Molinari E. (1980) in *Topics in Current Chemistry*, ed. Boschke F.L., Springer, Berlin Heidelberg, Vol. 90, p. 59.
58. Capitelli M., Gorse C. and Ricard A. (1981) J. Physique. Lettres **42**, 185
59. Akishev J.S., Demianov A.B., Kochetov I.B. et al. (1982) *Thermophys. High Temper.* **20**, 818 (*in Russian*)
60. Cacciatore M., Capitelli M., Gorse C., Massabieaux B. and Ricard A. (1982) Lett. Nuovo Cimento **34**. 417
61. Massabieaux B., Plain A., Ricard A., Capitelli M. and Gorse C. (1983) J. Phys. B: At. Mol. Phys. **16**, 1863
62. Plain A., Gorse C., Cacciatore M., Capitelli M., Massabieaux B. and Ricard A. (1985) J. Phys. B: At. Mol. Phys. **18**, 843
63. Cacciatore M., Capitelli M., De Benedictis S., Dilonardo M. and Gorse C. (1986) in *Non-equilibrium Vibrational Kinetics*, ed. M. Capitelli, Springer-Verlag, Berlin Heidelberg, p. 5
64. Gorse C., Billing G.D., Cacciatore M., Capitelli M. and De Benedictis S. (1987) Chem. Phys. **111**, 351
65. Loureiro J. and Ferreira C.M. (1986) J. Phys. D: Appl. Phys. **19**, 17
66. Loureiro J., Ferreira C.M., Capitelli M., Gorse C. and Cacciatore M. (1990) J. Phys. D. Appl. Phys. **23**, 1371
67. Loureiro J. (1991) Chem. Phys. **157**, 157.
68. Doroshenko V.M., Kudryavtsev N.N., Novikov S.S. and Smetanin V.V. (1990) High Temp. **28**, 82
69. Armenise I., Capitelli M., Colonna G., Kudriavtsev N. and Smetanin V. (1995) Plasma Chem. Plasma Process. **15**, 501

70. Armenise I., Capitelli M., Celiberto R., Colonna G., Gorse C. and Lagana A. (1994) Chem. Phys. Lett. **227**, 157
71. Armenise I., Capitelli M., Celiberto R., Colonna G. and Gorse C. (1994) AIAA Paper **94-1987**
72. Colonna G. and Capitelli M. (1995) AIAA Paper **95-2071**
73. Capitelli M. and Dilonardo M. (1978) Chem. Phys. **30**, 95
74. Capitelli M., Dilonardo M. and Gorse C. (1980) *Plasma Physik*, **20**, 83
75. Pinheiro M.J.G. (1993) Ph.D. thesis, IST, Lisbon Technical University
76. De Benedictis S., Capitelli M., Cramarossa F., and Gorse C. (1987) Chem. Phys. **111**, 387; De Benedictis S., Cramarossa F. and Achasov O. V. (1988) Chem. Phys. **124**, 91
77. Adamovich I., Saupe S., Grassi M. J., Schultz O., Macheret S., and Rich J. W. (1993) Chem. Phys. **173**, 491; Lee W., Chidley M., Leiweke R., Adamovich I., and Lempert W. R. (1999) AIAA Paper **99-3273**; Plones E., Palm P., Chernuklo A. P., Adamovich I., and Rich J. W. (1999) AIAA Paper **99-3479**; Yano R., Contini V., Plones E., Palm P., Merriman S., Aithal S., Adamovich I., Lempert W. R., Subramanian V., and Rich J. W. (1999) AIAA Paper **99-3725**
78. Porshnev P. I., Wallaart H. L., Perrin M.-Y. and Martin J.-P. (1996) Chem. Phys. **213**, 111; (1996) ibid **222**, 289; Wallaart H. L., Piar B., Perrin M.-Y. and Martin J.-P. (1995) Chem. Phys. **196**, 149; Wallaart H. L., Piar B., Perrin M.-Y. and Martin J.-P. (1995) Chem. Phys. Lett. **246**, 587
79. Gordiets B.F. and Mamedov Sh.S. (1975) Quantum Electronics, **2**, 1992 (*in Russian*)
80. Demyanov L.B., Kochetov I.B., Napartovich A.P., Pevgov B.G. and Starostin A.N. (1982) Chem. High Energy, **16**, 161 (*in Russian*)
81. Likalter A.A. (1975) Quantum Electronics **2**, 2399 (*in Russian*); (1976) J. Appl. Mechan. Techn. Phys. N **4**, 3 (*in Russian*); (1975) Thermophys. High Temper. **17**, 960 (*in Russian*); Likalter A.A. and Naidis G.B. (1981) in *Plasma Chemistry*, ed. B.M. Smirnov, Energoizdat, Moscow Vol. 8, p. 156 (*in Russian*)
82. Dolinina V.I., Oraevsky A.N., Suchkov A.F., Urin V.M. and Shebeko G.M. (1978) J. Techn. Phys. **48**, 983 (*in Russian*)
83. Akulintsev V.M., Gorshunov N.M. and Neschimenko Ju.P. (1979) Chem. High Energy **13**, 441 (*in Russian*)
84. Akulintsev V.M., Gorshunov N.M. and Neschimenko Ju.P. (1979) Chem. High Energy **15**, 165 (*in Russian*)
85. Akulintsev V.M., Gorshunov N.M. and Neschimenko Ju.P. (1983) J. Appl. Mechan. Techn. Phys. **24**, 1 (*in Russian*)
86. Bayadze K.V. and Vetsko V.M. (1983) Chemical Physics N **9**, 1185 (*in Russian*)
87. Rich J.W. and Bergman R.C. (1986) in *Non-equilibrium Vibrational Kinetics*, ed. M. Capitelli, Springer, Berlin Heidelberg, p. 271
88. Adamovich I.V., Borodin V.I., Chernukho A.P., Rich J.W. and Zhdanok S.A. (1995) AIAA Paper **95-1988**
89. Rusanov V.D., Fridman A.A. and Sholin G.V. (1981) Adv. Phys. Sci. **134**, 185 (*in Russian*)
90. Rusanov V.D., Fridman A.A. and Sholin G.V. (1986) in *Non-equilibrium Vibrational Kinetics*, ed. M. Capitelli, Springer, Berlin Heidelberg, p. 295
91. Bruno D., Capitelli M., and Longo S. (1998) Chem. Phys. Lett. **289**, 141
92. Vijyakumar P., Sun Q. and Boyd I. (1999) Phys. Fluids **11**, 2117

4. Kinetics of Free Electrons

4.1 The Boltzmann Equation

Free electrons are the most important component of a plasma, the energy balance and electrical conductivity of which are determined solely by these particles. Collisions of free electrons with heavy particles are the principal mechanism for excitation of internal degrees of freedom (for example, electronic and vibrational levels), ionization and dissociation. The kinetics of free electrons is closely connected with the vibrational and chemical kinetics, and the optical and electrical characteristics of the plasma.

The electron velocity distribution function, $f(c)$, provides the most complete information about the state of plasma free electrons. In the case of local thermodynamic equilibrium this distribution is the well-known Maxwell distribution, which can be characterized by a single parameter – the electron temperature T_e. This distribution, expressed in terms of the electron kinetic energy ϵ, takes the form

$$f(\epsilon) = A \exp\left(-\frac{\epsilon}{kT_e}\right), \tag{4.1}$$

where k is the Boltzmann's constant. The factor A is determined from the normalization condition

$$\int_0^\infty \sqrt{\epsilon} f(\epsilon) d\epsilon = 1, \tag{4.2}$$

which takes into account that the density of free electron states is proportional to $\sqrt{\epsilon}$. This normalization yields $A = \frac{2}{\sqrt{\pi}}(kT_e)^{-3/2}$.

Under real conditions the distribution function is often anisotropic in velocity space due to the presence of a directed flux of electrons in the plasma. To describe such a situation it is necessary to expand the function $f(c)$ in spherical harmonics (or, if necessary, in associated Legendre functions) in velocity space. In many cases the anisotropy is relatively small and it suffices to keep only the two first terms of this expansion (the so-called Lorentz approximation). The first term, which depends only on the absolute value of electron velocity $|c|$ or its kinetic energy $\epsilon = m_e c^2/2$, is the isotropic (spherically symmetrical) part of the distribution, while the second is the anisotropy. The anisotropies caused by applied electric fields or spatial gradients can be easily expressed as a function of the isotropic component itself. For this reason, the subsequent discussion will focus only on this isotropic component, usually termed the electron energy distribution function (EEDF).

Note that the EEDF generally deviates considerably from a Maxwellian distribution when any appreciable electric field is present. There may be a difference of several orders of magnitude between the two distributions in the energy range where electrons are able to make inelastic collisions with gas particles, that is, to excite vibrational or electronic levels, dissociate or ionize gas particles. In such a situation, the EEDF must be determined from the Boltzmann equation. When the plasma is subject to an external electric field and a beam of high-energy particles penetrates the gas (one way of producing gas ionization), this equation can be written

$$\sqrt{\epsilon}\frac{\partial f}{\partial t} = -\frac{\partial J_F}{\partial \epsilon} - \frac{\partial J_{el}}{\partial \epsilon} - \frac{\partial J_{rot}}{\partial \epsilon} + In + Sup - \frac{\partial J_{ee}}{\partial \epsilon} + I_{ion} + \frac{q-p}{N_e}. \quad (4.3)$$

Here the first three terms on the right-hand side describe the effects on the EEDF of the external field, elastic collisions with heavy particles, and excitation and deactivation of molecular rotational levels, respectively. In these processes the relative change in energy of an electron between or due to collisions is small, that is, $\Delta\epsilon/\epsilon \ll 1$. Hence, the effects of these processes on the EEDF can be described in differential form, as the changes in the fluxes J_F, J_{el} and J_{rot} per unit energy interval, respectively. These fluxes can be written as

$$J_F = -D_F\sqrt{\epsilon}\frac{\partial f}{\partial \epsilon} \quad (4.4a)$$

$$J_{el} = -D_{el}\sqrt{\epsilon}\frac{\partial f}{\partial \epsilon} - \mu_{el}\sqrt{\epsilon}f \quad (4.4b)$$

$$J_{rot} = -D_{rot}\sqrt{\epsilon}\frac{\partial f}{\partial \epsilon} - \mu_{rot}\sqrt{\epsilon}f, \quad (4.4c)$$

where D and μ are, respectively, the diffusion and the mobility coefficients of electrons in energy space defined as follows:

$$D_F = \frac{1}{3}N^2\left(\frac{eE}{N}\right)^2\frac{c^2}{\nu_m}, \quad (4.5)$$

$$D_{el} = \delta_{el}\nu_{el}\epsilon kT; \quad \mu_{el} = \frac{D_{el}}{kT}, \quad (4.6)$$

$$D_{rot} = \Delta\epsilon_{rot}\nu_{rot}kT_{rot}; \quad \mu_{rot} = \frac{D_{rot}}{kT_{rot}}. \quad (4.7)$$

Here, N is the total concentration of heavy particles, E is the applied electric field, T and T_{rot} are, respectively, the translational and the rotational temperature, ν_{el} and ν_m are collision frequencies for momentum transfer in, respectively, elastic collisions and collisions of all kinds of electrons with heavy particles, δ_{el} is the mean relative electron energy loss in elastic collisions, and $\Delta\epsilon_{rot}$ is the mean electron energy loss in excitation of rotational levels. One has

$$\nu_{el} = Nc \sum_s \alpha_s \sigma_{sel} \; ;$$

$$\nu_m = \nu_{el} + \nu_{rot} + \sum_{s,n,k}(\nu_{snk} + \nu^*_{snk}) + \sum_{s,i} \nu_{si} \; ; \qquad (4.8)$$

$$\delta_{el} = \left(\sum_s \frac{2m_e}{M_s}\alpha_s \sigma_{sel}\right)\left(\sum_s \alpha_s \sigma_{sel}\right)^{-1} \qquad (4.9a)$$

$$\Delta\epsilon_{rot} = \left(\sum_s \Delta\epsilon_{srot}\alpha_s \sigma_{srot}\right)\left(\sum_s \alpha_s \sigma_{srot}\right)^{-1}, \qquad (4.9b)$$

where α_s and M_s are, respectively, the relative concentration and the mass of heavy particles of type "s" in a mixture, σ_{sel} and σ_{srot} are the cross-sections of elastic scattering and of rotational excitation for such particles, ν_{rot} is the total frequency of rotational excitation, ν_{snk} and ν^*_{snk} are the frequencies of inelastic collisions of the first and second kind, respectively (corresponding to excitation or deactivation transitions $n \leftrightarrow k$ between two electronic or vibrational states n and k), for component "s", and ν_{si} is the ionization frequency for this component.

It should be noted that Maxwell-Boltzmann distributions for translational and rotational degrees of freedom of heavy particles (with temperatures T and T_{rot}, respectively) have been assumed in (4.6) and (4.7). For simplicity, it can usually also be assumed that $T_{rot} = T$ essentially at all times due to the high rate of rotational-translational relaxation (see Chap. 2).

It is convenient to express all the frequencies defined above in terms of the corresponding cross-sections and particle concentrations. One has

$$\nu_{rot} = Nc \sum_s \alpha_s \sigma_{srot} \; ; \qquad (4.10)$$

$$\nu_{snk}(\epsilon) = N_n^{(s)} c(\epsilon)\sigma_{snk}(\epsilon); \quad \nu^*_{snk}(\epsilon) = N_k^{(s)} c(\epsilon)\sigma^*_{skn}(\epsilon) \; , \qquad (4.11)$$

where $N_n^{(s)}$, $N_k^{(s)}$ are the concentrations of "s"-particles on levels n and k, and σ_{snk}, σ^*_{skn} are, respectively, the cross-sections of the first and second kind transitions $n \leftrightarrow k$. As is well known, these cross-sections are related by the detailed balance formula

$$\epsilon g_k \sigma^*_{kn}(\epsilon) = (\epsilon + \epsilon^*_{kn}) g_n \sigma_{nk}(\epsilon + \epsilon^*_{kn}) \; , \qquad (4.12)$$

where g_k and g_n are the statistcal weights of levels "k" and "n", and ϵ^*_{kn} is the energy of transition $k \leftrightarrow n$. In the subsequent discussion it is assumed that $\epsilon^*_{kn} \equiv \epsilon^*_{nk}$ are positive values.

Note that in most cases $\nu_{el} \gg \nu_{rot}$, ν_{nk}, ν_{kn}^*. In such cases, the assumption $\nu_m \simeq \nu_{el}$ is therefore correct to a good approximation.

The terms In and Sup on the right-hand side of the Boltzmann equation (4.3) account for the effects of inelastic collisions of the first or second kind respectively, resulting in the excitation or deactivation of vibrational and electronic levels of heavy particles. In such collisions the fraction of energy $\Delta\epsilon$ lost or acquired by the electron can be considerable ($\Delta\epsilon/\epsilon \sim 1$). For this reason, the terms In and Sup are usually expressed in the form:

$$In = \sum_{s,n,k} [\nu_{snk}(\epsilon + \epsilon_{snk}^*)\sqrt{\epsilon + \epsilon_{snk}^*} f(\epsilon + \epsilon_{snk}^*) - \nu_{snk}(\epsilon)\sqrt{\epsilon}f(\epsilon)] ; \quad (4.13)$$

$$Sup = \sum_{s,k,n} [\nu_{skn}^*(\epsilon - \epsilon_{skn}^*)\sqrt{\epsilon - \epsilon_{skn}^*} f(\epsilon - \epsilon_{skn}^*) - \nu_{skn}^*(\epsilon)\sqrt{\epsilon}f(\epsilon)] . (4.14)$$

Here $\nu_{snk}(\epsilon')$, $\nu_{snk}^*(\epsilon')$ represent the inelastic collision frequencies of the first and second kind, respectively, determined according to (4.11) at energy ϵ'. The expressions (4.13), (4.14) have a clear physical sense: the first terms in the square brackets account for the scattering of electrons into a small energy range around ϵ upon inelastic collisions of the first or the second kind occurring in the energy ranges around ϵ plus or minus the threshold energies $\epsilon_{skn}^* > 0$, respectively; and the second terms account for the electrons scattered out from the neighborhood of energy ϵ by both types of collisions.

The term $\partial J_{ee}/\partial\epsilon$ in (4.3) takes into account the effects of electron–electron collisions. The flux in energy space due to such collisions, J_{ee}, is given by the following expressions:

$$J_{ee} = -2\left[\mu_{ee}\sqrt{\epsilon}f + D_{ee}\sqrt{\epsilon}\frac{\partial f}{\partial \epsilon}\right] ; \quad (4.15)$$

$$D_{ee} = \frac{2}{3}\nu_{ee}\epsilon\left[\int_0^\epsilon x^{3/2} f(x)dx + \epsilon^{3/2}\int_\epsilon^\infty f(x)dx\right] ; \quad (4.16)$$

$$\mu_{ee} = \nu_{ee}\epsilon\int_0^\epsilon x^{1/2} f(x)dx , \quad (4.17)$$

where

$$\nu_{ee} = 4\pi\left(\frac{e^2}{4\pi\epsilon_0 m_e}\right)^2 \frac{\ln\Lambda}{c^3} N_e \quad (4.18)$$

is the electron–electron collision frequency. Here, ϵ_0 is the permittivity of free space and $\ln\Lambda$ is the well-known Coulomb logarithm ($\Lambda = 12\pi N_e \lambda_D^3$, where $\lambda_D = (\epsilon_0 k T_e/N_e e^2)^{0.5}$ is the Debye length, T_e denoting the electron kinetic temperature).

The term I_{ion} in equation (4.3) describes the effects of electron impact ionization on the EEDF. This term can be written as:

4.1 The Boltzmann Equation

$$I_{\text{ion}} = \sum_{s,i} \left[\int_{\epsilon+\epsilon_{si}^*}^{\infty} \nu_{sii}(\epsilon_1, \epsilon)\sqrt{\epsilon_1}f(\epsilon_1)\mathrm{d}\epsilon_1 - \nu_{si}(\epsilon)\sqrt{\epsilon}f(\epsilon) \right]. \quad (4.19\mathrm{a})$$

The second term on the right-hand side of (4.19a) is identical to the second term in square brackets on the right-hand side of (4.13), i.e., it accounts for the energy loss of the primary (ionizing) electron as for an ordinary inelastic process of the first kind. The first term, in which ν_{sii} is the differential ionization frequency, takes into account the creation of a new electron with energy ϵ by ionization of a heavy particle "s" from level i by a primary electron with energy ϵ_1. If the primary electron has energy ϵ_1, the two electrons resulting from the ionization process will share an energy $\epsilon_1 - \epsilon_{si}^*$. It is customary to call the electron with lower energy secondary electron and the one with higher energy the scattered electron. Therefore, the energy spectrum of the secondary electrons will extend from 0 to $\frac{1}{2}(\epsilon_1 - \epsilon_{si}^*)$, and the spectrum of scattered ones will range from $\epsilon_1 - \epsilon_{si}^* - \frac{1}{2}(\epsilon_1 - \epsilon_{si}^*)$ to $\epsilon_1 - \epsilon_{si}^*$. It is thus convenient to rewrite equation (4.19a) as:

$$I_{\text{ion}} = \sum_{s,i} \left[\int_{2\epsilon+\epsilon_{si}^*}^{\infty} \nu_{sii}(\epsilon_1, \epsilon)\sqrt{\epsilon_1}f(\epsilon_1)\mathrm{d}\epsilon_1 \right. \quad (4.19\mathrm{b})$$
$$\left. + \int_{\epsilon+\epsilon_{si}^*}^{2\epsilon+\epsilon_{si}^*} \nu_{sii}(\epsilon_1 - \epsilon_{si}^* - \epsilon, \epsilon)\sqrt{\epsilon_1}f(\epsilon_1)\mathrm{d}\epsilon_1 - \nu_{si}(\epsilon)\sqrt{\epsilon}f(\epsilon) \right].$$

Here, the first integral describes the process whereby a secondary electron appears with energy ϵ, while the second integral describes the appearance of a scattered electron with energy ϵ.

The differential ionization frequency ν_{sii} (usually expressed in s^{-1}eV^{-1}) is defined as

$$\nu_{sii}(\epsilon_1, \epsilon_2) = N_i^{(s)} c(\epsilon_1) \sigma_{sii}(\epsilon_1, \epsilon_2), \quad (4.20)$$

where $N_i^{(s)}$ is the concentration of heavy particles of type "s" on level i, on which ionization occurs, $c(\epsilon_1)$ is the velocity of the primary electron with energy ϵ_1, and $\sigma_{sii}(\epsilon_1, \epsilon_2)$ is the differential ionization cross-section, characterizing the probability of an electron appearing (either a secondary or a scattered electron) with energy ϵ_2 by ionization. The integral ionization cross-section (usually termed simply the ionization cross-section), $\sigma_{si}(\epsilon_1)$, is related to the differential one as follows:

$$\sigma_{si}(\epsilon_1) = \int_0^{(\epsilon_1-\epsilon_{si}^*)/2} \sigma_{sii}(\epsilon_1, \epsilon_2)\mathrm{d}\epsilon_2$$
$$= \int_0^{(\epsilon_1-\epsilon_{si}^*)/2} \sigma_{sii}(\epsilon_1, \min(\epsilon_2, \epsilon_1 - \epsilon_{si}^* - \epsilon_2))\mathrm{d}\epsilon_2$$
$$= \int_{(\epsilon_1-\epsilon_{si}^*)/2}^{\epsilon_1-\epsilon_{si}^*} \sigma_{sii}(\epsilon_1, \epsilon_1 - \epsilon_{si}^* - \epsilon_2)\mathrm{d}\epsilon_2. \quad (4.21)$$

Finally, the last term on the right of the Boltzmann equation (4.3) describes, for the sake of completeness, the creation of electrons at rate q through the action of an external source and their disappearance at rate p from the bulk plasma via such processes as electron-ion recombination and electron attachment. Losses by diffusion towards the wall can also be accounted for by such a term as a first approximation.

From the previous description, it may be concluded that the electron Boltzmann equation is a complex nonlinear integral-differential equation. In practice, the application of this equation to real situations requires detailed and accurate knowledge of a considerable amount of data characterizing the interaction of electrons with plasma particles (primarily the cross-sections of the various electron scattering processes). Moreover, the concentrations of numerous excited vibrational and electronic levels of heavy particles, free electrons and dissociated atoms need to be known or self-consistently determined. In general, these concentrations depend strongly on the rate of excitation, ionization and dissociation of heavy particles by electron impact, i.e., on the EEDF itself. This means that a rigorous formulation of the problem requires, in general, a self-consistent analysis of the electron, rotational, vibrational, electronic level and chemical kinetics, along with an analysis of the thermal balance of the medium. In other words, it is necessary to solve the Boltzmann equation (4.3) simultaneously with the equations for the vibrational and electronic level populations, the concentrations of neutral and charged components, and the equations determining the translational and rotational temperatures of heavy particles. It is clear that determination of the EEDF in the context of such a general formulation is an extremely difficult and laborious task that obviously requires complex numerical codes and adequate computing facilities.

In practice, some simplifications are always needed to solve the problem. For example, it is sufficient to take into account only the most important processes for the conditions under examination. Let us discuss some circumstances in which simplifications can be made.

1. *Invariant chemical composition of the main components.* When the chemical composition of the gas remains essentially unchanged by the discharge (with respect to its original composition prior to the discharge) the analysis can be considerably simplified. In this case, the concentrations of other neutral particles produced by the discharge, such as products of dissociation and of chemical reactions, can be assumed so small that they do not affect the electron kinetics or the EEDF. These are determined solely by collisions of the electrons with the pre-existing major components. Conversely, the EEDF of course affects the chemical composition of the minor components and therefore needs to be determined.
2. *Weakly ionized plasma.* When the degree of ionization is sufficiently small, electron–electron collisions play no significant role, and thus the term $\partial J_{ee}/\partial \epsilon$, describing such interactions, can be neglected in the Boltz-

mann equation (4.3). This considerably simplifies the problem, for then this equation becomes linear. From (4.4)–(4.7), (4.13) and (4.15)–(4.18), electron–electron interactions can be neglected provided that

$$\nu_{ee} \ll \delta_{el}\nu_{el}; \quad \frac{\epsilon^*_{skn}}{\epsilon}\nu_{skn}.$$

Under typical conditions of non-equilibrium plasmas in atmospheric gases, with mean electron energies well above thermal energy, the above conditions are usually met for degrees of ionization in the range $N_e/N \leq 10^{-4}$. Note that, since the ion concentration $N_{ion} \simeq N_e$, the inequality $\frac{2m_e}{M_{ion}}\nu_{eion} \ll \nu_{ee}$ (ν_{eion} denoting the electron-ion collision frequency and M_{ion} the ion mass) is always satisfied. This means that electron-ion momentum transfer collisions can always be disregarded in the analysis of the Boltzmann equation.

For degrees of ionization in the range $N_e/N \geq 10^{-4}$ electron- electron collisions can significantly affect the shape of the EEDF, especially its high energy tail, although they do not change the average electron energy. Therefore, the term $\partial J_{ee}\partial \epsilon$ must then be kept in (4.3), with the degree of ionization treated as an independent parameter or the electron concentration determined from appropriate balance equations (for charged particles, discharge current, or deposited power).

3. *Superelastic collisions.* Superelastic collisions are taken into account in the term Sup in (4.3) and the terms containing the derivative $\partial f/\partial \epsilon$ in (4.4) for J_{el} and J_{rot}. Neglect of such collisions is justified whenever the populations of excited vibrational and electronic levels involved in the term Sup are small, or the value of $\left(\frac{1}{f}\frac{\partial f}{\partial \epsilon}\right)^{-1}$ is considerably greater than the translational and the rotational temperatures. However, in many circumstances superelastic collisions cannot be neglected and, indeed, they significantly affect the EEDF, especially by enhancing the high energy tail of the distribution. In order to account approximately for the effects of superelastic collisions, without solving the problem self-consistently, the relative populations of excited electronic levels and the vibrational, rotational and translational temperatures must be treated as independent parameters (for example, taken from experiment) when solving the Boltzmann equation.

4. *Specific types of plasma.* Plasmas are often sustained by either an electric field or an external source of ionization, such as ultra-violet radiation or an electron beam, but seldom by both processes simultaneously. In the former case, the term q/N_e in (4.3) is absent, while in the latter the electric field term can be ignored. Furthermore, for plasmas sustained by an electric field, equation (4.19) for I_{ion} can often be simplified by neglecting the creation of secondary electrons or assuming they are created with zero energy.

4.2 EEDF in Discharges Sustained by an Electric Field

Electron energy distribution functions in discharges sustained by an electric field have been examined in many theoretical and experimental works. Former investigations have been summarized in various reviews [1,2], while more recent ones have been described in various reviews [3,4] and monographs [5–13]. In the present section we will summarize only some general principles needed for a theoretical analysis of the EEDF in the atmospheric gases and present some basic results of numerical solution methods.

First of all, let us briefly discuss the validity of the two-term Lorentz approximation which was used to obtain (4.3). For this approximation to apply, the anisotropy of the velocity distribution must be small compared to the isotropic component, and thus the following inequality must hold [2,3]:

$$\left|\frac{\partial J_F}{\partial \epsilon}\right| \equiv \frac{1}{3}N^2\left(\frac{eE}{N}\right)^2 \left|\frac{\partial}{\partial \epsilon}\left(\frac{v^2}{\nu_m}\sqrt{\epsilon}\frac{\partial f}{\partial \epsilon}\right)\right| \ll \sqrt{\epsilon}\nu_m f . \qquad (4.22)$$

For discharges in atmospheric gases, this inequality is usually satisfied for reduced maintaining fields up to $E/N \simeq 3\times 10^{-14}$ Volt cm^2. It is worth noting that, under such conditions, equation (4.3) applies to both constant and time-varying applied fields. In general, in the latter case the EEDF will also change in time and the unsteady Boltzmann equation will have to be solved to derive the time-dependent EEDF. However, for high-frequency electric fields $E = E_0\cos(\omega t)$, the EEDF cannot follow the changes of the field intensity in time if the frequency of the field is sufficiently high. In this case, the EEDF will be stationary and as given by (4.3) for steady-state conditions but with E replaced with an effective root mean square field given by [4,5,14–16]:

$$E_{ef} = \frac{E_0}{\sqrt{2}}\frac{\nu_m}{\sqrt{\omega^2 + \nu_m^2(\epsilon)}} . \qquad (4.23)$$

This is the so-called "effective field approximation". It is valid when the characteristic relaxation time of the EEDF, τ_ϵ, is considerably longer than the period of the field. The characteristic time τ_ϵ is the reciprocal of the electron energy relaxation frequency, which is given by

$$\nu_\epsilon = \delta_{el}\nu_{el} + \frac{\Delta\epsilon_{rot}}{\epsilon}\nu_{rot} + \sum_{s,n,k}(\nu_{snk} + \nu^*_{skn}) + \sum_{s,i}\nu_{si} . \qquad (4.24)$$

The frequencies ν_m and ν_ϵ for electrons in N_2 at 1 Torr pressure are shown in Fig. 4.1. It can be seen from this Fig. that the condition $\tau_\epsilon \equiv 1/\nu_\epsilon \gg \omega^{-1}$ is essentially met in nitrogen, over the electron energy range of interest, provided that

$$\omega \gg 0.1\nu_m . \qquad (4.25)$$

A similar inequality is also required in the case of oxygen plasmas for the effective field approximation to be valid.

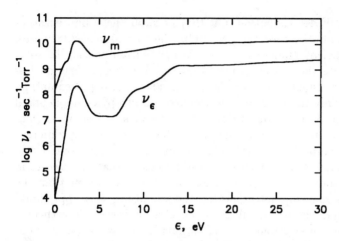

Fig. 4.1. Normalized electron collision frequencies for momentum transfer (ν_m) and energy relaxation (ν_ϵ) in N_2 as a function of the electron energy ϵ [16]

It should be noted that the effective field (4.23) usually varies with the electron energy through its dependence on the function $\nu_m(\epsilon)$. This is particularly true in the atmospheric gases N_2, O_2, in which ν_m varies significantly with ϵ (see Fig. 4.1, for nitrogen). Due to the dependence $E_{ef}(\epsilon)$, the EEDFs in HF and DC fields will generally differ substantially from each other. Further, in an HF field the EEDF will also change with field frequency, as can easily be understood from (4.23). A review of the kinetics of free electrons in HF discharges can be found in, for example, [4].

When the inequality (4.25) is reversed, the EEDF will instantaneously follow the field intensity. This means that the shape of the EEDF at any time is determined solely by the value of the field intensity at that instant of time. Therefore, this regime can be treated by considering the instantaneous values of the field intensity in (4.3) and assuming $\partial f/\partial t = 0$.

Analytical solutions of the kinetic equation (4.3) can easily be derived only in a few particular cases. One such a case is, for example, when only the first three terms on the right-hand side, given by (4.4)–(4.7), need to be retained. The steady-state solution is then

$$f(\epsilon) = f_0 \exp\left\{-\int_0^\epsilon \frac{\mu_{el} + \mu_{rot}}{D_{el} + D_{rot} + D_F} d\epsilon'\right\} . \tag{4.26}$$

This type of solution, in the absence of rotational excitation (i.e., without the terms μ_{rot} and D_{rot}) has long been known (see, e.g., [17]). The well-known Druyvesteyn distribution constitutes a particular case of (4.26). However, excitation of vibrational and electronic levels, and ionization most frequently need to be considered when solving (4.3). In this case, analytical solutions for the EEDF can only be derived using a number of crude assumptions and approximations (see, e.g., [17-22]. Such solutions provide physical insight into the problem of EEDF determination, but they generally have limited accuracy due to the assumptions used, especially in the case of molecular gases. For this reason, efficient numerical codes are now currently used to solve (4.3) and to calculate the EEDF in low temperature non-equilibrium plasmas. These codes are usually based on the so-called two-term approximation, that is considering only the isotropic and the first anisotropy components of the expansion of the velocity distribution function in spherical harmonics [23,24]. The validity of this approximation has been discussed by different authors [25–28] by comparing it with multiterm expansions and Monte Carlo simulations. The main result of these investigations is that the two-term approximation is valid when elastic collisions are more important than the inelastic ones. This may put into question the validity of this approximation in nitrogen due to the strong vibrational excitation losses. Nevertheless, superelastic collisions of electrons with vibrationally excited molecules contribute to decrease the effects of vibrational excitation on the distribution so that the two-term expansion can be considered a good approximation also in nitrogen plasmas under many circumstances.

Let us briefly present and discuss numerical calculations of the EEDF in discharges in the atmospheric gases N_2 and O_2, and in their mixtures with other gases. Such calculations have been performed, for example, in [16,29,30,31,33,40,42–78] for N_2, in [79–85] for O_2, in [55,86–93] for N_2-O_2 and N_2-O_2-O mixtures, and in [94–103] for mixtures of N_2 with other gases. In particular, the EEDF in the laser mixtures CO_2-N_2-He [94–97] and $CO-N_2$ [96], [103], and in the mixtures N_2-Ne [99–101], N_2-H_2 [102], and N_2-He [104] has been extensively investigated. In most cases, the solutions have been found under stationary conditions as a function of the reduced maintaining field E/N. In a number of works, solutions have also been found for applied high frequency fields [54,67,77,84,88,92] (see also the review [4]) using the effective field approximation, that is, for $\omega \gg \nu_\epsilon$ (see (4.23)). The time-dependent Boltzmann equation has also been solved in a number of works. For instance, in [16] the time evolution of the EEDF and related electron characteristics were calculated for a periodic field E of arbitrary frequency ω. The problem of the time-evolution of the EEDF when the applied field is either switched on or off has also been investigated by numerous authors [58–64,66,68–70,72,74,75,78,83].

In many of the above works, self-consistent sets of electron cross-sections have been used to solve the Boltzmann equation. Such sets have usually been derived from a fit of calculated electron transport coefficients (e.g., drift velocity, diffusion coefficients, Townsend ionization coefficient) to the corresponding experimental data obtained in drift tubes (see, e.g., [9] for an extensive discussion of the method). Using this method, the electron-molecule momentum transfer cross-sections and the electron inelastic cross-sections for excitation of rotational, vibrational and electronic levels have been determined for N_2 [42,43] and O_2 [79]. Such self-consistent cross-section sets enable calculations of relevant electron energy averaged properties, such as average energy, drift velocity, and the rate coefficients for excitation of vibrational and electronic levels and for ionization, as well as a detailed investigation of the discharge power balance as a function of the reduced maintaining field E/N. Some results of these calculations will be presented in Chap. 8.

Typical calculations of the EEDF in molecular and atomic nitrogen are shown in Fig. 4.2 as a function of E/N. In N_2, it can be seen that the EEDF decreases sharply in the energy range ~ 1.8 to ~ 3.5 eV, which is due to the excitation of the electronic ground state vibrational levels. The total cross-section for excitation of these levels first rises gently above the threshold, but suddenly rises sharply to form a narrow peak at an energy of about 2.5 eV. The excitation of these levels therefore constitutes a sharp barrier the electrons have to overcome in order to reach the higher energies for which excitation of electronic levels and ionization become possible. Of course, there is no such barrier in the case of atomic nitrogen and the EEDF falls off appreciably only for energies greater than the first electronic excitation threshold. In both cases, the high energy tail of the EEDF is considerably enhanced as E/N increases, because this increases the flux in energy space driven by the field. Figure 4.3 shows the time-evolution of the EEDF in N_2 when the applied field is suddenly increased. It can be seen that the EEDF reaches a steady state in the energy range 1.5 eV $\leq \epsilon \leq$ 2.5 eV in a time of about 10^{-9} s, whereas it needs a much longer time of about $\sim 10^{-8}$ s to reach such a condition at lower energies and in the tail. These times, which correspond to ~ 20 and ~ 200 electron-molecule collisions respectively, reflect the differences between the frequencies ν_m and ν_ϵ in the different energy ranges and the variation of ν_ϵ with electron energy shown in Fig. 4.1.

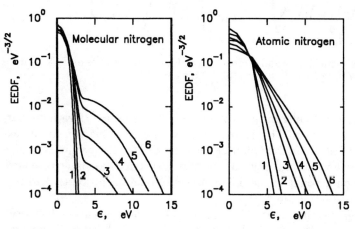

Fig. 4.2. EEDF in molecular and atomic nitrogen for different values of E/N. Curves 1–6 correspond to $E/N = 2;\ 3;\ 5;\ 6;\ 8;\ 10\ \times 10^{-16}$ Volt cm^2, respectively [58]

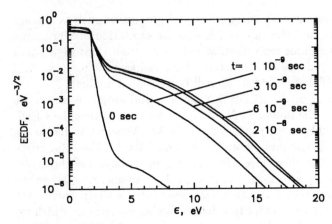

Fig. 4.3. Relaxation of the EEDF in an N_2 discharge when the reduced electric field E/N jumps instantaneously from 3×10^{-16} to 10×10^{-16} Volt cm^2, for $T = 500$ K and $p = 5$ Torr [60]

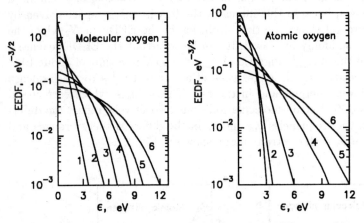

Fig. 4.4. EEDF in molecular and atomic oxygen for different values of E/N. Curves 1–6 are for $E/N = 2;\ 3;\ 5;\ 6;\ 8;\ 10 \times 10^{-16}$ Volt cm^2 [83]

The EEDF in molecular oxygen does not exhibit the same vibrational barrier effects as in nitrogen. This is due to the smallness of the vibrational excitation cross-sections for O_2. Qualitatively, the EEDFs have a similar shape in molecular and in atomic oxygen (see Fig. 4.4), but they differ of course quantitatively on account of the different sets of relevant processes and corresponding cross-sections in both cases.

Since the pioneering work by Nighan [44], much attention has been given to analyzing the effects of electron collisions with vibrationally excited molecules, in particular superelastic collisions, on the EEDF and various plasma properties. Such investigations have been carried out by numerous

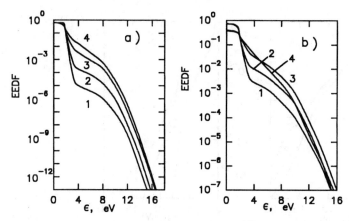

Fig. 4.5. Time evolution of the EEDF (eV$^{-3/2}$) in N_2 as the population of the N_2 vibrational levels evolves. Curves 1–4 correspond to the times 0; 6×10^{-4}; 5×10^{-3}; 5×10^{-2}; and 10^{-1} s, respectively, for $N_e = 10^{11}$ cm^{-3}; $T = 500$ K; $p = 5$ Torr; $E/N = 3 \times 10^{-16}$ (a); and 6×10^{-16} (b) Volt cm^2 [61]

authors for nitrogen [30,44,51,55,58–64,66,68,69,70–75,76–78], oxygen [83,85], and the mixtures N_2–O_2 [55,73], N_2–CO [73], N_2–CO_2–CO–He [103], N_2–He [104], and N_2–H_2 [105,106]. In some works [30,44,52,55,63,73,76–78,85,103], the degree of vibrational excitation was usually characterized by the vibrational temperature T_V which was taken as an independent parameter. However, in other works [58–62,64,66,68–70,74,75,83,105,106] both the EEDF and the vibrational level populations were calculated self-consistently by simultaneously solving the Boltzmann equation (4.3) (taking into account the term "Sup" for electron-vibrational interactions) and the rate balance equations for the vibrational level populations. Typical self-consistent calculations of this type are shown in Fig. 4.5. From a comparison of Figs. 4.5 and 4.3, it can be seen that time evolution of the self-consistent EEDF takes place over a time scale considerably larger than the own EEDF relaxation time. This longer time scale arises because the vibrational distribution function (VDF) is still evolving in time and affecting the EEDF. In fact, Figure 4.5 reflects the influence on the quasi-stationary EEDF of the excited vibrational level populations which, comparatively, change much more slowly in time.

Superelastic collisions of electrons with electronically excited molecules, especially those with metastable states, can also affect the EEDF in some circumstances. Indeed, the populations of some metastable states such as $N_2(A^3\Sigma_u^+;\ a'\Sigma_u^-)$ and $O_2(u^1\Delta_g;\ b^1\Sigma_g^+))$ can be sufficiently high for electron superelastic collisions with these states to become an important source of "hot" plasma electrons. The influence of such collisions on the EEDF has been numerically investigated in nitrogen [69,72,74,75], N_2–He mixtures (influence of He(2^3S) metastables) [104], and O_2 [85]. In some works [69,72,85,104] the concentrations of excited molecules were considered as independent parameters, while in others [74,75] they were calculated self-consistently with the

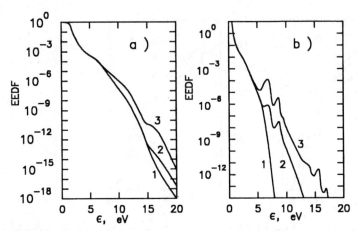

Fig. 4.6. Quasi-stationary EEDF (eV$^{-3/2}$) in an N_2 discharge, for $E/N = 3 \times 10^{-16}$ Volt cm^2 (**a**), and in the post-discharge (**b**) for a vibrational temperature of 3400 K and the following relative populations φ_A and φ_B of the electronically excited levels $A^3\Sigma_u^+$ and $B^3\Pi_g$: $\varphi_A = \varphi_B = 0$ (1); $\varphi_A = 10^{-6}$, $\varphi_B = 10^{-7}$ (2); $\varphi_A = 10^{-4}$, $\varphi_B = 10^{-5}$ (3) [69]

EEDF and the VDF. Figure 4.6a illustrates the influence of electron superelastic collisions with vibrationally and electronically excited N_2 molecules on the EEDF in pure nitrogen. The influence of such superelastic collisions is particularly strong in the post-discharge, i.e., after the applied field is switched off. In this situation, such collisions are the sole source of "hot" electrons and, for this reason, they can strongly influence or even determine the EEDF shape as illustrated in Fig. 4.6b. In gases like N_2 and CO, where the electron-vibration cross-sections are particularly high and the probabilities of V-T processes are small, the electron temperature in the post-discharge is strongly correlated with the vibrational "temperature". Both of these temperatures can far exceed the translational temperature over an extended period of time during the afterglow [68,70].

The influence of electron–electron collisions on the EEDF in plasmas of atmospheric gases has been investigated in a few works (see, for example, [29,30,60]). It was found that the effects of such collisions become important for degrees of ionization $N_e/N \geq 10^{-3}$.

Let us briefly refer to the behaviour of the EEDF in nitrogen in the presence of an HF field with frequency $\omega \sim \nu_\epsilon$, i.e., in conditions when the quasi-stationary (or the effective field) approximation does not apply. Figure 4.7 shows the EEDF at different instants of time within a cycle of the field oscillation, for a reduced angular frequency $\omega/p_0 = 10^7 \pi$ s^{-1} Torr^{-1}. Taking into account the values of the frequency ν_ϵ given in Fig. 4.1, one notes that $\omega < \nu_\epsilon$ in the electron energy ranges 1.8 eV $\leq \epsilon \leq$ 4 eV and $\epsilon \geq 8$ eV. This explains why the EEDF is significantly modulated in time within the above energy ranges.

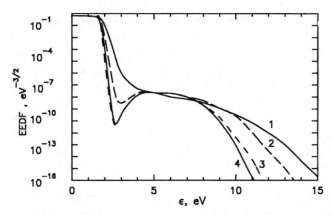

Fig. 4.7. Time evolution of the EEDF in an N_2 discharge with an alternating field $E/p = E_0/p_0 \cos(\omega t)$, for $E_0/p = 9.9$ Volt cm^{-1} Torr^{-1} and $\omega/p = \pi 10^7$s^{-1} Torr^{-1}, where p is the gas pressure. Curves 1-4 are for $(\omega t)/(2\pi)$=0.05; 0.16; 0.25; 0.34, respectively [16]

So far we have only discussed conditions in which the EEDF is in equilibrium with the local electric field. However, when a discharge is confined in a limited volume space at quite low pressures the energy transfer between the electric field and the electrons can be nonlocal, i.e., the EEDF can be influenced by electron diffusion from plasma regions where the electric field is different. As a result, it is not possible to calculate the EEDF by solving the Boltzmann equation in velocity space alone, using the local electric field strength (local approximation). These conditions have been studied by Monte Carlo [107] and fluid [108–110] methods and also by approximately solving the space-dependent Boltzmann equation [111,112]. The most comprehensive but substantially equivalent approaches to model this regime are the Particle-in-Cell/Monte Carlo collision (PIC/MCC model) and the Convective Scheme (CS) methods. Both methods solve the Vlasov/Poisson/Boltzmann problem exactly, with the aid of mathematical particles to propagate the solution (Green function). The main difference between the two methods is the following. In the PIC/MCC method [113–116], particles represent the distribution in the phase space at any time (Lagrangian view), collision processes are included by a Monte Carlo method, and the only mesh used is the one to sample the space charge and solve Poisson's equation. In the CS method [117], the distribution is represented on a mesh in the phase space, new particles are used at any time step to propagate the solution stored in the mesh under the effect of the electric field, and collision processes are included by a collision kernel on the grid, thus treating the solution as a fluid. As a rule, PIC/MCC is more convenient than CS if one is not interested in the very tail of the EEDF. The non local regime has also been studied analytically, taking advantage of the conservation of total energy [118,119].

4.3 EEDF in Discharges Excited by an Electron Beam

Discharges maintained by external ionization sources such as hard UV- and X-radiation or high-energy particle beams are now widely used for plasma creation. The Earth's ionosphere is also a low-temperature plasma formed by the action of these types of source under natural conditions. In many laboratory investigations, such sources (most often electron beams and particles resulting from nuclear reactions) are used to excite active laser media and to initiate plasma-chemical reactions [120]. For this reason, analysis of the electron kinetics under the action of external wave or corpuscular radiation is a problem of paramount importance.

When a beam of high energy electrons is injected into a gas, its energy can be dissipated mainly through the following two mechanisms: elementary collision processes and collective effects associated with the interaction of the plasma electrons with Langmuir oscillations which are excited by the beam. Beam-plasma discharges with generation of plasma oscillations can be achieved with powerful electron beams at relatively low gas pressures. The physical aspects of this type of discharge and its use in plasma chemistry have been discussed in detail in the literature [120]. This subject will not be dealt with in this book. Here, we will discuss only the problem of the formation of the EEDF in beam-plasma discharges in the absence of collective plasma effects.

Under the action of an ionization source such as a beam of high-energy electrons, with an energy ϵ_b largely exceeding the energy thresholds ϵ^*_{snk}, ϵ^*_{si} of all the excitation ($n \to k$) and ionization (i) processes, the EEDF is formed through the following mechanisms. The primary processes generated by the beam itself mostly consist in the ionization of gas particles (say, gas "s"). The secondary electrons so produced have considerable energy $\epsilon > \epsilon^*_{si}$, and thus they can further ionize and excite the gas particles. The EEDF, which is traditionally named the "degradation spectrum" in this case, spreads in energy space from zero energy to ϵ_b. It mostly results from the interactions of all the secondary electrons with the gas particles.

The significance of the degradation spectrum was first discussed by Fermi [121]. The mathematical bases for its calculation were formulated by a number of authors [122–124], and detailed calculations have subsequently been presented [125–132].

Let us outline a method for obtaining the degradation spectrum [130] from the Boltzmann equation (4.3), in the presence of a source q accounting for the rate of production of secondary electrons by a beam of electrons with energy $\epsilon_b \gg \epsilon^*_{snk}$, ϵ^*_{si}. In the electron energy range

$$\epsilon \gg \epsilon^*_{snk}, \ \epsilon^*_{si} \tag{4.27}$$

the terms In and Sup (see (4.13),(4.14)) can be described in the continuous approximation. In this case, one obtains

4.3 EEDF in Discharges Excited by an Electron Beam

$$(In + Sup)_{\text{evE}} \simeq -\frac{\partial J_{\text{evE}}}{\partial \epsilon}; \qquad J_{\text{evE}} = -D_{\text{evE}}\sqrt{\epsilon}\frac{\partial f}{\partial \epsilon} - \mu_{\text{evE}}\sqrt{\epsilon}f, \quad (4.28)$$

where

$$D_{\text{evE}} \simeq \frac{1}{2}\sum_{s,n,k}(\epsilon^*_{snk})^2 \nu_{snk}\left(1 + \frac{g_{sn}\varphi_{sn}}{g_{sk}\varphi_{sk}}\right) \quad (4.29a)$$

$$\mu_{\text{evE}} \simeq \sum_{s,n,k}\epsilon^*_{snk}\nu_{snk}\left(1 - \frac{g_{sn}\varphi_{sn}}{g_{sk}\varphi_{sk}}\right) \quad (4.29b)$$

are, respectively, the diffusion coefficient and the mobility of electrons in energy space due to excitation and deactivation of vibrational and electronic levels. In (4.29), g_{sn}, g_{sk} are the statistcal weights and φ_{sn}, φ_{sk} are the relative populations of the levels n and k of particle "s". Using (4.28) and (4.29) to describe In and Sup in (4.3), and integrating the latter from ϵ to ∞ with boundary conditions $f(\infty) = 0$, $\frac{df}{d\epsilon}|_\infty = 0$, one obtains

$$\sqrt{\epsilon}\left[D_\Sigma \frac{df}{d\epsilon} + \mu_\Sigma f\right] = \int_\epsilon^\infty I_{\text{ion}}(\epsilon')d\epsilon' + Q, \quad (4.30)$$

where

$$D_\Sigma = D_{\text{el}} + D_{\text{rot}} + D_{\text{F}} + D_{\text{evE}}; \qquad \mu_\Sigma = \mu_{\text{el}} + \mu_{\text{rot}} + \mu_{\text{evE}} \quad (4.31)$$

are the total diffusion coefficient and mobility accounting for all electron collision processes. In (4.30), $Q(\epsilon) \equiv \frac{1}{N_e}\int_\epsilon^\infty(q-p)d\epsilon'$ and the integral term $\int_\epsilon^\infty I_{\text{ion}}(\epsilon')d\epsilon'$ describes the flux due to ionization. In the energy range where the inequality (4.27) holds, taking into account (4.19a,b) and (4.21), this flux can be written as

$$\int_\epsilon^\infty I_{\text{ion}}(\epsilon')d\epsilon' = -\sqrt{\epsilon}\mu_{\text{ion}}f(\epsilon) + \sum_{s,i}\int_{\epsilon+\epsilon^*_{si}}^\infty \sqrt{\epsilon_1}fK_{si}(\epsilon_1,\epsilon)d\epsilon_1; \quad (4.32)$$

$$\mu_{\text{ion}} = \sum_{s,i}\epsilon^*_{si}\nu_{si}; \qquad K_{si}(\epsilon_1,\epsilon) = \int_\epsilon^{(\epsilon_1-\epsilon^*_{si})/2}\nu_{sii}(\epsilon_1,\epsilon_2)d\epsilon_2. \quad (4.33)$$

In this energy range, it is further possible to simplify equation (4.30) by neglecting the diffusion flux $D_\Sigma df/d\epsilon$. The solution of the integral equation so obtained can be expressed as a Neumann series. If $Q = \text{const.}$, one has

$$f(\epsilon) \simeq \frac{Q}{\sqrt{\epsilon}(\mu_\Sigma + \mu_{\text{ion}})}$$

$$\times \left[1 + \sum_{s,i}\int \frac{K_{si}}{\mu_\Sigma + \mu_{\text{ion}}}d\epsilon'\right.$$

$$\left. + \sum_{s,i}\int \frac{K_{si}}{\mu_\Sigma + \mu_{\text{ion}}}d\epsilon' \sum_{s,i}\int \frac{K_{si}}{\mu_\Sigma + \mu_{\text{ion}}}d\epsilon'' + \ldots\right]^{-1} \quad (4.34)$$

In order to calculate K_{si} in (4.34) from (4.20), (4.21) and (4.33), it is convenient to describe the differential ionization cross-sections σ_{sii} by the Breit–Wigner formula [129]

$$\sigma_{sii}(\epsilon_1, \epsilon_2) = \sigma_{si}(\epsilon_1) \frac{\Gamma_s^2}{(\epsilon_2 - \epsilon_s^0)^2 + \Gamma_s^2}$$
$$\times \left[\arctan\left(\frac{\epsilon_1 - \epsilon_{si}^* - 2\epsilon_s^0}{2\Gamma_s}\right) + \arctan\left(\frac{\epsilon_s^0}{\Gamma_s}\right) \right]^{-1}, \quad (4.35)$$

where Γ_s and ϵ_s^0 are parameters. Note that it is possible in some circumstances to use a simpler expression for σ_{sii} [133], assuming $\epsilon_s^0 = 0$ in (4.35). Using (4.35), one gets [130]

$$f(\epsilon) \simeq \frac{Q}{\sqrt{\epsilon}(\mu_\Sigma + \mu_{ion} + \mu^*)} ;$$
$$\mu^* = \sum_{s,i} \nu_{si} \left\{ \epsilon_s^0 + \Gamma_s \ln\left[\frac{\epsilon}{\sqrt{(\epsilon_s^0)^2 + \Gamma_s^2}}\right] \left[\frac{\pi}{2} + \arctan\left(\frac{\epsilon_s^0}{\Gamma_s}\right)\right]^{-1} \right\}. \quad (4.36)$$

Inspection of (4.34) and (4.36) reveals that the EEDF varies slowly with energy ϵ in the high energy range. It is worth noting that this type of slow variation is typical for any distribution (translational, rotational or vibrational distribution functions) in the presence of a particle source at energies considerably greater than the average energy of the distribution (see Chaps. 1–3).

Analytical methods as described above have been used [130–132,134,135] to analyze the degradation spectra and the energy losses of electrons in different gases.

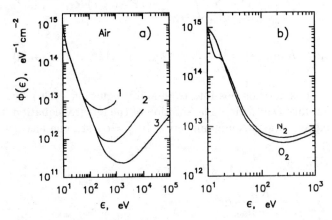

Fig. 4.8. Electron flux in air (a) and in nitrogen and oxygen (b) for high energies ϵ of secondary electrons. Curve 1 in (a) and the curves in (b) were obtained for a beam of electrons with energy $\epsilon_b = 1$ keV, and curves 2 and 3 of (a) for $\epsilon_b = 10$ keV and 100 keV [141]

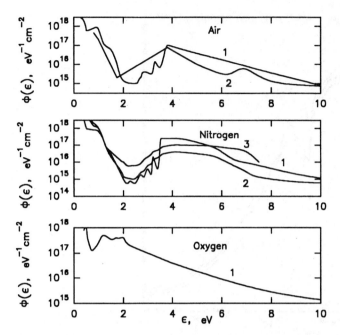

Fig. 4.9. Electron flux in air, nitrogen and oxygen in the energy range of inelastic thresholds [141]. Curves 1 [141] and 2 [144] are calculations, and curve 3 is from experiment [144]

To determine the EEDF at lower energies, when the inequality (4.27) no longer applies, numerical methods of solution must be used. Monte-Carlo [136–140] and iteration methods have mostly been used for this purpose.

When electron–electron and superelastic collisions are negligible, the equation (4.3) is such that the EEDF at a given energy depends on the collision processes taking place only at higher energies, and not on those at lower ones. For this reason, this equation has been solved by a backward prolongation technique, i.e., by starting the solution from the highest energy point, the energy ϵ_b of the primary beam electrons [141–143]. The combination of numerical calculations at low energies with analytical solutions at high energies, where the inequality (4.27) holds, has also proved very effective.

Specific calculations of degradation spectra have been made in the atmospheric gases N_2 [137,141,144], O_2 [141], and H_2 [146], and the mixtures N_2-O_2 [135,139,141], N_2-O_2-O [142,143], N_2-Ar [138], and O_2-Ar [147]. Measurements were also performed in the case of nitrogen [145]. Calculated electron fluxes in N_2, O_2 and the air mixture N_2-O_2 are shown in Figs. 4.8 and 4.9. These electron fluxes, $\Phi(\epsilon)$, are related with the EEDF $f(\epsilon)$ through the expression

$$\Phi(\epsilon) = \sqrt{\epsilon} f(\epsilon) \frac{c(\epsilon)}{Q} \left[1 + \frac{\epsilon_b}{\epsilon_{\text{ion}}} \right], \qquad (4.37)$$

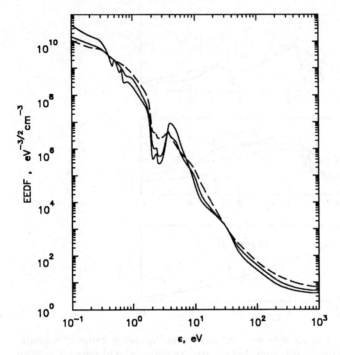

Fig. 4.10. EEDF (degradation spectrum) in typical ionospheric $N_2:O_2:O$ gas mixtures excited by a beam of electrons with energy 1 keV. Solid line – $N_2:O_2:O$ = 0.784:0.179:0.037; medium dash line – 0.459:0.096:0.495; long dash line – 0.13:0.007:0.863 [143]

where $\overline{\epsilon_{ion}}$ is the mean energy necessary for the fast electron to make an ionization, the so-called mean energy per electron-ion pair. For $\epsilon_b \geq 300$ eV, the value of $\overline{\epsilon_{ion}}$ is, respectively, 33, 34.4 and 34 eV, in nitrogen, oxygen, and the air mixture. Expression (4.37) is suitable for determining the energy cost of any process from the corresponding cross-section and the flux $\Phi(\epsilon)$.

Calculations of the EEDF in N_2-O_2-O mixtures, for typical ionospheric conditions at different altitudes, are shown in Fig. 4.10. As seen from Figs. 4.9 and 4.10, the EEDF exhibits a typical depletion in the low energy range of 1–3 eV. This depletion is due to electron energy losses associated with vibrational excitation of N_2 and O_2. Note that electron–electron and electron superelastic collisions have not been taken into account in the above calculations of the EEDF in plasmas excited by electron beams. However, such effects have been taken into account in multicusp magnetically confined hydrogen plasmas [148,149].

To conclude this chapter, let us recall that determination of the EEDF from the Boltzmann equation (4.3) is just an intermediate step in the kinetic modelling of plasmas. The ultimate aim is to determine from the EEDF such plasma properties as the electron drift velocity, the rate coefficients for excitation of different levels and for ionization and dissociation of atoms and molecules, and the plasma power balance. The determination of such properties will be addressed in Chap. 8.

References

1. Holstein T. (1946) Phys. Rev. **70**, 367
2. Ginzburg V.L. and Gurevich A.V. (1960) Adv. Phys. Sci. **70**, 201 (*in Russian*)
3. Aleksandrov N.L. and Son E.E. (1980) in *Plasma Chemistry*, ed. B.M. Smirnov, Atomizdat, Moscow **7**, p. 35 (*in Russian*)
4. Diatko N.A., Kochetov I.V. and Napartovich A.P. (1988), in *HF Discharge in Wave Fields*, ed. A.G. Litvak, Institute of Applied Physics, Gorky, p. 9 (*in Russian*)
5. Ginzburg V.L. (1967) *Propagation of Electromagnetic Waves in Plasmas*, Nauka, Moscow (*in Russian*)
6. Shkarovsky I., Jonston T., Bachinsky M. (1969) *Kinetics of Plasma Particles*, Atomizdat, Moscow (*in Russian*)
7. Granovsky V.L. (1971) *Electric Current in Gases*, Nauka, Moscow (*in Russian*)
8. Gurevich A.V. and Shwartsburg A.B. (1973) *Nonlinear Theory of Propagation of Radio Waves in the Ionosphere*, Nauka, Moscow (*in Russian*)
9. Crompton R. (1994) Adv. Atom. Mol. and Opt. Phys. **33**, 97
10. Cercignani C. (1988) *The Boltzmann Equation and its Applications*, Springer, Berlin Heidelberg
11. Biberman L.M., Vorobiev V.S. and Yakubov I.T. (1987) *Kinetics of Nonequilibrium Low Temperature Plasmas*, Colsultants Bureau, New York London (translated from Russian)
12. Raizer Yu.P. (1991) *Gas Discharge Physics*, Springer, Berlin Heidelberg
13. Velihov E.P., Kozlov A.S. and Rahimov A.T. (1987) *Physical Phenomena in Discharge Plasma*, Nauka, Moscow (*in Russian*)
14. Winkler R., Dilonardo M., Capitelli M., and Wilhelm J. (1987) Plasma Chem. Plasma Process. **7**, 125
15. Gurevich A.V. (1980) Adv. Phys. Sci. **132**, 685 (*in Russian*)
16. Capitelli M., Celiberto R., Gorse C., Winkler R. and Wilhelm J. (1988) J. Phys. D: Appl. Phys. **21**, 691
17. Davydov B.I. (1936) J. Exper. Theor. Phys. **6**, 463; 471 (*in Russian*)
18. Allis W.P. and Haus H.A. (1974) J. Appl. Phys. **45**, 781
19. Konovalov V.P. and Son E.E. (1982) Thermophys. High Temper. **20**, 412 (*in Russian*)
20. Lyaguschenko R.I. and Tendler M.B. (1975) Physics of Plasma **1**, 836 (*in Russian*)
21. Naidis G.V. (1977) J. Techn. Phys. **47**, 941 (*in Russian*)
22. Reshetniak S.A. and Shlepin L.A. (1976) Short Inf. Phys. **7**, 21 (*in Russian*); (1977) Physics of Plasma **3**, 859 (*in Russian*)
23. Rockwood S.D. (1973) Phys. Rev. **A8**, 2348
24. Morgan W.L. and Penetrante B.M. (1990) Comp. Phys. Commun. **58**, 127
25. Ness K.F. and Robson R.E. (1986) Phys. Rev. **A34**, 2185
26. Braglia G.L., Diligenti M., Wilhelm J. and Winkler R. (1985) Nuovo Cimento **12D**, 257
27. Braglia G.L., Wilhelm J. and Winkler R. (1982) Lett. Nuovo Cimento **44**, 365

28. Pitchford L.C. and Phelps A.V. (1982) Phys. Rev. **A25**, 540
29. Colonna G., Gorse C., Capitelli M., Winkler R., and Wilhelm J. (1993) Chem. Phys. Lett. **213**, 5
30. Mnatsakanian A.H. and Naidis G.V. (1976) Physics of Plasma **2**, 152 (*in Russian*)
31. Diatko N.A., Kochetov I.V., Napartovich A.P. and Taran M.D. (1983) Preprint of Kurchatov IAE **3842/12** (*in Russian*)
32. Itoh T. and Musha T. (1960) J. Phys. Soc. Japan **15**, 1675
33. Bell M.J. and Kostin M.D. (1968) Phys. Rev. **169**, 150
34. Kline L.E. and Siambis J.G. (1972) Phys. Rev. A **5**, 794
35. Friedland L. (1972) Phys. Fluids **20**, 1461
36. Kucuparpaci H.N. and Lucas J. (1979) J. Phys. D: Appl. Phys. **12**, 2123
37. Taniguchi T., Tagashira H. and Saki Y. (1978) J. Phys. D: Appl. Phys. **11**, 1757
38. Hunter S.R. (1977) Austral. J. Phys. **30**, 83
39. Pitchford L.C. and Phelps A.V. (1983) Phys. Rev. A **25**, 540
40. Pitchford L.C., O Neil S.V. and Rumble J.R.Jr. (1981) Phys. Rev. A **23**, 294
41. Rockwood S.D. (1973) Phys. Rev. A **8**, 2348
42. Frost L.S. and Phelps A.V. (1962) Phys. Rev. **127**, 1621
43. Engelhardt A.G., Phelps A.V. and Risk C.G. (1964) Phys. Rev. A **135**, 1566
44. Nighan W.L. (1970) Phys. Rev. A **2**, 1989
45. Lukovnikov A.I., Fetisov E.E. and Trehov E.S. (1980) J. Techn. Phys. **40**, 1916 (*in Russian*)
46. Latypova R.A., Lukovnikov A.I. and Fetisov E.P. (1972) J. Techn. Phys. **42**, 115 (*in Russian*)
47. Winkler R. and Pfau S. (1973) Beitr. Plasmaphys. **13**, 273
48. Winkler R. and Pfau S. (1974) Beitr. Plasmaphys. **14**, 169
49. Newman L.A. and De Temple T.A. (1976) J. Appl. Phys. **47**, 1912
50. Osipov A.P. and Rahimov A.T. (1977) Physics of Plasma **3**, 644 (*in Russian*)
51. Konev Yu.V., Kochetov I.V., Marchenko V.S. and Pevgov V.G. (1977) Quantum Electronics **4**, 1359 (*in Russian*)
52. Aleksandrov N.L., Konchakov A.M. and Son E.E. (1978) Physics of Plasma **4**, 169; 1182 (*in Russian*)
53. Taniguchi T., Tagashira H. and Sakai Y. (1978) J. Phys. D: Appl. Phys. **11**, 1757
54. Nikonov S.V., Osipov A.P. and Rahimov A.T. (1979) Quantum Electronics **6**, 1258 (*in Russian*)
55. Aleksadrov N.L., Konchakov A.M. and Son E.E. (1979) Thermophys. High Temper. **17**, 210 (*in Russian*)
56. Tachibana K. and Phelps A.V. (1979) J. Chem. Phys. **71**, 3544
57. Tagashira H., Taniguchi T. and Sakai Y. (1980) J. Phys. D: Appl. Phys. **13**, 235
58. Capitelli M., Dilonardo M. and Gorse C. (1981) Chem. Phys. **56**, 29
59. Capitelli M., Gorse C. and Ricard A. (1981) J. Physique–Lettres **42**, L185
60. Capitelli M., Gorse C., Wilhelm J. and Winkler J. (1981) Lettere al Nuovo Cimento **32**, 225
61. Cacciatore M., Capitelli M. and Gorse C. (1982) Chem. Phys. **66**, 141
62. Cacciatore M., Capitelli M., Gorse C., Massabieaux B. and Ricard A. (1982) Letters al Nuovo Cimento **34**, 417
63. Capitelli M., Gorse C. and Ricard A. (1982) J. Physique-Lettres **44**, 4251
64. Massabieaux B., Plain A., Ricard A. and Capitelli M. (1983) J. Phys. B: At. Mol. Phys. **16**, 1863
65. Yoshida S., Phelps A.V. and Pitchford L.C. (1983) Phys. Rev. A **27**, 2858

66. Capitelli M., Gorse C., Wilhelm J. and Winkler R. (1984) Ann. der Phys. **41**, 119
67. Diatko N.A., Kochetov I.V. and Napartovich A.P. (1985) Physics of Plasma **11**, 739 (*in Russian*)
68. Gorse C., Capitelli M. and Ricard A. (1985) Chem. Phys. **82**, 1900
69. Paniccia F., Gorse C., Bretagne J. and Capitelli M. (1986) J. Appl. Phys. **59**, 4004
70. Capitelli M., Gorse C. and Ricard A. (1986) in *Non-equilibrium Vibrational Kinetics*, ed. M. Capitelli, Springer, Berlin Heidelberg, p. 315
71. Loureiro J. and Ferreira C.M. (1986) J. Phys. D:Appl. Phys. **19**, 17
72. Paniccia F., Gorse C., Cacciatore M. and Capitelli M. (1987) J. Appl. Phys. **61**, 1323
73. Aleksandrov N.L. and Kochetov I.V. (1987) Thermophys. High Temper. **25**, 1062 (*in Russian*)
74. Gorse C. and Capitelli M. (1987) J. Appl. Phys. **62**, 4072
75. Gorse C., Cacciatore M., Capitelli M., de Benedictis S. and Dilecci G. (1988) Chem. Phys. **119**, 63
76. Loureiro J. and Ferreira C.M. (1988) J. Phys. D: Appl. Phys. **22**, 67
77. Ferreira C.M. and Loureiro J. (1989) J. Phys. D: Appl. Phys. **22**, 76
78. Loureiro J., Ferreira C.M., Capitelli M., Gorse C. and Cacciatore M. (1990) J. Phys. D: Appl. Phys. **23**, 1371
79. Nagpal R. and Ghosh P.K. (1990) J. Phys. D: Appl. Phys. **23**, 1663
80. Lucas J., Price D.A. and Moruzzi J.L. (1973) J. Phys. D: Appl. Phys. **6**, 1503
81. Masek K., Ruzicka T. and Laska L. (1977) Czechosl. J. Phys. **27**, 888
82. Lawton S.A. and Phelps A.V. (1978) J.Chem. Phys. **69**, 1055.
83. Capitelli M., Dilonardo M. and Gorse C. (1980) Plasma Physik. **20**, 83
84. Masek K. and Rohlena K. (1984) Czechosl. J. Phys. **B34**, 1227
85. Gousset G., Ferreira C.M., Pinheiro M., Sa P.A., Touzeau M., Vialle M. and Loureiro J. (1991) J. Phys. D: Appl. Phys. **24**, 290
86. Megill L.R. and Cahn J.H. (1967) J. Geophys. Res. **69**, 5041
87. Gurevich A.V., Milih G.M. and Shluger I.S. (1975) J. Techn. Phys. **69**, 1690 (*in Russian*)
88. Gurevich A.V. (1979) Geomagnetism and Aeronomy **19**, 633 (*in Russian*)
89. Fournier G., Bonnet J. and Pigache D. (1980) J. Physique Lettres **41**, L173
90. Aleksandrov N.L., Vysikailo F.I., Islamov. R.Sh., Kochetov I.V., Napartovich A.P. and Pevgov V.G. (1981) Thermophys. High Temper. **19**, 22 (*in Russian*)
91. Aleksandrov N.L, Vysikailo F.I., Islamov R.Sh., Kochetov I.V., Napartovich A.P. and Pevgov V.G. (1981) Thermophys. High Temper. **19**, 485 (*in Russian*)
92. Borisov I.D., Gurevich A.V. and Mileh G.M. (1986) *Artificial Ionized Area in Ionosphere*, Nauka, Moscow (*in Russian*)
93. Diatko N.A., Kochetov I.V., Mishin E.V. and Telegin V.A. (1989) Geomagnetism and Aeronomy **29**, 275 (*in Russian*)
94. Nighan W.L. and Bennet J.H. (1969) Appl. Phys. Lett. **14**, 240
95. Lowke J.J., Phelps A.V. and Irwin B.W. (1973) J. Appl. Phys. **44**, 4664
96. Rockwood S.D. (1974) J.Appl.Phys. **45**, 5229
97. Lobanov A.N. and Suchkov A.F. (1974) Quantum Electronics **1**, 1527 (*in Russian*)
98. Judd O.P. (1974) J. Appl. Phys. **45**, 4572
99. Klagge S. et al. (1977) Beitr. Plasmaphys. **17**, 237
100. Winkler R. and Pfau S. (1977) Beitr. Plasmaphys. **17**, 317
101. Pfau S. and Winkler R. (1977) Beitr. Plasmaphys. **17**, 397
102. Michel R., Pfau S. and Winkler R. (1978) Beitr. Plasmaphys. **18**, 131

103. Capitelli M., Gorse C., Berardini M. and Braglia G.L. (1981) Lettere al Nuovo Cimento **31**, 231
104. Slavik J. and Colonna G. (1997) Plasma Chem. Plasma Process. **17**, 305
105. Garscadden A. and Nagpal R. (1995) Plasma Sources Sci. Technol. **4**, 266
106. Gorse C., De Benedictis S., Dilecce G. and Capitelli M. (1990) Spectrochemica Acta **41 B**, 521
107. Kushner M.J. (1983) J. Appl. Phys. **54**, 4958; (1987) J. Appl. Phys. **61**, 2784; Kushner M.J. (1986) IEEE Trans.Plasma Science **PS14**, 188
108. Boeuf J.P. and Belenguer Ph. (1990) in *Non Equilibrium Processes in Partially Ionized Gases*, eds. M. Capitelli and J. N. Barsdsley, Plenum Press, Ney York
109. Capriati G., Boeuf J.P. and Capitelli M. (1993) Plasma Chem. Plasma Process. **13**, 499
110. Nienhuis G. J. and Goedheer W. (1999) Plasma Sources Sci.Technol. **8**, 295
111. Sigeneger F. and Winkler R. (1997) Plasma Chem.Plasma Process. **17**, 281; Petrov G. and Winkler R. (1997) J. Phys. D: Appl. Phys. **30**, 53
112. Alves L.L., Gousset G. and Ferreira C.M. Phys. Rev. E **55**, 890
113. Hockney R.W. and Eastwood J.W. (1991) *Computer Simulation Using Particles*, Adam Hilger, Bristol
114. Birdsall C.K. and Langdon A.B. (1985) *Plasma Physics via Computer Simulations*, Mc Graw Hill, New York
115. Birdsall C.K. (1991) IEEE Trans. Plasma Science **19**, 68; Vahedi V., Birdsall C.K., Liebermann M.A., Dipeso G., and Rognlien T.D. (1993) Plasma Sources Sci. Technol. **2**, 273; Cooperberg D.J. and Birdsall C.K. (1998) Plasma Sources Sci.Technol. **7**, 41
116. Liebermann M.A. and Lichtenberg A.J. (1994) *Principles of Plasma Discharges and Material Processing*, J. Wiley and Sons, New York
117. Hitchon W.N.G., Parker G.J. and Lawler J.E. (1993) IEEE Trans. Plasma Science **21**, 228
118. Kortshagen U. (1995) Plasma Sources Sci. Technol. **4**, 172
119. Tsendin L.D. (1995) *Plasma Sources Sci. Technol.* **4**, 200
120. Eletsky A.V. (1978) Adv. Phys. Sci. **125**, 279 (*in Russian*); Bychkov V.L. and Eletsky A.V. (1985) in *Plasma Chemistry*, ed. B.M. Smirnov, Energoizdat, Moscow **12**, p. 119 (*in Russian*); Rusanov V.D. and Fridman A.A. (1984) *Physics of Chemically Active Plasma*, Nauka, Moscow (*in Russian*)
121. Fermi E. (1950) *Nuclear Physics*, Chicago Press, Chicago
122. Fano U. (1953) Phys. Rev. **92**, 328
123. Spencer L.V. and Fano U. (1954) Phys. Rev. **93**, 1172
124. Fano U. and Spencer L.V. (1975) Int. J. Radiat. Phys. Chem. **7**, 63
125. Inokuti M. (1975) Radiat. Res. **64**, 6; Kimura M., Inokuti M. and Dillon M. A. (1993) Adv. Chem. Phys. **84**, 192
126. Kim Y.K. (1975) Radiat. Res. **64**, 96
127. Khare S.P. (1975) Radiat. Res. **64**,106
128. Green A.E. (1975) Radiat. Res. **64**, 119
129. Donthat D.A. (1975) Radiat. Res. **64**, 141
130. Konovalov V.P. and Son E.E. (1987) in *Plasma Chemistry*, ed. B.M.Smirnov, Energoizdat, Moscow **17**, p. 194 (*in Russian*); Son E. (1992) in *Plasma Technology: Fundamentals and Applications*, eds. M. Capitelli and C. Gorse, Plenum Press, New York
131. Krinberg I.A. (1978) *Electron Kinetics in Earth's Ionosphere and Plasmasphere*, Nauka, Moscow (*in Russian*)
132. Nikerov V.A. and Sholin G.V. (1985) *Kinetics of Degradation Processes*, Energoizdat, Moscow (*in Russian*)
133. Opal C.B., Peterson W.K. and Beaty E.C. (1971) J. Chem. Phys. **55**, 4100

134. Krinberg I.A., Garifullina L.A. and Akatova L.A. (1974) J. Atm. Terr. Phys. **36**, 1727
135. Medvedev Yu.A. and Hohlov V.D. (1979) J. Techn. Phys. **49**, 309 (*in Russian*)
136. Heaps M.G. and Green A.E.S. (1974) J. Appl. Phys. **45**, 3183
137. Ryjov V.V. and Yastremsky A.G. (1978) Physics of Plasma **4**, 1262 (*in Russian*)
138. Ryjov V.V. and Yastremsky A.G. (1979) J. Techn. Phys. **49**, 2141 (*in Russian*)
139. Lappo G.B., Prudnikov M.M. and Checherin V.G. (1980) Thermophys. High Temper. **18**, 677 (*in Russian*)
140. Sergienko T.I., Ivanov V.E. and Ivanov G.A. (1988) Geomagnetism and Aeronomy **28**, 984 (*in Russian*)
141. Konovalov V.P. and Son E.E. (1980) J. Techn. Phys. **50**, 300 (*in Russian*)
142. Gordiets B.F. and Konovalov V.P. (1990), in *Plasma Jets in the Development of New Materials Technology*, ed. O.P. Solonenko and A.I. Fedorehenko, VSP, Utrecht Tokyo, p. 617
143. Gordiets B.F. and Konovalov V.P. (1991) Geomagnetism and Aeronomy **31**, 649 (*in Russian*)
144. Punkevich B.S., Stal N.L., Stepanov B.M. and Hohlov V.D. (1984) Inf. Universities. Radio Physics **27**, 174 (*in Russian*)
145. Suhre D.R. and Verdeen J.T. (1976) J. Appl. Phys. **47**, 4484.
146. Oks E.A., Rusanov V.D. and Sholin G.V. (1979) Physics of Plasma **5**, 211 (*in Russian*)
147. Keto J.W. (1981) J. Chem. Phys. **74**, 4445
148. Bretagne J., Delouya G., Gorse C., and Capitelli M. (1985) J. Phys. D: Appl. Phys. **18**, 811
149. Gorse C., Celiberto R., Cacciatore M., Lagana A., and Capitelli M. (1992) Chem. Phys. **161**, 221

5. Energetic and Spectroscopic Parameters of Atmospheric Species

The study of kinetic processes in gases requires a detailed knowledge of the internal states of atoms and molecules and of the associated energetic and spectroscopic constants. Such knowledge is essential to understand the basic behaviour of atoms and molecules in non-equilibrium conditions. Much attention has been given to finding and examining these parameters, and many works on the subject are available in the literature, for example, under the form of reviews and monographs [1–9]. For this reason, this chapter only presents some basic information needed for the study of kinetic processes in atmospheric atomic and molecular species.

First of all, data on the ionization potential I_{ion}, electron affinity I_{aff}, and (for molecules or molecular ions) dissociation energy D_0 of the species are needed for the modelling of plasmachemical processes. Values of these parameters for some atmospheric species, taken from different sources, are given in Tables 5.1–5.3. The values given in the literature for some parameters are scattered, therefore the accuracy of some of the data presented in these tables is not high, as is the case of the dissociation energy of polyatomic molecules and ions.

Some molecules indicated in Table 5.1, as, for example, N_3, are not stable at $T \simeq 300$ K. For such cases, the ionization potential is to be regarded, in fact, as the recombination energy of the parent ions (for example, N_3^+). Note also that the ionization potentials I_{ion} and dissociation energies D_0 can be estimated using the following relationship

$$D_0(A - B^+) = D_0(A - B) + I_{ion}(B) - I_{ion}(AB) \,. \tag{5.1}$$

Some values of D_0 given in Table 5.4 have, in fact, been obtained from (5.1).

Table 5.1. Ionization potentials I_{ion}

Species	N	O	N_2	O_2	NO	N_3
I_{ion} (eV)	14.53	13.62	15.58	12.08	9.26	11.5
Species	O_3	N_2O	NO_2	N_4	O_4	NON_2
I_{ion} (eV)	12.52	12.89	9.78	14.78	11.64	9.06
Species	H	H_2	OH	H_2O	HO_2	H_2O_2
I_{ion} (eV)	13.59	15.44	13.2	12.61	11.5	10.9
Species	NH	NH_2	NH_3	N_2H_2	N_2H_3	CO_2
I_{ion} (eV)	13.1	11.4	10.17	9.85	7.9	13.77

Table 5.2. Electron affinities I_{aff}

Species	O	O_2	NO	O_3	NO_2	NO_3	H
I_{aff} (eV)	1.47	0.44	0.031	2.89	2.43	3.7	0.75

Table 5.3. Dissociation energies D_0 of nitrogen-oxygen molecules and molecular ions

Species	N_2	O_2	NO	O_3	N_2O	N_2O
Bond	N–N	O–O	N–O	O_2–O	N_2–O	N–NO
D_0 (eV)	9.76	5.12	6.5	1.04	1.67	4.93
Species	NO_2	NO_2	N_2^+	O_2^+	NO^+	NO^+
Bond	NO–O	N–O_2	N–N^+	O–O^+	N–O^+	N^+–O
D_0 (eV)	3.11	4.5	8.71	6.66	10.85	11.77
Species	N_3^+	N_3^+	N_2O^+	N_2O^+	N_2O^+	NO_2^+
Bond	N_2–N^+	N_2^+–N	NO–N^+	N_2–O^+	NO–N^+	N–O_2^+
D_0 (eV)	2.6	3.65	6.57	2.4	1.3	6.8
Species	NO_3^+	NO_3^+	NO_3^+	N_4^+	O_4^+	$N_2O_2^+$
Bond	N^+–O_2	NO^+–O	NO–O^+	N_2–N_2^+	O_2–O_2^+	N_2–O_2^+
D_0 (eV)	9.25	2.59	6.95	0.8	0.44	0.24
Species	N_2NO^+	$N_2O_3^+$	$O_2O_4^+$	O_2^-	O_3^-	O_3^-
Bond	N_2–NO^+	N_2O–O_2^+	O_2–O_4^+	O–O^-	O_2–O^-	O–O_2^-
D_0 (eV)	0.2	0.6	0.28	3.22	2.47	3.06

Table 5.4. Dissociation energies D_0 of hydrogen-oxygen-nitrogen molecules and molecular ions

Species	H_2	OH	HO_2	HO_2	H_2O	H_2O_2
Bond	H–H	O–H	H–O_2	OH–O	OH–H	OH–OH
D_0 (eV)	4.48	4.39	2.04	2.78	5.12	2.15
Species	HNO	HNO_2	HNO_3	HNO_3	NH	NH_2
Bond	H–NO	OH–NO	OH–NO_2	H–NO_3	N–H	NH–H
D_0 (eV)	2.11	2.24	1.41	1.41	3.6	3.9
Species	NH_3	NH_3	N_2H	N_2H	N_2H_2	N_2H_3
Bond	NH_2–H	NH–H_2	NH–N	N_2–H	NH–NH	NH–NH_2
D_0 (eV)	4.07	4.05	3.1	0.86	5.17	2.07
Species	N_2H_3	N_2H_4	N_2H_4	H_2^+	OH^+	OH^+
Bond	N_2H_2–H	N_2H_3–H	NH_2–NH_2	H–H^+	O–H^+	O^+–H
D_0 (eV)	2.24	2.76	1.78	2.63	4.78	4.81

Other important parameters for the modelling of kinetic processes are the energies of quantum states and terms of atomic and molecular species. The energies of some low terms of N and O atoms are given in Table 5.5.

The total energy T of a molecule in a given electronic-vibrational-rotational state can be determined from the expressions

$$T = T_e + E_v(v) + F_r(J) \tag{5.2a}$$

$$E_v(v) = \omega_e \left(v + \frac{1}{2}\right) - x_e \omega_e \left(v + \frac{1}{2}\right)^2 + y_e \omega_e \left(v + \frac{1}{2}\right)^3 + ... \tag{5.2b}$$

$$F_r(J) = B_v J(J+1) - D_e J^2(J+1)^2 + ... \tag{5.2c}$$

5. Energetic and Spectroscopic Parameters of Atmospheric Species

Table 5.5. The energies of low terms of N and O atoms

	N atom			O atom	
Term	Energy (cm^{-1})	Energy (eV)	Term	Energy (cm^{-1})	Energy (eV)
$^2S_{3/2}$	0	0	$^3P_2$	0	0
$^2D_{5/2}$	19223	2.385	$^3P_1$	158.56	0.0197
$^2D_{3/2}$	19231	2.384	$^3P_0$	226.5	0.0281
$^2P_{1/2,3/2}$	28840	3.576	$^1D_2$	15867.7	1.967
$^4P_{1/2}$	83285.5	10.326	$^1S_0$	33792.4	4.19
$^4P_{3/2}$	83319	10.33	^5S	73767.8	9.146
$^4P_{5/2}$	83336	10.332	^3S	76794.7	9.521

$$B_v = B_e - \alpha_e \left(v + \frac{1}{2}\right) + ..., \tag{5.2d}$$

where T_e is the electronic energy, $E_v(v)$ is the vibrational energy, and $F_r(J)$ is the rotational energy of the molecule, v and J denoting the vibrational and the rotational quantum numbers, respectively. The parameters T_e, ω_e, $x_e\omega_e$, $y_e\omega_e$, B_e, α_e, and D_e are the spectroscopic constants (usually expressed in cm^{-1}) determining the energy. Data for N$_2$ and O$_2$ are given in Tables 5.6 and 5.7, respectively, in the above usual units [4]. The parameter r_e is the equilibrium internuclear distance (expressed in Å). The potential energy curves for the most important electronic states of these molecules are shown in Figs.5.1 and 5.2, respectively.

Table 5.6. Spectroscopic constants of ^{14}N$_2$ [4]

	D_0 = 9.759 ev;		I_{ion} = 15.5808 ev			
State	T_e	ω_e	$\omega_e x_e$	B_e	α_e	r_e
$X^1\Sigma_g^+$	0	2358.57	14.324	1.99824	0.017318	1.09768
$A^3\Sigma_u^+$	50203.6	1460.64	13.87	1.4546	0.0180	1.2866
$B^3\Pi_g$	59619.3	1733.39	14.122	1.6374	0.0179	1.2126
$W^3\Delta_u$	59808.	1501.	11.6	-	-	-
$B'^3\Sigma_u^-$	66272.4	1516.88	12.18	1.473	0.0166	1.278
$a'^1\Sigma_u^-$	68152.66	1530.25	12.0747	1.4799	0.01657	1.2755
$a^1\Pi_g$	69283.06	1694.20	13.949	1.6169	0.01793	1.2203
$w^1\Delta_u$	72097.4	1559.36	11.63	1.498	0.0166	1.268
$G^3\Delta_g$	87900.	742.49	11.85	0.9280	0.0161	1.6107
$C^3\Pi_u$	89136.88	2047.17	28.445	1.8247	0.01868	1.1486
$E^3\Sigma_g^+$	95856.	2185.	-	1.927	6.0	1.117
$C'^3\Pi_u$	98351.	791.	33.5	1.0496	-	1.5414
$a''^1\Sigma_g^+$	100016.	-	-	1.9133	-	1.1218
$b^1\Pi_u$	101675.6	-	-	1.961	-	1.108
$D^3\Sigma_u^+$	104746.	-	-	1.961	-	1.108
$b'^1\Sigma_u^+$	104498.	760.08	4.418	1.1549	0.007387	1.4439
$c_3^1\Pi_u$	104476.	2192.20	14.70	1.9320	0.0395	1.1163
$c_4^{\prime 1}\Sigma_u^+$	104519.	2201.78	25.199	1.9612	0.0436	1.1080

Table 5.7. Spectroscopic constants of $^{16}O_2$ [4]

	$D_0 = 5.115$ ev;		$I_{ion} = 12.071$ ev			
State	T_e	ω_e	$\omega_e x_e$	B_e	α_e	r_e
$X^3\Sigma_g^-$	0	1580.19	11.98	1.44563	0.0159	1.20752
$a^1\Delta g$	7918.1	1483.5	12.	1.4264	0.0171	1.2156
$b^1\Sigma_g^+$	13195.1	1432.77	14.00	14.0037	0.01820	1.22688
$c^1\Sigma_u^-$	33057.	794.2	12.73	0.915	0.0139	1.517
$A'^3\Delta u$	34690.	850.	20.	0.96	0.026	1.48
$A^3\Sigma_u^+$	35397.8	799.07	12.16	0.9106	0.0141	1.5215
$B^3\Sigma_u^-$	49793,28	709.31	10.65	0.8190	0.01206	1.6042

The expressions (5.2b) and (5.2c) can be readily obtained from a series expansion of the potential energy of the nuclei in powers of $(r - r_e)$, the deviation of internuclear distance r from equilibrium. It is clear that calculations of $E_v(v)$ and $F_r(J)$ using expansions (5.2b) and (5.2c) with a limited number of terms become increasingly inaccurate as the vibrational and the rotational quantum numbers increase. In fact, to calculate vibrational level energies, only one or two terms of expansion (5.2b) are often considered. The one-term approximation corresponds to the harmonic oscillator model (equidistant vibrational levels), while the two-term one corresponds to an anharmonic oscillator with level energies E_n and energy gaps given by, respectively,

$$E_n = nE_{10} - n(n-1)\Delta E; \quad E_n - E_{n-1} = E_{10} - 2(n-1)\Delta E, \quad (5.3)$$

where E_{10} and ΔE are determined by the spectroscopic constants ω_e and x_e, specifically, $E_{10} = \omega_e - 2\omega_e x_e$. In (5.3) E_n is the energy of level n with respect to the ground level, whose energy in absolute value is $E_0 = \frac{1}{2}\omega_e - \frac{1}{4}\omega_e x_e$.

The internuclear potential that corresponds to this anharmonic oscillator is the well-known Morse potential

$$U(r) = D_0\{1 - \exp[(r - r_e)/L]\}^2. \quad (5.4)$$

This potential is widely used in calculations of vibrational level energies, matrix elements of vibrational transitions, Franck–Condon factors, and other quantities of interest. For this potential, the squares of the matrix elements for transitions between neighboring levels follow the law [10]

$$\frac{|V_{n,n-1}|^2}{|V_{10}|^2} \approx \frac{(1-x_e)n}{1-nx_e}. \quad (5.5)$$

The total number of bound vibrational levels, n_d, and the dissociation energy, D_0, for the Morse anharmonic oscillator are given by

$$n_d = \frac{E_{10}}{2\Delta E} + 1; \quad D_0 \approx \frac{E_{10}^2}{4\Delta E}. \quad (5.6)$$

However, the values of D_0 predicted by (5.6) often differ significantly from experiment. For example, for N_2 equation (5.6) yields $D_0 \sim 9.47 \; 10^4 \; cm^{-1}$,

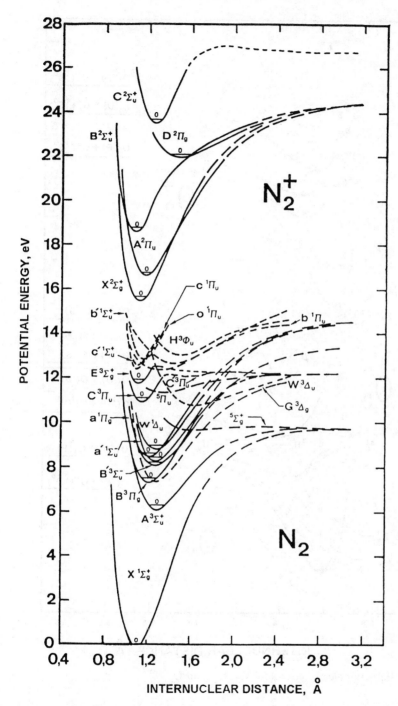

Fig. 5.1. Potential energy curves for N_2, N_2^- and N_2^+ [9]

90   5. Energetic and Spectroscopic Parameters of Atmospheric Species

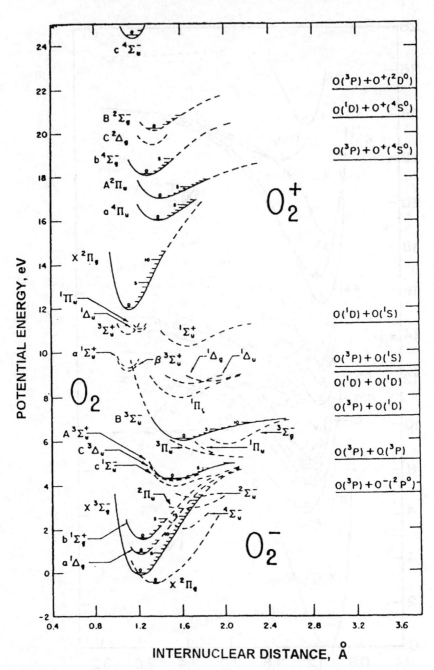

Fig. 5.2. Potential energy curves for O_2, O_2^- and O_2^+ [8]

which significantly exceeds the widely accepted value $\sim 7.87 \cdot 10^4$ cm^{-1} (see Table (5.6)). This indicates that the Morse potential deviates considerably from the real one close to the dissociation limit.

In some vibrational and chemical kinetic problems (for example, calculation of dissociation rates) the populations of vibrational levels close to the dissociation limit need to be determined. Although the Morse oscillator model is often used to investigate such problems, the total number of bound levels n_b cannot be calculated in this case from (5.6). This number has instead to be calculated from the condition that the energy of the highest bound level, according to (5.3), be equal to the experimental dissociation energy. Thus, a truncated Morse oscillator has to be used. This yields, for example, $n_b \approx 46$ for N_2. Although this procedure is a crude approximation, it is justified in view of the present lack of knowledge on the transition probabilities for vibrational levels close to the dissociation limit.

To improve quantitatively the description of kinetic processes involving upper vibrational levels, more exact models of internuclear potentials must be used [11–14]. One example is the potential [13]

$$U(y) = \begin{cases} a_0 y^2 \left(1 - \frac{a_1}{a_2} y\right)^{a_2} & \text{for} \quad y \leq y_m \quad (5.7a) \\ D_0 - \frac{C}{r_e^6 (1+y)^6} & \text{for} \quad y > y_m \quad (5.7b) \end{cases}$$

Here, $y = (r - r_e)/r_e$, D_0 is as before the dissociation energy, and the parameters a_0, a_1 and a_2 are given by

$$a_0 = \frac{\omega_e^2}{4B_e}; \quad a_1 = -1 - \frac{\alpha_e \omega_e}{6B_e^2}; \quad a_2 = \frac{(6B_e^2 + \alpha_e \omega_e)^2}{\frac{3}{2}(6B_e^2 + \alpha_e \omega_e)^2 - 48 B_e^3 \omega_e x_e}.$$

(5.8)

The constant C and the value of y_m are determined from the requirement of continuity both of $U(y)$ and its derivative at the point y_m.

The potential (5.7) yields energies of lower vibrational levels and dissociation energies in agreement with experiment. The Schrodinger equation for such a potential was numerically solved in [15] for N_2 molecules. Both the vibrational level energies and the matrix elements of single-quantum transitions were calculated. According to this study, there exist a total of 65 bound vibrational levels in the electronic ground state of N_2, and the energy distance between levels close to the dissociation limit is $E_{65} - E_{64} \approx 2$ cm^{-1}. This value is much smaller and more realistic than the value $E_{46} - E_{45} \approx 1040$ cm^{-1} that follows from the truncated Morse oscillator model with 46 levels. Moreover, for the potential (5.7) the square matrix elements for single-quantum transitions increase faster with vibrational quantum number than in the case of the Morse potential (see (5.5))

The Franck–Condon factors, $q_{v'v''}$, are important molecular parameters which connect different electronic terms. They are given by the square modulus of the overlap integral of two vibrational wave functions belonging to different electronic states, that is,

$$q_{v'v''} = \left|\int \Psi_{v'}(r)\Psi_{v''}(r)dr\right|^2 \equiv |<v'|v''>|^2 , \qquad (5.9)$$

where Ψ_n stands for the normalized vibrational wave function of vibrational state n. The importance of Franck–Condon factors stems from the fact that the probability of an optically allowed transition between two given electronic-vibrational states is proportional to the corresponding $q_{v'v''}$ factor. The magnitudes of cross-sections and rate constants of many collision processes are also often determined by the Franck-Condon factors.

To calculate the Franck–Condon factors, the vibrational wave functions have to be determined from the Schrodinger equation for the potential $U(r)$ of the relevant electronic term. RKR-potentials, constructed on the basis of experimental spectroscopic data, and also harmonic, Morse, Lennard-Jones and other potentials, expressed by analytical formulas, are often used. Detailed information on the magnitudes of the Franck-Condon factors and probabilities of optical transitions for N_2 is presented in [9,16,17], and for O_2, in [8].

For low vibrational levels, Franck–Condon factors can be calculated with sufficient accuracy supposing harmonic (parabolic) potentials for the electronic terms. Analytical expressions for $q_{v'v''}$, obtained for these potentials, are presented in [18–20]. For example, an approximate formula, which allows $q_{v'v''}$ to be calculated directly for transitions between levels $0 \leq v' \leq 5$, $0 \leq v'' \leq 6$, was proposed in [20]. It reads

$$q_{v'v''} = \frac{R^{v''-v'}\exp(-R)}{v'!v''!}$$
$$\times \left\{(R-v'')^{v'} - H(v'-2)v''\left[v'(R-v'')+2\right]^{v'-2}\right\}^2, \qquad (5.10)$$

where $H(v'-2)$ is the step function, and R is determined by the expressions:

$$R = \frac{S^2}{2}; \quad S = 0.1722 \times (r'_e - r''_e)\sqrt{\mu \overline{\omega_e}}; \quad \sqrt{\overline{\omega_e}} = \frac{2\sqrt{\omega'_e \omega''_e}}{\sqrt{\omega'_e} + \sqrt{\omega''_e}}. \qquad (5.11)$$

Here, $\mu = M_1 M_2/(M_1 + M_2)$ is the reduced mass of the oscillator in a.u., M_1 and M_2 denoting the masses of the nuclei. The equilibrium internuclear distances r'_e, r''_e and the spectroscopic constants ω'_e, ω''_e are given in Å and cm^{-1}, respectively.

A comparison of the Franck–Condon factors calculated from (5.10) with more exact calculations using RKR or Morse-potentials was carried out in [17] for different molecules. This comparison revealed that equation (5.10) yields values considerably different from the exact ones (more than one order of magnitude difference) for only 29 transitions, out of more than 5,000 transitions examined. No such differences were found for N_2 and N_2^+. For O_2 and O_2^+, differences were found only for the factors $q_{5,5}$ of the transitions $O_2(A^3\Sigma_u^+ - X^3\Sigma_g^-)$ and $O_2(b^4\Sigma_g^- - a^4\Pi_u))$, respectively. Therefore, equation (5.10) can be used for rapid, but fairly accurate calculations of Franck-

Condon factors of transitions between lower vibrational levels of electronic terms.

References

1. Herzberg G. (1950) *Spectra of Diatomic Molecules*, John Wiley and Sons, New York
2. Barrow R.F., ed. (1973) *Diatomic molecules: A Critical Bibliography of Spectroscopic Data*, Editions du CNRS, Paris
3. Radtsig A.A. (1975) in *Plasma Chemistry*, ed. Smirnov B.M., Atomizdat, Moscow **1**, 3 (*in Russian*)
4. Huber K.P. and Herzberg G. (1978) *Molecular Spectra and Molecular Structure. Constants of Diatomic Molecules*, Van Nostrand Reinhold Company, New York London
5. Rosen B., ed. (1970) *Spectroscopic Data Relative to Diatomic Molecules. International Tables of Selected Constants* Pergamon Press, Oxford
6. Suchard S.N., ed. (1975) *Spectroscopic Data: Heteronuclear Diatomic Molecules* Plenum Press, New York
7. Suchard S.N. (1976) *Spectroscopic Data: Homonuclear Diatomic Molecules*, Plenum Press, New York
8. Krupenie P.H. (1972) J. Phys. Chem. Ref. Data **1**, 423
9. Lufthus A. and Krupenie P.H. (1977) J. Phys. Chem. Ref. Data **6**, 113
10. Nikitin E.E. (1974) *Theory of Elementary Atomic and Molecular Processes in Gases*, Clarendon Press, Oxford
11. Murrell J.N. and Sorbie K.S. (1974), J. Chem. Soc. Faraday **1170**, 1552
12. Varandas A.J.C. (1980) J. Chem. Soc. Faraday **1176**, 129
13. Dmitrieva I.K.and Zenevich V.A. (1983) Chem. Phys. Lett. **96**, 228
14. Dmitrieva I.K., Zenevich V.A. and Plindov G.I. (1985) Chem. Phys. Lett. **121**, 485
15. Porshev P.I. (1989) *Electron and Vibrational Kinetics of Non-equilibrium Low Temperature Plasma of Molecular Nitrogen*, PhD Thesis, Lykov Institute TMO, Minsk (*in Russian*)
16. Gilmore F.R., Laher R.R. and Espy P.J. (1992) J. Phys. Chem. Ref. Data **21**, 1005
17. Kuznetsova L.A., Kuzmenko N.E, Kuzyakov Yu.Ya. and Plastinin Yu.A. (1980) *Probabilities of Optical Transitions for Diatomic Molecules*, Nauka, Moscow, (*in Russian*); Gurvich A.V., Kuznetsova L.A., Kuzmenko N.E., Kuzyakov Y.Y., Mikirov A.E. and Smirnov D.Y. (1985) Proceedings Institute of Applied Geophysics, Gidrometeoizdat, Moscow, No. 64 (*in Russian*)
18. Lin C.S. (1975) Canad. J. Phys. **53**, 310
19. Nicholls R.W. (1973) JQSRT **13**, 1059
20. Nicholls R.W. (1981) J. Chem. Phys. **74**, 6980

6. Rates of Translational and Rotational Relaxation

6.1 Translational Relaxation and Diffusion Coefficients

The translational relaxation time of atoms, molecules and ions is determined by the mean free time τ_0 or, equivalently, by its reciprocal, the mean collision frequency $\nu = 1/\tau_0$. In the case of binary collisions between hard-sphere particles A and B, one has:

$$\nu_{AB} = \frac{\pi d_{AB}^2}{N_A} \int\int f_A(c_1) f_B(c_2) c \, dc_1 dc_2 \,, \tag{6.1}$$

where d is the collision diameter, $d_{AB} = (d_A + d_B)/2$, $c = |c_1 - c_2|$, N_A is the density of particles A, $f_A(c_1)$ and $f_B(c_2)$ are the non-equilibrium velocity distribution functions of particles A and B, respectively. Since, as a rule, these functions are unknown beforehand, equilibrium (Maxwellian) distributions at a given temperature T can used to estimate ν_{AB} in order of magnitude. In this case one has

$$\nu_{AB} = 4 d_{AB}^2 \left(\frac{\pi k T}{\mu}\right)^{1/2} N_B \equiv \sqrt{2} \pi d_{AB}^2 \bar{c} N_B \,, \tag{6.2}$$

where $\bar{c} = 2 \left(\frac{2kT}{\pi \mu}\right)^{1/2}$, μ is the reduced mass of the colliding particles A and B, and N_B is the density of particles B. By defining the mean free path of particles A in gas B as

$$l = (\sqrt{2} \pi d_{AB}^2 N_B)^{-1} \,, \tag{6.3}$$

then τ_0 is related to l through the simple expression

$$\tau_0 = 1/\nu = l/\bar{c} \,. \tag{6.4}$$

According to the rigorous theory of hard-sphere gases [1], the viscosity coefficient η, for $d_A \simeq d_B$, is related to the translational relaxation parameters l and τ_0 through the expression

$$\eta = \xi_\eta \frac{1}{3} \rho_B \bar{c} l = \xi_\eta \frac{1}{3} \rho_B (\bar{c})^2 \tau_0 \,, \tag{6.5}$$

where $\xi_\eta = 15\pi/32$ and ρ_B is the mass density of the gas B (in gram/cm^3). Note that the factor ξ_η is unity in the framework of simple gas kinetic theory.

Table 6.1 provides values of l, $\bar{c}$, τ_0, and of the rate coefficients of translational relaxation $k_{TT} = \nu_{BB}/N_B \equiv 1/(N_B \tau_0)$, for various pure gases (A–A collisions; A=B). All these data are for gas temperature T=273 K and pressure p=1 atm.

Table 6.1. Translational relaxation data for pure gases under standard conditions

Gas	$\bar{c}$ (10^4 cm/s)	l (10^{-6} cm)	τ_0 (10^{-10} s)	k_{TT} (10^{-10} cm^3/s)
N_2	4.53	5.88	1.30	2.87
O_2	4.25	6.33	1.49	2.50
H_2	16.94	11.06	0.65	5.75
He	12.02	17.36	1.44	2.59
Ar	3.80	6.26	1.65	2.26
CO_2	3.51	3.98	1.13	3.30

Extension of these data to other temperatures and pressures can be carried out using (6.2) and (6.3).

Experimental values of other transport coefficients such as the heat conductivity λ and the diffusion coefficient $\mathcal{D}$, can also be used to determine τ_0. The values of τ_0 obtained in this way are usually slightly different from those derived from the viscosity coefficient. For example, the effective molecular diameter d of the N_2 molecule as obtained from viscosity data is 3.7 Å, while it is 3.61 Å and 3.69 Å, respectively, from heat conductivity and diffusion data [2]. These discrepancies can be explained by the different accuracies of the experimental determinations of the various transport coefficients (the viscosity coefficients are usually measured with greater accuracy), on one side hand, and of the theoretical approximations used to obtain l, τ_0 from transport data, on the other.

The modelling of numerous phenomena in low pressure gases and plasmas requires accurate knowledge of the diffusion coefficients $\mathcal{D}$. For example, the values of $\mathcal{D}$ are required to calculate the frequencies of wall losses of different gas phase species, which are transported to the wall by diffusion (see (11.13)). For pure gases (self-diffusion), these coefficients are connected with l and τ_0 by the relationship

$$\mathcal{D} = \xi_D \frac{1}{3} \bar{c} l = \xi_D \frac{1}{3} (\bar{c})^2 \tau_0 , \qquad (6.6)$$

where $\xi_D = 9\pi/16$. Note that $\xi_D = 1$ in simple gas kinetic theory. The diffusion coefficients are inversely proportional to gas density and increase with gas temperature T. The values $\mathcal{D}$ are often presented as a function of gas pressure p. In this case one can write

$$\mathcal{D} = \frac{\mathcal{D}^0}{p} \left(\frac{T}{273}\right)^\beta , \qquad (6.7)$$

where p is in atm, T is in K, and $\mathcal{D}^0$ is the diffusion coefficient under standard conditions ($T = 273$ K, $p = 1$ atm). According to the kinetic theory for hard spheres, $\beta = 1.5$. However, for real intermolecular interactions the collision diameter d depends on the molecular velocity, and thus the value of β may actually differ from 1.5. Values of $\mathcal{D}^0$ and β for diffusion in atmospheric gases and binary mixtures of these gases are given in Table 6.2.

Table 6.2. Diffusion data for the atmospheric gases and various mixtures

Gases	$\mathcal{D}^0$ (cm^2/s)	β	Ref.	Gases	$\mathcal{D}^0$ (cm^2/s)	β	Ref.
N_2-N_2	0.17	1.90	[3,4]	O_2-CO_2	0.174	1.661	[3,5]
O_2-O_2	0.18	1.92	[3,4]	H_2-CO	0.66	1.548	[3,5]
H_2-H_2	1.28	1.95	[3,4]	H_2-CO_2	0.56	1.75	[3,5]
H_2O-H_2O	0.277	2.10	[3,4]	H_2-air	0.66	1.75	[3,5]
$CO-CO$	0.175	1.90	[3,4]	$CO-CO_2$	0.142	1.803	[3,5]
CO_2-CO_2	0.097	1.859	[3,4]	$CO-$air	0.182	1.73	[3,5]
N_2-O_2	0.182	1.724	[3,5]	$N_2(A)-N_2$	0.179	1.90	[6]
N_2-H_2	0.66	1.548	[3,5]	$N(^4S)-N_2$	0.28	1.50	[7]
N_2-H_2O	0.204	2.072	[3,5]	$N(^2D)-N_2$	0.227	1.50	[8]
N_2-CO	0.175	1.576	[3,5]	$N(^2P)-N_2$	0.185	1.50	[8]
N_2-CO_2	0.208	1.57	[3,5]	$O_2(b)-O_2$	0.187	1.92	[9]
O_2-H_2	0.69	1.732	[3,5]	$O(^3P)-O_2$	0.292	1.50	[7]
O_2-H_2O	0.207	2.072	[3,5]	$O(^1D)-O_2$	0.25	1.50	[10]
O_2-CO	0.175	1.724	[3,5]	$H-H_2$	1.932	1.72	[11]

Most of the data in Table 6.2 are valid in the temperature range $T \simeq 270$–1000 K. Note that the diffusion coefficients for atoms (except H–H_2) and metastable molecules in their parent molecular gas have been determined for one temperature only (usually ~ 300 K). This is why in Table 6.2 the values β for atoms are assumed to be 1.5 (as in the hard-sphere collision model) and for metastable molecules they are taken to be the same as for ground state molecules.

Ion diffusion is also an important transport process in plasmas. If the ion and the gas temperatures are the same, the dependence on gas pressure p and temperature T of the diffusion coefficient $\mathcal{D}_\pm$ of positive or negative ions is

$$\mathcal{D}_\pm = \frac{\mathcal{D}_\pm^0}{p}\left(\frac{T}{273}\right)^2, \tag{6.8}$$

where p is in atm, T is in K, and $\mathcal{D}_\pm^0$ is the diffusion coefficient under standard conditions ($T = 273$ K, $p = 1$ atm). Note that the ion diffusion coefficients are smaller in magnitude than those of hard spheres. Their temperature dependence ($\mathcal{D}_\pm \sim T^2$) also differs from that of hard spheres ($\mathcal{D} \sim T^{1.5}$). This is explained by the important role of polarization interaction between an ion and a neutral particle (atom or molecule), which increases the effective collision cross-section σ and at the same time results in a dependence of the cross section on the relative collision velocity c of the form $\sigma \sim 1/c$.

Table 6.3. Reduced diffusion coefficients of atmospheric ions in atmospheric gases

Ion–gas	$\mathcal{D}_+^0$ (10^{-2} cm^2/s)	Ion–gas	$\mathcal{D}_\pm^0$ (10^{-2} cm^2/s)
N^+–N_2	6.98	H_3^+–H_2	26.1
N_2^+–N_2	4.39	H_3O^+–H_2	29.6
N_3^+–N_2	5.31	NO^+–NO	4.49
N_4^+–N_2	5.48	O^-–O_2	7.52
O_2^+–O_2	5.26	O_2^-–O_2	5.08
O_4^+–O_2	5.08	O_3^-–O_2	5.99
H^+–H_2	36.9	O_4^-–O_2	5.03

Values of reduced diffusion coefficients $\mathcal{D}_\pm^0$ for atmospheric ions in atmospheric gases are given in Table 6.3 [12].

When experimental data for the diffusion or mobility coefficients of ions are lacking, the following expression can be used to estimate $\mathcal{D}_\pm^0$ [13]:

$$\mathcal{D}_\pm^0 \left(\frac{\text{cm}^2}{\text{s}}\right) \simeq \frac{0.325}{(\alpha\mu)^{1/2}}, \tag{6.9}$$

where α is the polarizability of the neutral particle (in Å^3) and μ is the reduced mass of the colliding particles (in a.e.).

The diffusion of both electrons and positive ions in a quasi-neutral plasma leads to the creation of a space-charge field of such a polarity as to retard the loss of electrons (which, being lighter, diffuse faster and tend to leave the plasma first) and accelerate the loss of ions. As is well known, the transport of electrons and positive ions under such circumstances can be described as a diffusion process with a new diffusion coefficient called the *ambipolar diffusion coefficient*, $\mathcal{D}_a$. This is connected with $\mathcal{D}_+$ by the relationship

$$\mathcal{D}_a \simeq \mathcal{D}_+(1 + \epsilon_k/T_+), \tag{6.10}$$

where ϵ_k is the electron characteristic energy, that is, the ratio of the electron free diffusion coefficient to the electron mobility. For a Maxwellian electron energy distribution, ϵ_k is equal to the electron temperature. Therefore, in thermal equilibrium, that is, when $\epsilon_k = T_+ = T$, we get from (6.10) $\mathcal{D}_a \simeq 2\mathcal{D}_+$.

The diffusion coefficient $\mathcal{D}_{M-A-B}$ of any neutral or charged species M (atoms, molecules, or ions) present as a trace in a binary mixture of gases A and B can be calculated from Blanc's law as follows:

$$\mathcal{D}_{M-A-B}^{-1} = X_A/\mathcal{D}_{M-A} + X_B/\mathcal{D}_{M-B}, \tag{6.11}$$

where X_A and X_B are the relative concentrations of gases A and B ($X_A + X_B = 1$) and $\mathcal{D}_{M-A}$ and $\mathcal{D}_{M-B}$ are the diffusion coefficients of species M in the gases A and B, respectively, at the same pressure as the mixture A+B.

The values of the frequencies ν_{AB} for translational relaxation of O^+ ions in the gases N_2, O_2 and O are important for the investigation of ionospheric plasma kinetics. These frequencies (hereafter denoted as ν_{in} and expressed in s^{-1}) are given by the following expressions [14,15]:

$$\nu_{in} \simeq 5.4 \times 10^{-11} T_{eff}^{1/2}(1 - 6.3 \times 10^{-2} \lg T_{eff})^2 [O] \quad \text{for} \quad O^+ - O \quad (6.12a)$$

$$\nu_{in} \simeq 6.9 \times 10^{-10} [N_2] \quad \text{for} \quad O^+ - N_2 \quad (6.12b)$$

$$\nu_{in} \simeq 6.7 \times 10^{-10} [O_2] \quad \text{for} \quad O^+ - O_2 . \quad (6.12c)$$

Here, the concentrations of [O], [N_2], [O_2] are expressed in cm^{-3} and T_{eff} in (6.12a) denotes an effective temperature (in K) which may be different from the gas temperature. In general, this effective temperature must be calculated from expression (10.21) given in Chap. 10.

The diffusion coefficient of O^+ ions in O, N_2 and O_2 and in the mixtures of these gases can be determined from (6.4), (6.6), (6.11), and (6.12).

Recent calculations of collision integrals of high–temperature air species have been recently performed [16] including resonant charge transfer for ion–parent atom systems. These collision integrals have been used to calculate transport coefficients (thermal and electrical conductivity, and viscosity) for equilibrium N_2-O_2 atmospheric plasmas in the temperature range 50–100,000 K and for two-temperature nitrogen and oxygen plasmas [16].

6.2 Rotational Relaxation

The data on rotational relaxation can be divided into three categories. The first category includes rotational relaxation times. These data are the most plentiful, but they have limited applicability, since they characterize the rotational relaxation of macroscopic variables (for example, the rotational energy density, ϵ_{rot}). In practical terms, rotational relaxation times are used in hydrodynamics of non-equilibrium gas flows.

The second category includes the probabilities (or rate coefficients) of rotational transitions between levels. These data characterize rotational relaxation in more detail because they allow rotational kinetics and the time evolution of level populations to be described. The relaxation times can be obtained by a proper averaging of these probabilities. Therefore, the availability of data on the transition probabilities considerably widens the knowledge of the subject and allows the investigation of important subjects such as the action of lasers on vibrlational–rotational transitions. Unfortunately, these data are rather scarce.

Finally, the third category includes the most detailed characteristics – the differential cross-sections of inelastic scattering associated with rotational transitions. These data can as a rule be obtained by molecular beam methods. The transition probabilities, included in the previous group of data, can be calculated by averaging the differential cross-sections over the scattering angles and relative energies of the colliding particles.

The experimental methods used to investigate rotational relaxation can be also divided into three groups in accordance with the above three categories of data. Methods connected with the determination of relaxation times make up the first group. These include methods based on the examination of transport phenomena, for example, in ultrasonic flows, and gas kinetic methods (shock tubes, expanding flows, etc.). Methods involving analyses of level population distributions belong to the second group. All types of molecular beam methods are included in the third group. Only the first group will be briefly discussed here.

Historically, transport phenomena were the first type of processes for which rotational relaxation phenomena were taken into account. The transport coefficients derived from kinetic theory of polyatomic gases (based on the solution of the Boltzmann gas-kinetic equation) depend, in fact, on the rotational relaxation time [1,5]. The physical reason for this dependence can be illustrated by the example of heat conductivity. In gases with internal degrees of freedom, the heat flow is determined by the flow of both internal and translational energies. The energy flow associated with internal (rotational) degrees of freedom depends on whether or not the translational and the rotational degrees of freedom are in equilibrium with each other. If the relaxation times for translational and rotational energies are different, the rotational temperature is not equal to the translational one, and the effective heat conductivity coefficient will depend, therefore, on the rotational relaxation time.

The internal (rotational) energy relaxation time also influences the other transport coefficients. For example, this time strongly affects the bulk viscosity coefficient. It is thus possible to determine rotational relaxation times from measurements of transport coefficients.

Ultrasonic methods, currently used to investigate ultrasound absorption and dispersion in molecular gases, are also employed to determine relaxation times [17]. This method was first used to measure vibrational relaxation times only. However, progress in the use of high-frequency sound and low gas pressures has allowed also the study of rotational relaxation by ultrasonic methods. The physical basis for application of this method is the following. As is well known, the sound speed can be expressed as $c_s = \sqrt{\text{elasticity/density}}$, where the gas elasticity is determined by the energy of translational degrees of freedom. With increasing frequency of the sound wave, the translational energy of the molecules, changed by the wave, has no time to partly transfer to and reach equilibrium with rotational energy. This increases the gas elasticity and, therefore, the sound velocity. As this dispersion of the wave takes place, sound absorption rises due to the irreversibility of the energy redistribution processes. For this reason, both the sound velocity and the absorption coefficient depend on the relaxation time for energy redistribution, which therefore enables the relaxation time to be determined experimentally.

Shock tube methods have been widely used since the 1950s [18]. In a shock tube, fast compression, leading to sharp increases in density and translational temperature, takes place in a thin layer behind the wave front as a shock wave moves through a gas. Subsequently, the translational energy is transferred into rotational and vibrational degrees of freedom, so translational energy decreases and density increases (the pressure remains approximately constant). The relaxation time is determined by the density profile or by other measured parameters behind the shock front.

Supersonic gas expansion is another useful method for the study of rotational relaxation at low temperatures [19]. The rapid fall in gas density and temperature beyond the nozzle results in a slowing-down of the relaxation processes. Therefore, the rotational energy has no time to follow the decrease of translational temperature. Information on relaxation times can be obtained from investigation of the gas state at different distances from the nozzle.

At present, most data on rotational relaxation times have been obtained for diatomic molecules (see [20] (review) and [21]). Hydrogen and its isotopes have been most studied. They have the largest rotational constants of all molecules ($B_e = 60.85$ and 30.44 cm^{-1}, for H_2 and D_2, respectively). For this case, the rotation time is less than the duration of a molecular collision at room temperature (adiabatic Massey parameter > 1). Thus, the probabilities of rotational transitions are small and energy exchanges between translational and rotational degrees of freedom are inhibited.

Rotational relaxation times τ_{RT} are usually expressed in dimensionless form, as $Z_{RT} = \tau_{RT}/\tau_0$, where τ_0 is the mean time between collisions (mean free time), or in the form $\tau_{RT} p$, where p is the pressure. Z_{RT} is the number of collisions needed for rotational relaxation. For example, for H_2–H_2 collisions at room temperature, $Z_{RT} \simeq 10^2$. This number decreases with increasing temperature.

Values of Z_{RT} for pure N_2 and O_2, and for collisions between N_2 and rare gas atoms, are shown in Table 6.4.

The rotational constant B_e of most diatomic molecules is small. For example, for N_2, $B_e = 1.998$ cm^{-1}=2.9 K. Therefore, the duration of a collision is short in comparison with the rotation time (Massey parameter < 1). In this case, the collision is impulsive, and the transferred energy, which is approximately equally distributed between translational and rotational degrees of freedom (see Chap. 2), is of the order of kT. In this case, $Z_{RT} \sim 1$ and multi-quantum transitions between rotational levels are highly probable.

The experimental values of τ_{RT} for N_2–N_2 collisions, in the temperature range 6–78 K, can be fitted by the expression [36]

$$\tau_{RT}[N_2] = 4.9 \times 10^8 T^{0.78}, \tag{6.13}$$

where the time τ_{RT} is in seconds, the density $[N_2]$ in cm^{-3} and the gas temperature T in K.

For N_2–Ar collisions in the temperature range 15–200 K, the following fitting expression has also been obtained:

$$\tau_{RT}[Ar] = 2.3 \times 10^9 T^{0.66} \tag{6.14}$$

using the same units as before.

Table 6.4. Rotational relaxation data for N_2 and O_2

T (K)	Z_{RT}	Ref.	T (K)	Z_{RT}	Ref.
	N_2-N_2			N_2-He	
77.1	1.8	[23]	309	3.8–8.6	[31]
80	4.1	[24]		N_2-Ne	
86	4.1	[24]	309	1.9–3.1	[31]
110	3.4	[24]		N_2-Ar	
150	4.0	[24]	309	2.5–3.6	[31]
180	4.3	[24]		N_2-Xe	
223	4.6	[24]	309	3.8–5.8	[31]
273	6.2	[25]		O_2-O_2	
293	5.6	[23]	162	0.7–2.6	[23]
300	4.3	[26]	196	32	[32]
366	3.8	[27]	229	23	[32]
398	5.0	[27]	239	2	[33]
444	6.1	[27]	273	16	[25]
500	3.8	[28]	293	3	[33]
580	13.8	[29]	300	4.1	[34]
676	10.6	[29]	366	3.1	[27]
700	8.1	[26]	400	7.0	[35]
830	5.5	[30]	495	5.6	[29]
1000	9.8	[26]	700	6.7	[26]
1250	6.0	[31]	1000	7.6	[26]
1300	11.4	[26]	1300	8.5	[26]

Values of Z_{RT} for N_2-N_2 collisions at high temperatures (1500–6000 K), obtained in shock tube experiments, are also available [37].

As seen from equations (6.13) and (6.14), the values of Z_{RT} and τ_{RT} for N_2 at very low gas temperatures increase with this temperature. This can be explained by the influence of the attractive part of the intermolecular potential. At very low temperatures, this attractive potential increases the kinetic energy of the colliding particles, and thus it also increases the rotational-translational energy exchange (see Chap. 2). As the temperature increases, the influence of the attractive potential becomes less and less important and τ_{RT} (or Z_{RT}), therefore, decrease. The dependence of Z_{RT} on the gas temperature for nonpolar molecules was theoretically investigated in [37,38].

Values of Z_{RT} for CO_2-CO_2 and H_2O-H_2O collisions are given in Table 6.5.

It should be noted that Z_{RT}, for the polar H_2O molecules, somewhat decreases as the temperature rises, unlike the situation for the nonpolar CO_2 molecules.

The above analysis of experimental data shows that $Z_{RT} \sim 5$–10 for most molecules. However, light molecules (H_2, D_2) or molecules excited in upper rotational levels, which are separated by large energy gaps from neighboring levels, are exceptions to this. These results are relevant for determining the space and time scales of rotational non-equilibrium phenomena.

Rotational relaxation has received new attention in recent years due to its effects on transport coefficients. A variety of semiclassical and quantum

Table 6.5. Rotational relaxation data for pure CO_2 and H_2O

T (K)	Z_{RT}	Ref.	T (K)	Z_{RT}	Ref.
CO_2-CO_2			H_2O-H_2O		
284	2.4	[39]	350.0	2.2	[44]
298	0.6	[40]	381.2	2.7	[28]
308	2.8	[41]	400.0	1.4	[44]
475	2.8	[29]	426.1	2.5	[28]
500	3.9	[42]	450.0	0.9	[44]
504	1.2	[43]	478.0	2.3	[28]
580	3.3	[29]	500.0	0.5	[44]
676	4.1	[29]	525.6	2.0	[28]

calculations have been made to obtain accurate values of rotational relaxation times [45–49]. At the same time DSMC (Direct Simulation Monte Carlo) is now being extensively used to calculate rotational relaxation times and rotational distribution functions [50–52].

References

1. Hirschfelder J.O., Curtiss C.F. and Bird R.B. (1964) *Molecular Theory of Gases and Liquids*, John Wiley and Sons, New York
2. Kaye G.W. and Laby T.H. (1959) *Tables of Physical and Chemical Constants*, Longmans, London
3. Eletsky A.V. (1966) in *Physical Values*, ed. I.S.Grigoriev and E.Z.Melikhov, Energoizdat, Moscow, Vol. 375 (*in Russian*)
4. *Chemist's Handbook* (1964) ed. B.P.Nikolsky, Nauka, Moscow, Vol. 3 (*in Russian*)
5. Marrero T.A. and Mason E.A. (1972) J. Phys. Chem. Ref. Data **1**, 3
6. Zipf E.C. Jr. (1963) J. Chem. Phys. **38**, 2034; Haydon S.C. Fewell M.P., Ernest E.D., and Baldwin M.J. (1996) Chem. Phys. **206**, 246; Fewell M.P. and Haydon S.C. (1997) J. Phys. D: Appl. Phys. **30**, 1778
7. Morgan J.E. and Schiff H.I. (1964) Can. J. Chem. **42**, 2300
8. Cernogora G. and Sadeghi N. (1980) Chem. Phys. Lett. **74**, 417
9. Lawton S.A. and Phelps A.V. (1978) J. Chem. Phys. **69**, 1055
10. Pinheiro M. (1993) PhD Thesis, Instituto Superior Tecnico, Lisbon
11. Lede J. and Villermaux J. (1974) J. Chimie Phys. **71**, 85
12. McDaniel E.W. and Mason E.A. (1973) *The Mobility and Diffusion of Ions in Gases*, John Wiley and Sons, New York
13. Dalgarno A., McDowell M.R.C. and Williams A. (1958) Philos. Trans. **A250**, 411
14. Banks P.M. and Kockarth G. (1973) Aeronomy, Part B, No. 4, Academic Press, London
15. Schunk R.W. and Nagy A.F. (1978) Rev. Geophys. Space Phys. **16**, 355
16. Capitelli M., Gorse C., Longo S., and Giordano D. (2000) J. Termophys. and Heat Transfer **14**, p. 259; Capitelli M., Celiberto R., Gorse C., and Giordano D. (1996) Plasma Chem. Plasma Process. **16**, 267S; Capitelli M., Colonna G., Gorse C., and D'Angola A. (2000) European Journal of Physics D (in press); Casavola A., Cascarano C., Milella A., Minelli P., Mininni R., Pagano D., Sardella E., Capitelli M., Colonna G., and Gorse C. (1999) ESA **SP-426**, 357

17. Mihailov I.G., Soloviev B.A. and Syrnikov Yu.P. (1964) *Basis of Molecular Acoustics*, ed. I.G. Mihailov, Nauka, Moscow (*in Russian*)
18. Stupochenko J.V., Losev S.A. and Osipov A.I. (1967) *Relaxation Processes in Shock Waves*, Springer, Berlin Heidelberg
19. Sharafutdinov R.G. (1983) in *High Temperature Gas Dynamics, Shock Tubes and Shock Waves*, ITMO, Minsk, Vol. 13 (*in Russian*)
20. Osipov A.I. (1990) in *Plasma Chemistry*, ed. B.M. Smirnov, Energoatomizdat, Moscow **16**, 3 (*in Russian*)
21. Bogdanov A.V., Dubrovsky J.V., Osipov A.I. and Strelchenya V.M. (1991) *Rotational Relaxation in Gases and Plasmas*, Energoatomizdat, Moscow (*in Russian*)
22. Eletsky A.V., Palkina L.A. and Smirnov B.M. (1975) *Transfer Phenomena in Weakly Ionized Plasma*, Atomizdat, Moscow (*in Russian*)
23. Prangsma G.J., Alberga A.H. and Becnakker J.J. (1973) Physica **64**, 278
24. Sakharov N.P. (1983) Inf. VINITI **N 5683-83** (*in Russian*)
25. Boyer R.A. (1951) J. Acoust. Soc. America **23**, 176
26. Carnevale E.H., Carey C. and Larson G. (1967) J. Chem. Phys. **47**, 2829
27. Healy R.N. and Storovick T.S. (1969) J. Chem. Phys. **50**, 1419
28. Baker C.E. and Brokaw R.S. (1964) J. Chem. Phys. **40**, 1523
29. Annis B.K. and Malinauskas A.P. (1971) J. Chem. Phys. **54**, 4763
30. Butherus T.F. and Storvick T.S. (1974) J. Chem. Phys. **60**, 521
31. Kistemaker P.G. and De Vries A.E. (1975) Chem. Phys. **7**, 371
32. Zink H., Van Jtterbeck A. and Bose T.K. (1965) Phys. Lett. **16**, 34
33. Parker J.G., Adams C.E. and Slavseth R.M. (1958) J. Acoust. Soc. America **25**, 253
34. Creenspan M. (1959) J. Acoust. Soc. America **31**, 155
35. Green E.F. and Hornig D.F. (1953) J. Chem. Phys. **21**, 617
36. Belikov A.J. (1986) PhD Thesis, Novosibirsk (*in Russian*)
37. Brau C.A. and Jonkman R.M. (1970) J. Chem. Phys. **52**, 477
38. Parker J.G. (1959) Phys. Fluids **2**, 449
39. O'Neal C.J. and Brokaw R. (1963) Phys. Fluids **6**, 1675
40. Bogdanov A.V. and Gorbachev I.E. (1982) Preprint Phys. Techn. Inst. Leningrad, **70** (*in Russian*)
41. Kistemaker P.G., Tom A. and De Vries A.E. (1970) Physica **48**, 414
42. Tip A., Los J. and De Vries A.E. (1967) Physica **35**, 489
43. Malinauskas A.P., Gooch J.W. and Annis B.K. (1970) J. Chem. Phys. **53**, 1317
44. Zeleznik B.J. and Svehla R.A. (1970) J. Chem. Phys. **53**, 632
45. Parker G.A. and Pack R.T. (1978) J. Chem. Phys. **68**, 1585
46. Mc Court F. R. W., Beenaker J. J. M., Kohler W. E., and Kuscer I. (1990) *Nonequilibrium Phenomena in Polyatomic Gases*, Clarendon Press, Oxford
47. Dichinson A.S. and Lee M.S. (1986) J. Phys. B: At. Mol. Phys. **19**, 3091
48. Billing G.D. and Wang L. (1992) J. Phys. Chem. **96**, 2572
49. Nyeland C. (1996) in *Molecular Physics and Hypersonic Flows* ed. M. Capitelli, Kluwer, Dordrecht, NATO ASI Series Vol. 482, p. 393
50. Koura K. (1999) in *Rarefied Gas Dynamics*, eds. R. Brun, R. Campargue, R. Gatignol, and J.C. Lengrand, Cépadues Editions, Toulouse, Vol. 2, p. 23
51. Koura K. (1997) Phys. Fluids **9**, 3543
52. Wysong I.J. and Wadsworth D.C. (1999) in *Rarefied Gas Dynamics*, eds. R. Brun, R. Campargue, R. Gatignol, and J.C. Lengrand, Cépadues Editions, Toulouse, Vol. 2, p. 321

7. Rate Coefficients for Vibrational Relaxation

The experimental data on vibrational relaxation, like that on rotational relaxation, can be divided into three categories. The first category includes the characteristic times of vibrational energy relaxation. The second covers the probabilities – or rate coefficients – of the transitions between different levels. The third category includes the differential cross-sections for inelastic collisions leading to vibrational transitions. The probabilities P_{mn} used in kinetic calculations are related to the average of the differential cross-sections over the scattering angles and relative energies of the colliding particles.

The first two categories of experimental results referred to above represent the bulk of the available data. The experimental methods used to obtain these data are based on the study of different phenomena related to vibrational non-equilibrium. Analysis of the relaxation zone behind a shock wave in shock wave tubes is the principal method for investigation of vibrational relaxation at high temperatures (≥ 1000 K). Most results at temperatures $T \leq 1000$ K are obtained from ultrasound dispersion and absorption characteristics connected with vibrational non-equilibrium. Optical methods (in particular, laser pumping) for creation of non-equilibrium are also frequently used to investigate vibrational kinetics.

7.1 Rate Coefficients of V-T Relaxation

Vibrational-translational (V-T) relaxation processes for small deviations from equilibrium can be fairly well described by the harmonic oscillator model. In this case, as shown in Chap. 3, the vibrational relaxation time τ_{VT} of diatomic molecules is connected with the probability P_{10} of $1 \to 0$ transitions by the Landau–Teller relation (3.17):

$$\tau_{VT} = \frac{1}{ZP_{10}\left(1 - e^{-\hbar\omega/kT}\right)}, \tag{7.1}$$

where Z is the collision frequency, ω is the frequency of the transition $1 \to 0$, and T is the gas temperature. In this case, the probability P_{10} is connected with the probabilities of $n \to n \pm 1$ transitions by expressions (3.24) and (3.23). Generally speaking, relation (7.1) is not correct for polyatomic molecules or for diatomic molecules under strong vibrational non-equilibrium.

Vibrational relaxation processes of the atmospheric molecules N_2, O_2, CO, CO_2, N_2O, NO_2, and H_2O have now been experimentally investigated in detail. These results can usually be described by simple and well theoretically founded analytical approximations.

The rate coefficient k_{10} (cm^3/s) for deactivation of the lowest excited vibrational level, as well as the dimensionless value P_{10}, are generally used to describe the experimental data on vibrational relaxation. Note that P_{10} and k_{10} are related through the expression

$$k_{10} = P_{10} \frac{Z}{N}, \qquad (7.2)$$

where $Z/N = 2d_{12}^2 (2\pi kT/\mu)^{1/2}$ is the number of collisions per unit time of a relaxing molecule with particles of the surrounding gas at unit density, d_{12} is the collision diameter ($d_{12} = \frac{1}{2}(d_1 + d_2)$, where d_1 and d_2 are the hard-sphere diameters of the colliding particles), and μ is the reduced mass of the colliding particles.

An empirical formula for k_{10}, based on the SSH theory [1], was proposed in [2]:

$$k_{10} \text{ (cm}^3\text{/s)} = 0.92 \times 10^{-6} \left(\frac{\mu}{10}\right)^{2.06} \left(\frac{\omega}{2000}\right)^{2.66}$$

$$\exp\left\{-22.37 \left(\frac{\omega}{2000}\right)^{0.681} \left(\frac{\mu}{10}\right)^{0.302} \left(\frac{T}{1000}\right)^{-1/3}\right\}. \qquad (7.3)$$

Here the reduced collision mass μ is given in a.u., the vibrational frequency ω for transition $1-0$ is in cm^{-1}, and the gas temperature T is in Kelvin.

Expression (7.3) describes within an accuracy of a factor of ~ 3 [2] much of the experimental V-T relaxation data for: O_2 in an atmosphere of Ar, He, or H_2; N_2 in an atmosphere of N_2; and CO in an atmosphere of Ar, CO, He, or H_2, for $T \geq 1000$ K.

More exact approximations of experimental data exist for each specific pair of colliding particles. The following approximations (which are also founded by SSH theory [1]) can also be used to describe experimental data:

$$k_{10} \text{ (cm}^3\text{/s)}$$
$$= AT^n \exp\left\{-\frac{B}{T^{1/3}} + \frac{C}{T^m}\right\} \times \left[1 - D \times \exp\left\{-\frac{E_{10}}{T}\right\}\right]^{-1}. \qquad (7.4)$$

Here the energy of transition $1 \to 0$ is measured in Kelvin. The parameter D equals 1 or 0. Taking $D = 1$ enables the factor in the curved brackets in (7.1) to be taken into account. In some cases, this also enables more precise values of k_{10} to be obtained from data on relaxation times τ_{VT}, whenever such data are available. This can be important for high temperatures $T \geq E_{10}$.

Recommended values of the parameters n, m, A, B, C, D in (7.4) are given in Tables 7.1–7.8 for V-T relaxation of N_2, O_2, H_2, CO, CO_2, N_2O,

NO_2, and H_2O molecules in collisions with various species. In Tables 7.1–7.8 the temperature T is in Kelvin, A is in cm^3/s, B is in K$^{1/3}$, and C is in K^m.

Note that the data in Tables 7.1, 7.2, 7.4–7.8 are taken from reviews [3,10] where experimental results from different authors have been used to obtain the fitting coefficients n, m, A, B, C, D in (7.4). Most of these fitting values apply in the temperature range $T \geq 300$ K.

For the modelling of vibrational kinetic processes, the rate coefficients $k_{v+1,v}$ for transitions between upper vibrational levels $v \geq 1$ need often to be known. These coefficients can be calculated from the expression

$$k_{v+1,v} = k_{10} G(v+1), \qquad (7.5)$$

where $G(v+1)$ is an appropriate scaling function. Using SSH theory [1] and some approximations [11] for the Morse oscillator model, the following expression for $G(v+1)$ is obtained:

$$G(v+1) \simeq Z_{v+1} \exp\{v\delta_{VT}\}, \qquad (7.6)$$

$$Z_{v+1} \simeq \frac{(v+1)(1-x_e)}{1-(v+1)x_e}, \qquad (7.7)$$

$$\delta_{VT} \simeq \begin{cases} \frac{4\Delta E}{3E_{10}}\gamma_0 \equiv 0.427\Delta E L\sqrt{\frac{\mu}{T}} & \text{for } \gamma_v < 20 \\ \frac{4\Delta E}{E_{10}}\gamma_0^{2/3} \equiv \frac{1.87\Delta E L^{2/3}\mu^{1/3}}{E_{10}^{1/3}T^{1/3}} & \text{for } \gamma_v \geq 20 \end{cases} \qquad (7.8)$$

$$\gamma_v = 0.32 E_{v+1,v} L \sqrt{\frac{\mu}{T}}, \qquad (7.9)$$

Table 7.1. Values of the parameters in (7.4) for rate coefficients of the processes $N_2(v=1)+M \rightarrow N_2(v=0)+M$

M	n	m	A	B	C	D	Ref.
N_2	1	1	7.8×10^{-12}	218	690	1	[3]
H_2	1	2/3	4.9×10^{-12}	167.1	394	1	[3]
He	1/3	1	3×10^{-8}	196	1680	0	[3]
H_2O	1	0	2.5×10^{-15}	21.18	0	0	[3]
CO_2	1	1	1.1×10^{-12}	218	690	1	[3]

Table 7.2. Values of the parameters in (7.4) for rate coefficients of the processes $O_2(v=1)+M \rightarrow O_2(v=0)+M$

M	n	m	A	B	C	D	Ref.
O_2	1	0	1.35×10^{-12}	137.9	0	1	[3]
H_2	1	0	2.69×10^{-10}	91.5	0	1	[3]
He	1	0	4.54×10^{-14}	60.85	0	1	[3]
Ar	1	2	3.14×10^{-12}	173.1	6.2×10^5	1	[3]

7. Rate Coefficients for Vibrational Relaxation

Table 7.3. Values of the parameters in (7.4) for rate coefficients of the processes $H_2(v=1)+M \to H_2(v=0)+M$

M	n	m	A	B	C	D	Ref.
H_2	1/2	0	7.47×10^{-12}	93.87	0	1	[4–7]
He	1	0	1.31×10^{-13}	95.2	0	1	[5]
Ne	1	0	1.69×10^{-14}	65.5	0	1	[5]
Ar	1	0	1.23×10^{-13}	103.8	0	1	[5]
Kr	1	0	1.11×10^{-14}	84	0	1	[5]

Table 7.4. Values of the parameters in (7.4) for rate coefficients of the processes $CO(v=1)+M \to CO(v=0)+M$

M	n	m	A	B	C	D	Ref.
CO	1	0	$1. \times 10^{-12}$	150.7	0	0	[3]
He	1/3	1	3.02×10^{-8}	188.7	1542	0	[3,9]
Ar	0	0	3.1×10^{-6}	273	0	0	[9]
Kr	1	0	3.07×10^{-11}	187	0	1	[3,9]

Table 7.5. Values of the parameters in (7.4) for rate coefficients of the processes $CO_2(01^10)+M \to CO_2(00^00)+M$

M	n	m	A	B	C	D	Ref.
CO_2	1	2/3	2.89×10^{-12}	132	360.8	1	[10]
N_2	1	2/3	1.4×10^{-11}	158.4	434	1	[10]
O_2	1	2/3	7.05×10^{-12}	147.6	392.3	1	[10]
H_2	1	2/3	1.09×10^{-13}	65.5	331.6	1	[10]
NO	1	0	3.66×10^{-16}	6.45	0	1	[10]
CO	1	2/3	1.24×10^{-12}	118.9	280.5	1	[10]
H_2O	1	0	1.86×10^{-13}	4835.4	0	1	[10]

Table 7.6. Values of the parameters in (7.4) for rate coefficients of the processes $N_2O(01^10)+M \to N_2O(00^00)+M$

M	n	m	A	B	C	D	Ref.
N_2O	1	2/3	1.25×10^{-13}	75.7	204.8	0	[10]
N_2	1	0	3.26×10^{-14}	38.5	0	0	[10]
O_2	1	0	3.88×10^{-14}	40.9	0	0	[10]
H_2	1	0	9.66×10^{-15}	0	0	0	[10]
NO	1	0	1.64×10^{-15}	3.7	0	0	[10]
CO	1	2/3	3.41×10^{-11}	150.5	452.4	0	[10]

Table 7.7. Values of the parameters in (7.4) for rate coefficients of the processes $NO_2(010)+M \rightarrow NO_2(000)+M$

M	n	m	A	B	C	D	Ref.
NO_2	1	0	7.25×10^{-22}	-9.1	0	0	[10]
N_2	1	0	4.11×10^{-13}	65.2	0	0	[10]
O_2	1	0	5.25×10^{-13}	64.9	0	0	[10]
CO	1	0	3.66×10^{-12}	64.9	0	0	[10]

Table 7.8. Values of the parameters in (7.4) for rate coefficients of the process $H_2O(010)+M \rightarrow H_2O(000)+M$

M	n	m	A	B	C	D	Ref.
H_2O	1	2/3	3.21×10^{-8}	228.2	930.5	0	[10]
N_2	1	2/3	5.36×10^{-16}	18.6	47.2	0	[10]
O_2	1	2/3	5.67×10^{-19}	96.8	268.3	0	[10]

where: x_e is the anharmonicity of the molecule; ΔE is the anharmonicity (in K); $E_{v+1,v}$ is the energy of the vibrational transition $v+1 \rightarrow v$ (in K); μ is the reduced collision mass (in a.u.); and L (in Å) is the characteristic parameter of the short-range repulsive potential $U \sim \exp(-r/L)$. The values of L are usually in the range 0.16– 0.4 Å (typically, ~ 0.2 Å).

The rate coefficients $k_{v+1,v}$ increase with v and, for upper levels, their magnitudes strongly depend on the chosen values of the parameters δ_{VT} or L. It is important to note that these parameters must be chosen in such a way that the values of $k_{v+1,v}$ for upper levels do not exceed the value of k_{TT}. This should be carefully checked when a choice of parameters is made for given modelling purposes. Further, SSH theory should be used with caution for transitions involving high-lying vibrational levels [12,13].

Besides vibrational relaxation by collisions with molecules and atoms of inert gases, V-T relaxation by collisions with dissociated atoms (such as N, O, H) can be important in high temperature molecular gases and plasmas, in which the degree of dissociation can be quite high. These chemically active atoms are generally more effective for vibrational deactivation than molecules and atoms of inert gases. For example, the efficiency of vibrational relaxation by O atoms is explained by their electron-vibrational non-adiabacity in collisions [14]. Let us briefly discuss the data for V-T relaxation by collisions with O, N, and H atoms.

We consider first processes of the type

$$N_2(v) + M \rightarrow N_2(v' < v) + M \quad M = O, N, H. \tag{7.10}$$

The V-T relaxation (7.10) is fast in the case M=O, owing to its chemical nature and the nonadiabatic, vibrational-electronic interactions involving $O(^3P_{2,1,0})$ atoms [14]. The experimental data [15,16] have in some cases been approximated by the Landau–Teller expression:

$$k_{10}\,(\text{cm}^3/\text{s}) = 1.2 \times 10^{-13} \exp\left(-27.6 T^{-1/3}\right). \tag{7.11}$$

However, this temperature dependence, following from SSH theory [1] for adiabatic transitions, is not typical of nonadiabatic processes. An Arrhenius-type approximation of the form

$$k_{10}\,(\text{cm}^3/\text{s})$$
$$= 2.3 \times 10^{-13} \exp\left(-\frac{1280}{T}\right) + 2.7 \times 10^{-11} \exp\left(-\frac{10840}{T}\right) \tag{7.12}$$

can be used instead of approximation (7.11).

The experimental estimations [17] for V-T relaxation of N_2 by nitrogen atoms show that V-T relaxation for the level $v = 1$ is very slow. Numerical calculations [18,19] confirm this and show that the rate coefficients increase sharply with the vibrational level number and gas temperature. The numerical calculations of the total rate coefficients $k_v(N_2-N)$ for transitions from level v to all lower levels $v - \Delta v$, $1 \leq \Delta v \leq v$ can be approximated by the analytical expression

$$k_v = \begin{cases} k^0 \exp\left\{-\frac{E_a}{T} + \frac{\beta E_v}{T}\right\} & \text{if } \beta E_v \leq E_a \\ k^0 & \text{if } \beta E_v \leq E_a. \end{cases} \tag{7.13}$$

Here, E_v is the vibrational energy of level v (in K); k^0, E_a, β are process parameters. Note that equation (7.13) has sometimes been used [20,21] to describe interaction of chemically active atoms with vibrationally excited molecules when there is an energy barrier for the interaction (see Chap. 10). In (7.13), E_a is the energy of this barrier (the activation energy); $\beta \leq 1$ is a factor accounting for a possible decrease of the activation energy due to the vibrational energy, E_v, of the molecule.

Multi-quantum V-T relaxation in collisions of molecules with chemically active atoms usually has a chemical nature. For this reason equation (7.13) can be used for an approximate description of this type of V-T relaxation.

For simplicity, it can also be assumed that the rate coefficients $k_{v,v-\Delta v}$ for V-T transitions $v \to v - \Delta v$ are connected with the total rate coefficient (7.13) by the relationship:

$$k_{v,v-\Delta v} \simeq \begin{cases} \frac{1}{v} k_v & \text{for } 1 \leq v \leq \Delta v_m \\ \frac{1}{\Delta v_m} k_v & \text{for } v > \Delta v_m;\, \Delta v \leq \Delta v_m \\ 0 & \text{for } v > \Delta v_m;\, \Delta v > \Delta v_m. \end{cases} \tag{7.14}$$

For $N_2(v)+N$ collisions the value $k_0 \simeq 4 \times 10^{-10}(T/300)^{0.5}$ (cm^3/s) has been assumed and E_a, β have been obtained [22] by fitting (7.13) to numerical calculations [18,19] for $T = 500$ K. It has been found that the best fit was obtained for $E_a \simeq 7280$ K; $\beta \simeq 0.065$. The numerical results [18,19] show that the main channels of V-T relaxation are the transitions to the five lower levels. Therefore $\Delta v_m = 5$ can be used in (7.14).

7.1 Rate Coefficients of V-T Relaxation

The process (7.10) with M=H is important for the kinetics of upper N_2 vibrational levels. The rate coefficients for this process have been estimated using equations (7.13) and (7.14). According to ab initio calculations [23], the collision N_2–H has an energy barrier $E_a \simeq 7500$ K for the production of short-lived N_2H species, which subsequently dissociate to yield N_2–H again. If we assume that the V-T processes (7.10) for M=H take place through the formation of N_2H species as an intermediate step, the results in [23] can be used to estimate the rate coefficients. For these processes, the values $E_a = 7500$ K, $k^0 \simeq 4 \times 10^{-10}(T/300)^{0.5}$ (cm^3/s), and $\Delta v_m = v$ in equations (7.13), (7.14) have been proposed [22]. The parameter β in (7.13) was obtained by fitting the calculated relative changes in the electric field E and the $N_2^+(B)$ emission to measurements in an N_2 discharge with a small admixture of H_2. Due to dissociation in the discharge, this admixture provides H atoms which decrease the vibrational level population of N_2 via V-T process (7.10). Consequently, the rate of associative ionization of N_2 and $N_2^+(B)$ pumping also decreases (see Chap. 12). The best fit was found to occur for $\beta \simeq 0.105$.

We next consider the process

$$O_2(v) + O \rightarrow O_2(v' < v) + O . \tag{7.15}$$

This process has been experimentally investigated for the transition $v = 1 \rightarrow v = 0$ in the temperature range $T = 1000$–3400 K [24-26]. It is a very fast process, with a rate coefficient

$$k_{10}(O_2 - O) \, (\text{cm}^3/\text{s}) \simeq 4.5 \times 10^{-15} T , \tag{7.16}$$

where T is in K.

The relaxation

$$NO(v) + O \rightarrow NO(v' < v) + O \tag{7.17}$$

has a chemical nature. The rate coefficients for the transitions $1 \rightarrow 0$, $2 \rightarrow 1$, $2 \rightarrow 0$ have been measured [27,28] at $T = 2700$ K and $T = 300$ K, respectively. Calculations have also been carried out in this temperature range [29]. These investigations have shown that the rate coefficient k_{10} depends very weakly on the gas temperature. Its value is

$$k_{10}(NO - O) \, (\text{cm}^3/\text{s}) \simeq 6.5 \times 10^{-11} . \tag{7.18}$$

The rate coefficients of the transitions $2 \rightarrow 1$ and $2 \rightarrow 0$ are $0.6 \times k_{10}$ and $0.8 \times k_{10}$, respectively. Note that the excitation of $NO(v = 1)$ by atomic oxygen (the reverse process of (7.17)) followed by the infrared emission $NO(v = 1) \rightarrow NO(v = 0) + h\nu$ is an important energy balance mechanism of the Earth's thermosphere at altitudes of $120 - 200$ km (see also Chap. 14).

The rate coefficient of the process

$$CO(v) + O \rightarrow CO(v' < v) + O \tag{7.19}$$

for the transition $1 \rightarrow 0$ has been measured in the range $T = 1800$–4000 K [30] and $T = 265$–389 K [31]. The results obtained can formally be described to within an accuracy of a factor of ~ 2 assuming a Landau–Teller temperature dependence $k_{10} \sim \exp\{-B/T^{1/3}\}$ [30]. However, as for process (7.10), it is preferable to assume here an Arrhenius- type temperature dependence, in which case the data can be approximated as follows:

$$k_{10}(\mathrm{CO-O})\,(\mathrm{cm}^3/\mathrm{s}) \simeq 5.3 \times 10^{-13} T^{0.5} \exp\left\{\frac{-1600}{T}\right\}, \qquad (7.20)$$

where T is in K.

For the process

$$\mathrm{CO}_2(01^10) + \mathrm{O} \to \mathrm{CO}_2(00^00) + \mathrm{O}, \qquad (7.21)$$

the rate coefficient has been estimated in [32–34]. Although the estimations of these works are similar in magnitude in the range $T = 200$–300 K, the proposed temperature dependences differ as follows

$$k_{10}(\mathrm{CO}_2-\mathrm{O})\,(\mathrm{cm}^3/\mathrm{s}) \simeq \begin{cases} 1.5 \times 10^{-11} \exp\{-800/T\} & [32,33] \\ 5 \times 10^{-13} & [34], \end{cases} \qquad (7.22)$$

where T is in K.

For the process

$$\mathrm{O}_3(010) + \mathrm{O} \to \mathrm{O}_3(000) + \mathrm{O} \qquad (7.23)$$

the measured rate coefficient at room temperature is [35]

$$k_{10}(\mathrm{O}_3-\mathrm{O})\,(\mathrm{cm}^3/\mathrm{s}) \simeq 3.1 \times 10^{-12}. \qquad (7.24)$$

Generally speaking, the deactivation of $\mathrm{O}_3(010)$ by O atoms may occur via the chemical channel $\mathrm{O}_3(010) + \mathrm{O} \to 2\mathrm{O}_2$. With such an assumption, the measurements and estimations [35] yield a value for $k_{10}(\mathrm{O}_3-\mathrm{O})$ approximately two times smaller than (7.24).

7.2 Rate Coefficients of V-V and V-V′ Processes

Let us briefly discuss the available data on the rate coefficients for the following V-V energy exchange processes:

$$A(v) + A(w) \to A(v-1) + A(w+1). \qquad (7.25)$$

The rate coefficient k_{10}^{01} of process (7.25) for N_2 molecules has been both experimentally estimated [36–38] and calculated by the trajectory method [39] and SSH theory [40]. The obtained results can be summarized as follows:

$$k_{10}^{01}(\mathrm{N}_2-\mathrm{N}_2)\left(\frac{\mathrm{cm}^3}{\mathrm{s}}\right) \simeq \begin{cases} 0.9 \times 10^{-14}(T/300)^{1.5} & [39] & (7.26\mathrm{a}) \\ 2.5 \times 10^{-14}(T/300)^{1.5} & [36] & (7.26\mathrm{b}) \\ 10.9 \times 10^{-14}(T/300)^{1.5} & [37,38,40]. & (7.26\mathrm{c}) \end{cases}$$

The reason for these different results is unclear. The disagreement between (7.26b) and (7.26c) may be connected with the choice of the parameter δ_{VV}

7.2 Rate Coefficients of V-V and V-V' Processes

(see below), which influences the values of $k_{v,v-1}^{w,w+1}$ for transitions between upper levels and, together with k_{10}^{01}, determines the V-V relaxation kinetics. For example, in a quasi-stationary regime the ratio $k_{10}^{01}/\delta_{VV}^3$ determines the energy flux to the upper vibrational levels (see (3.95) and (3.108)).

Calculations based on SSH theory [40] and the trajectory method [41,42] yield the following values of k_{10}^{01} for O_2 and H_2 molecules:

$$k_{10}^{01}(O_2 - O_2)\,(\text{cm}^3/\text{s}) \simeq \begin{cases} 0.9 \times 10^{-14}(T/300)^{1.5} & [40] \quad (7.27a) \\ 2 \times 10^{-14}(T/300)^{1.04} & [41] \quad (7.27b) \end{cases}$$

$$k_{10}^{01}(H_2 - H_2)\,(\text{cm}^3/\text{s}) \sim 4.23 \times 10^{-15}(300/T)^{1/3} \quad [42] . \quad (7.28)$$

Equation (7.28) is a fitting expression [7] providing agreement with measurements of $k_{21}^{01}(H_2-H_2)$ for $T = 300$ K [43]. For the analysis of the experimental data [43], the values of $k_{21}^{01}(H_2-H_2)$, $k_{10}(H_2-H_2)$, and $k_{21}(H_2-H_2)$ have been taken from calculations [42] which yield values of $k_{10}(H_2-H_2)$ in agreement with experiment [4].

Since the CO molecule has a dipole moment, the long distance dipole-dipole interaction can influence the rate coefficients for vibrational transitions. Analyses of available experimental data along with theoretical considerations taking into account both short- and long-range interactions yields [8,44,45]:

$$k_{10}^{01}(CO - CO)\,(\text{cm}^3/\text{s}) \simeq 3.4 \times 10^{-10}(T/300)^{0.5}\left[S_{10}^{01} + L_{10}^{01}\right]; \quad (7.29)$$

$$S_{10}^{01} = 1.64 \times 10^{-6}T; \qquad L_{10}^{01} = 1.61/T , \quad (7.30)$$

where T is in Kelvin. The quantities S_{10}^{01} and L_{10}^{01} account for the contributions of the short- and the long-range interactions, respectively.

The V-V' processes

$$A(v) + B(w) \to A(v-1) + B(w + \Delta w) \quad (7.31)$$

can also be important for vibrational kinetics. In the processes (7.31) the molecules B and A can be either of different types or of the same type but excited in different vibrational modes, in the case of polyatomic molecules. Processes (7.31) include multiquantum transitions ($\Delta w \geq 1$). Cases in which $\Delta w \geq 2$ may be important when resonant or quasi-resonant conditions are met.

The rate coefficient k_{10}^{01} of the exothermic process (7.31) for $\Delta w = 1$, A=N_2, B=O_2 has been experimentally estimated as [17]:

$$k_{10}^{01}(N_2 - O_2)\,(\text{cm}^3/\text{s}) \simeq 3.69 \times 10^{-12}\frac{T}{300}\exp\left(-\frac{104}{T^{1/3}}\right) . \quad (7.32)$$

According to [22], the rate coefficient k_{10}^{01} of the exothermic process (7.31) for $\Delta w = 1$, A=H_2, B=N_2 is:

$$k_{10}^{01}(H_2 - N_2)\,(\text{cm}^3/\text{s}) \simeq 1.9 \times 10^{-13}\left(\frac{T}{300}\right)^{3/2} F\left(\frac{144}{T^{1/2}}\right) , \quad (7.33)$$

where $F(y)$ is given by

$$F(y) \simeq \begin{cases} 0.5\left[3 - \exp\left(-\frac{2y}{3}\right)\right]\exp\left(-\frac{2y}{3}\right) & \text{for } y \leq 20 \\ 8\left(\frac{\pi}{3}\right)^{1/2} y^{7/3} \exp\left(-3y^{2/3}\right) & \text{for } y > 20 \end{cases} \qquad (7.34)$$

The function $F(y)$ with $y = \gamma_0$ (see (7.9)) is a well known approximation of the adiabacity function in the SSH theory for the calculation of vibrational rate coefficients [1,11]. Equation (7.33) yields agreement with experiment for $T = 295$ K [46]. The temperature dependence in this expression is obtained assuming that $L = 0.25$ Å.

The function $F(y)$ can also be useful to calculate the rate coefficients $k_{v+1,v}^{w,w+1}$ for V-V and V-V' exchanges between upper levels due to short-range interactions:

$$k_{v+1,v}^{w,w+1} = k_{10}^{01} G_S(v+1, w+1), \qquad (7.35)$$

where $G_S(v+1, w+1)$ is the scaling function

$$G_S(v+1, w+1) = Z_{v+1}^A Z_{w+1}^B \frac{F(y_{v+1,v}^{w,w+1})}{F(y_{10}^{01})}. \qquad (7.36)$$

Here, Z_{v+1}^A and Z_{w+1}^B are the same functions as in (7.7), but for the anharmonicities x_e^A and x_e^B of the molecules A and B involved in process (7.31). In the case of process (7.25), B=A. The variables $y_{v+1,v}^{w,w+1}$ are defined as

$$y_{v+1,v}^{w,w+1} = 0.32(E_{v+1,v} - E_{w+1,w})L\sqrt{\frac{\mu}{T}}, \qquad (7.37)$$

where $E_{i+1,i}$ is the energy of the transition $i+1 \to i$ (in Kelvin); $E_{v+1,v} - E_{w+1,w}$ is the total change in vibrational energy due to the transition $v+1, w \to v, w+1$; L is in Å; the reduced collision mass μ is in a.u.; and T is in Kelvin.

For the V-V processes (7.25) the function $G_S(v+1, w+1)$ can be simplified as

$$G_S(v+1, w+1) \simeq Z_{v+1}^A Z_{w+1}^B \exp\left(\delta_{VV}|v-w|\right), \qquad (7.38)$$

where

$$\delta_{VV} \simeq 0.427 \Delta E L \sqrt{\frac{\mu}{T}}. \qquad (7.39)$$

In fact, $\delta_{VV} = \delta_{VT}$ (see (7.8)) for collisions between molecules of the same type. Therefore, the actual value of $k_{v+1,v}^{w,w+1}$ is determined by the choice of k_{10}^{01} and δ_{VV} (or L). For example, the values (7.27a) and $L \simeq 0.22$ Å have been recommended [40] for calculating the rate coefficients $k_{v+1,v}^{w,w+1}(O_2-O_2)$ using SSH theory. On the other hand, the fitting formula of [47] applies to trajectory calculations of $k_{v+1,v}^{w,w+1}(O_2-O_2)$ with the values (7.27b) and $L \simeq 0.18$ Å. Both types of calculations provide agreement with experimental values of $k^{01}_{v+1,v}(O_2-O_2)$, for $v \simeq 10-16$. However, the SSH calculations are preferable for $v \simeq 8$.

It is important to note that the scaling functions $G_S(v+1, w+1)$ calculated from semiclassical trajectory methods may differ from (7.36) and (7.38), which are derived from SSH theory. For example, the calculated rate coefficients $k^{w-1,w}_{v,v-1}$ for N_2-N_2 collisions [39] can be approximated by the expression [47]

$$k^{w-1,w}_{v,v-1}(N_2-N_2) \simeq k^{01}_{10} \times G_S(v,w) \times g(v,w), \tag{7.40}$$

where k^{01}_{10} is determined by (7.26a), $G_S(v,w)$ is the function (7.36) for $L = 0.2$ Å, and g(v,w) is a correction given by

$$g(v,w) = \begin{cases} 1 - 0.04 \times \text{Max}\{v,w\} & \text{for Max}\{v,w\} < 10 \\ 0.645 + 0.617 \times \left(\frac{\text{Max}\{v,w\}-10}{30}\right)^3 & \text{for Max}\{v,w\} \geq 10 \end{cases}$$
(7.41)

For H_2-H_2 collisions, the calculated rate coefficients of V-V processes [42] can also be approximated by using (7.35), (7.28), (7.38), $\delta_{VV} = 0.21\sqrt{T/300}$, and a correction function $g(v,w)$ given by [7]

$$g(v,w) = \exp[\Delta_1(v-w) - \Delta_2(v-w)^2] \quad \text{for } v > w;$$

$$\Delta_1 \simeq 0.236(T/300)^{1/4}; \quad \Delta_2 \simeq 0.0572(300/T)^{1/3}. \tag{7.42}$$

Finally, let us consider the calculations of the rate coefficients $k^{w,w+1}_{v+1,v}$ for collisions between dipole molecules (like CO). In this case

$$k^{w,w+1}_{v+1,v} = 3.4 \times \left(\frac{T}{300}\right)^{1/2}$$
$$\times [S^{01}_{10} \times G_S(v+1,w+1) + L^{01}_{10} \times G_L(v+1,w+1)]. \tag{7.43}$$

Here, S^{01}_{10}, L^{01}_{10} account for short- and long-range interactions, respectively (see (7.30)). The scaling function $G_L(v,w)$ due to long-range interactions is given by [8,44,45]

$$G_L(v+1,w+1) \simeq Z_{v+1}Z_{w+1} \exp\left[-\frac{(\Delta E)^2(v-w)^2}{bT}\right], \tag{7.44}$$

where b is a constant determined by the dipole moment of the molecule. For CO-CO collisions, $b = 39.9$ K [8,45], when ΔE and T are expressed in Kelvin. These data are in satisfactory agreement with the semiclassical calculations of V-V and V-T rates given in [48].

Finally, we want to emphasize that a complete set of V-V and V-T rates for the CO-N_2 system has been obtained by using semiclassical dynamics [49].

References

1. Nikitin E.E. (1974) *Theory of Elementary Atomic and Molecular Processes in Gases*, Clarendon Press, Oxford
2. Lifshitz A. (1974) J. Chem. Phys. **61**, 2478
3. Nikitin E.E., Osipov A.I. and Umansky C.Y.(1994) in *Reviews of Plasma Chemistry*, ed. B.M. Smirnov, Consultants Bureau, New York London p. 1
4. Audibert M.M., Vilaseca R., Lukasik J. and Ducuing J. (1976) Chem. Phys. Lett. **37**, 408
5. Dove J.L. and Teitelbaum H. (1974) Chem. Phys. **6**, 431
6. Cacciatore M., Capitelli M., and Billing G. D. (1989) Chem. Phys. Lett. **157**, 305
7. Matveev A.A. and Silakov V. P. (1995) Plasma Sources Sci. Technol. **4**, 606
8. Flament C., George T., Meister K.A., Tufts J.C., Rich J.W. and Subramanian V.V. (1992) Chem. Phys. **163**, 241
9. G.G. Chorniy and S.A. Losev, eds. *Physical-Chemical Processes in Gas Dynamics* (1998) Moscow State University, Moscow, Vol. 1 (*in Russian*)
10. Zuev A.P., Losev S.A., Osipov A.I. and Starik A.M. (1992) Chemical Physics **11**, 4 (*in Russian*)
11. Keck J. and Carrier G. (1965) J. Chem. Phys. **43**, 2284
12. Billing G. D. (1984) Comp. Phys. Rep. **1**, 237
13. Cacciatore M. (1996) in *Molecular Physics and Hypersonic Flows*, ed. M. Capitelli, Kluwer, Dordrecht, NATO ASI Series, Vol. 482, p. 21
14. Andreev E.A. and Nikitin E.E. (1976) in *Plasma Chemistry*, ed. B.M. Smirnov, Energoizdat, Moscow, Vol. 3, 28 (*in Russian*)
15. Breschears W.D. and Bird P.F. (1968) J. Chem. Phys. **48**, 4768; Eckstrom D.J. (1973) J. Chem. Phys. **59**, 2787; McNeal R.J., Whilson M.E. and Cook G.R. (1972) Chem. Phys. Lett. **16**, 507
16. Zaslonko I.S., Mukoseev Yu.K. and Smirnov V.N. (1982) Chemical Physics No 5, 662 (*in Russian*)
17. Kozlov P.V., Makarov V.N., Pavlov V.A., Uvarov A.V. and Shatalov O.P. (1996) J. Tech. Phys. **66**, 43 (*in Russian*)
18. Lagana A., Garcia E. and Ciccarelli L. (1987) J. Phys. Chem. **91**, 312; Lagana A. and Garcia E. (1996) *Quasiclassical and Quantum Rate Coefficients for the $N+N_2$ Reaction*, ed. A. Lagana, University of Perugia; Garcia E. and Lagana A. (1998) Plasma Sources Sci. Technol. **7**, 359; Lagana A., Crocchianti S., Ochoa de Aspuru G., Riganelli A., and Garcia E. (1997) Plasma Sources Sci. Technol. **6**, 270
19. Armenise I., Capitelli M., Garcia E., Gorse C., Lagana A., and Longo S. (1992) Chem. Phys. Lett. **200**, 597
20. Gordiets B.F., Osipov A.I. and Shelepin L.A. (1986) *Kinetic Processes in Gases and Molecular Lasers*, Gordon and Breach, New York (translated from Russian)
21. Rusanov V.D. and Fridman A.A. (1984) *Physics of Chemically Active Plasma*, Nauka, Moscow (*in Russian*)
22. Gordiets B.F., Pinheiro M., Ferreira C.M. and Ricard A. (1998) Plasma Sources Sci. Technol. **7**, 363
23. Walch S.P., Duchovic R.J. and Rohlfild C.M. (1984) Chem. Phys. Lett. **103**, 437
24. Kiefer J.H. and Lutz V.W. (1967) Proceed. 11th Symp. Combustion University of California, Berkeley, p. 67
25. Breen J.E., Quy R.B. and Glass G.D. (1973) J. Chem. Phys. **59**, 556
26. Dushin V.K., Zabelinsky I.E. and Shatalov O.P. (1988) Chemical Physics **7**, 1320 (*in Russian*)

27. Glanzer A. and Troe J. (1975) J. Chem. Phys. **63**, 4352
28. Fernando R.P. and Swith W.M. (1979) Chem. Phys. Lett. **66**, 218
29. Quack M. and Troe J. (1975) Ber. Bunsenges. Phys. Chem. **79**, 170
30. Center R.E. (1973) J. Chem. Phys. **58**, 5230
31. Lewittes M.E., Davis C.C. and McFarlane R.A. (1978) J. Chem. Phys. **69**, 1952
32. Gordiets B.F. and Kulikov Yu.N. (1981) Space Res. **19**, 249 (*in Russian*)
33. Gordiets B.F., Kulikov Yu.N., Markov M.N. and Marov M.Ja. (1982) J. Geophys. Res. **A87**, 4504
34. Caledonia G.E. and Kennealy J.P. (1982) Planet. Space Sci. **30**, 1043
35. West G.A. and Weston R.E. (1976) Chem. Phys. Lett. **42**, 488
36. Akishev Yu.S., Demianov A.V., Kochetov I.V. at al. (1982) Thermophys. High Temper. **20**, 818 (*in Russian*)
37. Suchkov A.F. and Shebeko Yu.N. (1981) Chem. High Energy **15**, 279 (*in Russian*)
38. Valiansky S.I., Vereschagin K.A., Volkov A.Yu. et al. (1984) Quant. Electron. **11**, 1836 (*in Russian*)
39. Billing G.D. and Fisher E.R. (1979) Chem. Phys. **43**, 395
40. Kirillov A.S. (1997) Space Res. **35**, 155 (*in Russian*)
41. Billing G.D. and Kolesnick R.E. (1992) Chem. Phys. Lett. **200**, 382
42. Cacciatore M. and Billing G.D. (1992) J. Phys. Chem. **96**, 217
43. Kreutz T.G., Gelfand J., Miles R.B. and Rabitz H. (1988) Chem. Phys. **124**, 359
44. Sharma R.D. and Brau C.A. (1969) J. Chem. Phys. **50**, 924
45. Deleon R.L. and Rich J.W. (1986) Chem. Phys. **107**, 283
46. Bott J.F. (1976) J. Chem. Phys. **65**, 3921
47. Guerra V.A. (1998) PhD Thesis, Instituto Superior Tecnico, Lisbon
48. Cacciatore M. and Billing G.D. (1981) Chem. Phys. **58**, 395
49. Cacciatore M., Billing G.D. and Capitelli M. (1984) Chem. Phys. **89**, 17

8. Electron Rate Coefficients

8.1 Integral Scattering Cross-Sections

Knowledge of the cross-sections for inelastic scattering of electrons by atoms and molecules is essential in the investigation of low temperature plasma physics, chemistry and optics. As follows from Chap. 4, the totality of scattering cross-sections determines the form of the electron energy distribution function (EEDF) and the rate coefficients for the various processes involving free electrons. There is now an extensive literature in this area (for example [1–8]). We will accordingly present here only the data on integral scattering cross-sections that is required for studying and modelling the kinetic processes in low-temperature plasmas of N_2 and O_2 and their mixtures.

Sets of cross-sections for inelastic scattering of electrons by the N_2 molecule are given in [8–15], and by the O_2 molecule in [8,16,17]. In [8], where different experimental data are presented and compared, the following analytical approximation for the dependence of most cross-sections $\sigma(\epsilon)$ on the electron energy ϵ was proposed:

$$\sigma(\epsilon) = \begin{cases} a_0^2 \left(1 - \frac{\epsilon_{\text{th}}^*}{\epsilon}\right) \exp\left[\sum_{k=0}^{4} b_k (\ln \epsilon)^k\right] & \text{for } \epsilon_{\text{th}}^* \leq \epsilon \leq \epsilon_0 \\ a_0^2 A \ln(B\epsilon)/\epsilon & \text{for } \epsilon > \epsilon_0 \quad \text{or} \\ a_0^2 C \epsilon^{-m} & \text{for } \epsilon > \epsilon_0 \,. \end{cases} \quad (8.1)$$

Here, $a_0^2 = 2.8 \times 10^{-17}$ cm^2, ϵ_{th}^* is the threshold energy, and b_k, ϵ_0, A, B, C are fitting parameters. In this expression, the energy ϵ is given in eV, and the cross-section $\sigma(\epsilon)$ is in cm^2.

The values of the parameters in equation (8.1) for the electron cross-sections of different processes in N_2 and O_2 are given in Tables 8.1 and 8.2.

The corresponding values of ϵ_0, which are not included in these tables, are as follows. For the majority of processes, this value is 100 eV. However, it equals 500 eV for elastic scattering in N_2 and O_2, and equals 40, 50, 500, 80 and 1.6, 1.7, 2.0, 3.0 eV for the excitation of, respectively, $N_2(C'^3\Pi_u)$, $N_2(a'^1\Sigma_u^-)$, $N_2(a^1\Pi_g)$, $N_2(b^1\Pi_u)$ and $v = 1, 2, 3, 4$ vibrational levels of O_2.

Note that, concerning the elastic cross-section in N_2, the fitting equation (8.1) is suitable only for energies $\epsilon \geq 4$ ev. This is explained by the major contribution to this cross-section in the region $\epsilon \sim 1.8 - 3$ eV of a resonant scattering mechanism, via the creation of a short-lived $N_2^-(^2\Pi_g)$ ion [18,19].

The cross-section of this mechanism cannot be described by equation (8.1). This mechanism is also responsible for the excitation of N_2 vibrational and rotational levels. For this reason, data on the cross-sections of vibrational and rotational excitations are not given in Table 8.1. A resonant scattering mechanism with the creation of a short-lived $O_2^-(^2\Pi_g)$ ion also occurs in electron collisions with $O_2(X^3\Sigma_g^-)$ molecules. This mechanism contributes to the cross-sections of elastic scattering and vibrational excitation, excitation of the electronic terms $O_2(a^1\Delta g)$, $O_2(b^1\Sigma_g^+)$, and dissociative attachment. It has an especially strong influence on vibrational excitation and dissociative attachment.

The expression

$$\sigma(\epsilon) = \sigma_{nr}(\epsilon) + \sigma_r^0 \left[1 + q\frac{\epsilon - \epsilon_r}{\Gamma/2}\right]^2 \left[1 + \left(\frac{\epsilon - \epsilon_r}{\Gamma/2}\right)^2\right]^{-1} \qquad (8.2)$$

can be used to describe approximately the integral cross-sections including the effect of resonant scattering [18]. Here, $\sigma_{nr}(\epsilon)$ is the nonresonant part of the scattering cross-section, and σ_r^0, q, ϵ_r and Γ are the resonant scattering cross-section parameters; σ_r^0 characterizes the cross-section magnitude and q its dependence on energy; ϵ_r is the resonance energy; and Γ is the energy width of the resonance. For $q = 0$, the resonant cross-section is symmetrical with respect to the point ϵ_r, which corresponds to the Breit–Wigner formula [20]. The actual resonant scattering cross-sections derived from experiment are generally the sum of a number of resonant cross-sections which peak at different energies, because the intermediate negative ions $N_2^-(^2\Pi_g)$ or $O_2^-(^2\Pi_g)$ can be excited in different vibrational states. For example, the cross-sections for vibrational excitation of the states $N_2(X^1\Sigma_g^+, v)$ have 3 to 6 maxima with widths of 0.2-0.5 eV [19], and those for excitation of $O_2(X^3\Sigma_g^-, v)$ states have 6 to 9 maxima with (theoretical) widths of $\simeq 10^{-3}$ eV [15,18,19]. These cross-sections are usually smoothed out by averaging over the larger experimental energy interval $\Delta\epsilon$. For example, this has been done for O_2 [8]. The smoothed cross-sections for excitation of the lowest four vibrational levels are described by (8.1) with the parameter values given in Table 8.2.

The cross-sections of electron scattering by atoms have frequently been described by approximate expressions other than (8.1). Examples include the expressions proposed by Gryzinski [21] and by Dravin [22], or the following expressions used in [8,23]:

a) for allowed transitions

$$\sigma(\epsilon) = \frac{q_0 F}{\epsilon \epsilon_{th}^*}\left[1 - \left(\frac{\epsilon_{th}^*}{\epsilon}\right)^\alpha\right]^\beta \ln\left(4\gamma\frac{\epsilon}{\epsilon_{th}^*} + e\right) ; \qquad (8.3)$$

b) for forbidden transitions

$$\sigma(\epsilon) = \frac{q_0 F}{(\epsilon_{th}^*)^2}\left[1 - \left(\frac{\epsilon_{th}^*}{\epsilon}\right)^\alpha\right]^\beta \left(\frac{\epsilon_{th}^*}{\epsilon}\right)^\delta . \qquad (8.4)$$

Table 8.1. Parameters of (8.1) for cross-sections of inelastic scattering of electrons by N_2 in the ground electronic state [8]

State	ϵ^*_{th} (ev)	b_0	b_1	b_2	b_3	b_4	C	m
elastic	0	2.172	1.799	−0.6725	0.0954	−0.00565	3186	1
Triplets								
$A^3\Sigma^+_u$	6.17	−17.406	18.706	−6.8067	1.0079	−0.05905	22510	3
$B^3\Pi_g$	7.35	−7.945	11.815	−51714	0.8866	−0.06001	25000	3
$W^3\Delta_u$	7.36	−53.767	60.284	−23.625	3.8971	−0.23895	16290	3
$B'^3\Sigma^-_u$	8.16	−45.582	53.652	−22.679	4.0674	−0.27095	15200	3
$C^3\Pi_u$	11.03	30.802	−22.856	6.5256	−0.8897	0.04204	15590	3
$E^3\Sigma^+_g$	11.88	−145.716	143.377	−51.6455	7.926	−0.44828	230.8	3
$C'^3\Pi_u$	12.08	−239.81	227.6	−72.775	7.9625	−0.07846	210	3
$F^3\Pi_u$	12.75	30.312	−22.856	6.5256	−0.8897	0.04204	9371	3
$G^3\Pi_u$	12.80	30.912	−22.856	6.5256	−0.8897	0.04204	17060	3
$D^3\Sigma^+_u$	12.85	43.826	−41.491	14.9628	−2.5092	0.15385	2691	3
M1(2)	13.15	29.642	−22.856	6.5256	−0.8897	0.04204	4752	3
Singlets								
$a'^1\Sigma^-_u$	8.4	−316.303	401.865	−187.851	38.2932	−2.88681	3.105	1
$a^1\Pi_g$	8.55	−11.365	12.291	−4.2994	0.5924	−0.02933	36.5	1
$w^1\Delta_u$	8.89	−56.661	67.038	−27.9411	4.8355	−0.30303	0.982	1
$a''^1\Sigma^+_g$	12.25	−113.067	123.523	−48.9832	8.3074	−0.51592	2.084	1

State	ϵ^*_{th}	b_0	b_1	b_2	b_3	b_4	A	B
$b^1\Pi_u$	12.5	−20.848	19.564	−5.9184	0.6868	−0.02351	53.8	0.075
$b'^1\Sigma^+_u$	12.85	−21.039	19.564	−5.9184	0.6868	−0.02351	45.3	0.075
$c^1\Pi_u$	12.91	−21.393	19.564	−5.9184	0.6868	−0.02351	31.7	0.075
$c'^1\Sigma^+_u$	12.94	−20.708	19.564	−5.9184	0.6868	−0.02351	63.1	0.075
$o^1\Pi_u$	13.1	−22.305	19.564	−5.9184	0.6868	−0.02351	12.7	0.075
$e^1\Pi_u$	14.33	−23.185	19.584	−5.9184	0.6868	−0.02351	5.2	0.075

N_2 ionization to different N_2^+ states

State or process	ϵ^*_{th} ev	b_0	b_1	b_2	b_3	b_4	A	B
$X^2\Sigma^+_g$	15.58	−16.888	12.05	−2.0623	0.2048	−0.00297	260.6	0.26
$A^2\Pi_u$	16.73	−23.623	18.05	−4.6185	0.4586	−0.01187	237.7	0.26
$B^2\Sigma^+_u$	18.75	−56.843	42.983	−11.0907	1.0507	−0.01842	106.5	0.26
$D^2\Pi_g$	22.00	0.453	−3.931	1.7057	−0.2055	0.00171	37.5	0.26
$C^2\Sigma^+_u$	23.60	−4.738	0.197	0.6301	−0.1108	0.00114	36.7	0.26
Common ionization	15.58	−20.631	16.19	−4.1815	0.4759	−0.0208	945.7	0.26
Production N_2^+	15.58	−19.178	15.1	−3.6644	0.3274	−0.0055	757.2	0.26
Dissociative ionization	26.00	66.256	−58.382	16.7515	−1.580	0.00553	240	0.26

Here, $q_0 = 6.513 \times 10^{-16}$ cm^{2}eV^2, F, α, β, γ, δ are fitting parameters, $e = 2.71828....$, and ϵ^*_{th} is the excitation threshold. The energies ϵ and ϵ^*_{th} are given in eV, and the cross-section $\sigma(\epsilon)$ is in cm^2.

Another fitting formula for the cross-sections of allowed transitions has been obtained [6] using numerical calculations, the Born approximation, and other theoretical methods. This formula contains only three fitting parameters and is as follows:

$$\sigma(\epsilon) = \sigma_0 \left(1 - \frac{\epsilon^*_{th}}{\epsilon}\right)^{1/2} \left(\frac{\epsilon^*_{th}}{\epsilon} + \phi\right). \tag{8.5}$$

122   8. Electron Rate Coefficients

Table 8.2. Parameters of (8.1) for cross-sections of inelastic scattering of electrons by O_2 in the ground electronic state [8]

State	ϵ^*_{th} (ev)	b_0	b_1	b_2	b_3	b_4	C	m
elastic	0	2.465	0.825	−0.2325	0.0194	−0.00094	3232	1
vibrational levels								
v=1	0.19	−2.49	−5.95	4.15	1.85	1.16	-	-
v=2	0.37	−2.18	−4.57	−10.39	−0.56	1.91	-	-
v=3	0.56	−2.85	1.45	−20.31	11.64	−1.73	-	-
v=4	0.74	−6.01	16.08	−44.43	39.31	−13.02	-	-
$a^1\Delta_g$	0.98	−3.51	2.45	−0.7619	0.0618	−0.00547	8054	3
$b^1\Sigma_g^+$	1.63	−2.638	−0.044	0.258	−0.107	0.00414	2536	3
$c^1\Sigma_u^- + A'^3\Delta_u + A^3\Sigma_u^+$	4.5	−28.096	34.199	−14.403	2.4836	−0.15457	4.718	1

							A	B
$B^3\Sigma_u^-$	7.1	−8.976	12.232	−4.7158	0.7368	−0.04197	42	0.294
9.7 - 12.1 ev	9.7	20.061	−18.003	6.2235	−0.9854	0.05783	10.3	0.156

O_2 ionization to different O_2^+ states

State or process	ϵ^*_{th} (ev)	b_0	b_1	b_2	b_3	b_4	A	B
$X^3\Pi_g$	12.1	−7.416	1.772	0.5184	−0.1283	0.00136	64.49	0.0365
$a^4\Pi_u$	16.1	−46.558	31.71	−7.1676	0.5703	−0.0055	169.6	0.027
$A^2\Pi_u$	16.9	−234.237	209.278	−69.6548	10.262	−0.56458	167.5	0.026
$b^4\Sigma_g^-$	18.2	−47.67	31.121	−6.8005	0.532	−0.00657	145.4	0.024
≥ 23 ev	23	80.926	−62.282	15.5622	−1.2904	0.00344	71.28	0.019
$O^+(^4S)+O(^3D)$	18.9	−35.7	20.972	−4.1277	0.3069	−0.00599	152	0.034
$O^+(^4S)+O(^1D)+$								
$O^+(^2D)+O(^3P)$	21.3	−36.106	20.972	−4.1277	0.3069	−0.00599	101.3	0.034
$O^+(^2D)+O(^1D)+$								
$O^+(^2P)+O(^3P)$	23.7	−36.799	20.972	−4.1277	0.3069	−0.00599	50.6	0.034
common ionization	12.1	23.037	−31.907	15.0781	−2.8861	0.19649	1240	0.0214

8.2 Electron Rate Coefficients in Plasmas Maintained by an Electric Field

Table 8.3. Values of the parameters in expressions (8.1), (8.3), (8.4) for the cross-sections of inelastic scattering of electrons by ground state O atoms [8]

State	ϵ^*_{th} (ev)	b_0	b_1	b_2	b_3	b_4	ϵ_0 (ev)	C	m
elastic	0	2.8	0.39	-0.078	-0.0102	0.00095	10000	2600	1
$2p^1 D^0$	1.96	0.0063	1.672	-1.512	0.4532	-0.05237	100	25440	3
$2p^1 S^0$	4.19	-0.147	-0.389	-0.5896	0.271	-0.03943	100	3176	3

State	ϵ^*_{th} (ev)	F	α	β	γ	δ
$3s^5 S^0$	9.14	0.0217	3	1	-	3
$3s^3 S^0$	9.52	0.056	0.86	1.44	0.32	-
$3p^5 P$	10.7	0.122	0.23	0.65	-	3
$3p^3 P$	11	0.081	0.88	0.41	-	1
$3d^3 D^0$	12.03	0.031	1.26	0.49	0.61	-
$3s^3 D^0$	12.48	0.056	0.26	1.44	0.32	-

The values of the parameters in (8.1), (8.3), (8.4) for elastic and inelastic electron scattering by O atoms are given in Table 8.3.

The fitting expressions (8.1), (8.3), (8.4) have not been used for the scattering cross-sections by N atoms. In this case, for allowed transitions the approximation (8.5) can be used with the values of σ_0 and ϕ determined in [6].

It should be noted that the Gryzinski's formulae have been used to get a complete set of state-to-state cross sections for e–H_2, e–N_2, and e–O_2 systems [24–26]

8.2 Electron Rate Coefficients and Power Balance in Plasmas Maintained by an Electric Field

The electron cross-sections for the atmospheric gases reported above can be inserted into the Boltzmann equation to obtain the electron energy distribution function (EEDF) in plasmas of such gases or of their mixtures. Once the EEDF is determined, one can calculate the most important plasma kinetic parameters, such as the electron mobility μ_e, diffusion coefficient D_e, average ($\bar{\epsilon}$) and characteristic (ϵ_k) energy, and the rate coefficients k_{skn}, k^*_{snk} of inelastic processes $k \to n$, $n \to k$, for the gas component "s". Using the normalization (4.2), these quantities are given by the following expressions:

$$\mu_e = -\frac{1}{3}\frac{2}{m_e} \int_0^\infty \frac{\epsilon^{3/2}}{\nu_m} \frac{df}{d\epsilon} d\epsilon ; \qquad (8.6)$$

$$D_e = \frac{1}{3}\frac{2}{m_e} \int_0^\infty \frac{\epsilon^{3/2}}{\nu_m} f d\epsilon ; \qquad (8.7)$$

$$\langle \epsilon \rangle \equiv \bar{\epsilon} = \int_0^\infty \epsilon^{3/2} f d\epsilon ; \qquad (8.8)$$

$$\epsilon_k = \frac{D_e}{\mu_e} = -\int_0^\infty \frac{\epsilon^{3/2}}{\nu_m} f d\epsilon \left(\int_0^\infty \frac{\epsilon^{3/2}}{\nu_m} \frac{df}{d\epsilon} d\epsilon \right)^{-1} ; \qquad (8.9)$$

$$k_{snk} = \sqrt{\frac{2}{m_e}} \int_0^\infty \sigma_{snk} \epsilon f d\epsilon ; \qquad (8.10)$$

$$k_{skn^*} = \sqrt{\frac{2}{m_e}} \int_0^\infty \sigma_{skn} \epsilon f(\epsilon - \epsilon^*_{skn}) d\epsilon . \qquad (8.11)$$

The ionization rate coefficient k_{sion} can be also calculated using an equation similar to (8.10), but with the corresponding ionization cross-section σ_{si}. The mobility μ_e determines the electron drift velocity v_d and the plasma electron current density J_e:

$$v_d = \mu_e E; \qquad J_e = \mu_e E N_e \qquad (8.12)$$

The electron power balance equation is another important plasma characteristic, which can be obtained by multiplying the Boltzmann equation (4.3) by ϵ and integrating from 0 to ∞. The resulting equation can be symbolically represented as

$$\frac{d\bar{\epsilon}}{dt} = \left(\frac{d\bar{\epsilon}}{dt}\right)_F - \left(\frac{d\bar{\epsilon}}{dt}\right)_{el} - \left(\frac{d\bar{\epsilon}}{dt}\right)_{rot} - \left(\frac{d\bar{\epsilon}}{dt}\right)_{in} - \left(\frac{d\bar{\epsilon}}{dt}\right)_{ion} + \left(\frac{d\bar{\epsilon}}{dt}\right)_{q,p} . \qquad (8.13)$$

The terms on the right of this equation describe the rates of change in electron energy due to different processes, namely the power gain from the electric field and from external sources (the first and the last terms) and the losses or gains due to elastic collisions $((d\bar{\epsilon}/dt)_{el})$, excitation and deactivation of rotational $((d\bar{\epsilon}/dt)_{rot})$, vibrational and electronic levels $((d\bar{\epsilon}/dt)_{in})$, and ionization $((d\bar{\epsilon}/dt)_{ion})$. Taking into account (4.4)–(4.20), (4.49), (8.6)–(8.12), these terms can be written as

$$\left(\frac{d\bar{\epsilon}}{dt}\right)_F = \mu_e E^2 \qquad (8.14a)$$

$$\left(\frac{d\bar{\epsilon}}{dt}\right)_{el} = \delta_{el} \int_0^\infty \nu_{el} \epsilon^{3/2} \left(f + kT \frac{df}{d\epsilon}\right) d\epsilon \qquad (8.14b)$$

$$\left(\frac{d\bar{\epsilon}}{dt}\right)_{rot} = \int_0^\infty \Delta\epsilon_r \nu_r \epsilon^{1/2} \left(f + kT_r \frac{df}{d\epsilon}\right) d\epsilon \qquad (8.14c)$$

$$\left(\frac{d\bar{\epsilon}}{dt}\right)_{in} = \sum_{s,n,k} \epsilon^*_{skn} N_n^{(s)} \left(k_{snk} - k^*_{skn} \frac{g_{sn}}{g_{sk}} \frac{\varphi_{sk}}{\varphi_{sn}}\right) \quad (d) \qquad (8.14d)$$

$$\left(\frac{d\bar{\epsilon}}{dt}\right)_{ion} = \sum_{s,i} \epsilon^*_{si} N_i^{(s)} k_{si} \qquad (8.14e)$$

$$\left(\frac{d\bar{\epsilon}}{dt}\right)_{q,p} = \int_0^\infty \epsilon(q - p) d\epsilon \qquad (8.14f)$$

8.2 Electron Rate Coefficients in Plasmas Maintained by an Electric Field

The calculation and experimental determination of the integral characteristics (8.6)–(8.14) are of great importance, because they determine the macroscopic properties of the plasma. This is why the main theoretical task in the solution of the Boltzmann kinetic equation (4.3) is not to obtain the EEDF itself, but rather to obtain the integral quantities (8.6)–(8.14).

Theoretical and experimental research with the aim of obtaining parameters (8.6)–(8.14) for low-temperature plasmas of atmospheric molecular gases has been intensive since the end of the 1960s as a result of the development of molecular discharge lasers and of investigations into plasma chemical and ionosphere processes. Results of these investigations have been presented in various monographs and reviews [27–37].

The kinetic coefficients for nitrogen plasmas maintained by an electric field have been calculated in [38–41] with the cross-sections from [42,43]. The average electron energy and the energy balance in the discharge were calculated as functions of the electric field [38]. In [44,45] the kinetic coefficients were calculated with cross-sections from [46]. In [47] the average electron energy, the fractional power transfer to vibrational degrees of freedom, and the rate coefficients for excitation of vibrational levels and electronic state $A^3\Sigma$ were calculated with other cross-sections. The authors of [47] considered the influence on these characteristics of vibrationally excited molecules and electron–electron collisions. Rotational excitation cross-sections [48] significantly different from those generally used [49] were employed in [50] to analyze the energy balance in laser plasma mixtures of N_2, CO, CO_2 and He. It was found that in this case the energy balance in the plasma is significantly changed and the contribution of rotational excitation to gas heating plays an essential role for electric fields $E/N \sim 10^{-16}$ Volt cm^2.

Calculations of different kinetic coefficients and energy balances were carried out in [50–56] using appropriate selections of cross-section sets, and good agreement with experimental data was obtained. For example, the calculated N_2 ionization rate coefficients [50,51,53,54,57] agree with the measurements [59–63] for $E/N \geq 6 \times 10^{-16}$ Volt cm^2. A good agreement between theory and experiment could also usually be reached without difficulty for the transport coefficients (8.6)–(8.9) over a wide range of E/N values.

References [14,55,57–65] provide experimental data on the rate coefficients for excitation and ionization of nitrogen by electrons. For such processes, agreement between theory and measurements usually falls off for small E/N and high energy thresholds (since such processes involve the high energy tail of the EEDF). This tail is very sensitive both to the choice of the cross-sections for the calculations and to the influence of other processes, such as electron collisions with excited particles. The time evolution of the EEDF tail also determines the time dependence of the rate coefficients for processes with high thresholds. Time-dependent rate coefficients for excitation of the vibrational level $v = 1$ and the $N_2(A^3\Sigma_u^+)$ electronic state have been calculated in [66] in order to analyze the evolution of the EEDF towards quasi-stationary conditions after the electric field is switched on. It was found that the rate coefficients for $v = 1$ and $A^3 \Sigma_u^+$ excitation become quasi-stationary after a time $\sim 10/\nu_m$ and $\sim 200/\nu_m$, respectively, where ν_m is the collision frequency for electron momentum transfer.

The influence of vibrationally excited molecules on the electron kinetic coefficients and energy balance in nitrogen plasmas has been investigated in [24,26,41,47,51,53,54,65,67–80]. In some of these works the degree of vibrational excitation was considered as an independent parameter and characterized by the vibrational temperature T_V of lower levels. In this situation, the vibrational distribution function was assumed as either a Boltzmann or a Treanor distribution (see Chap. 3). The self-consistent problem was solved in [67,69,70, 72,75,76], i.e., the kinetic coefficients were calculated simultaneously with the vibrational distribution function (VDF). In particular, the changes in time of these coefficients due to the evolution of the VDF in discharge or post-discharge conditions have been investigated. It follows from all these works that N_2 vibrational excitation can significantly influence the electron kinetic coefficients. This influence is most important for reduced electric fields $E/N \leq 5 \times 10^{-16}$ Volt cm^2, for all inelastic processes with excitation thresholds ≥ 2 ev. In this case, the increase in vibrational temperature from 300 to 5000 K increases the rate coefficients by several orders of magnitude, due to the great enhancement of the EEDF tail.

Electronically excited states (principally metastable states, the concentrations of which can be quite significant) can also strongly influence the excitation and ionization rate coefficients. These effects for nitrogen plasmas were theoretically investigated in [71,73,75,76].

The calculated mean and characteristic energies, drift velocities as well as rate coefficients for excitation of different states and for ionization of N_2 in nitrogen plasmas are shown in Figs. 8.1–8.5 as a function of the reduced electric field E/N. These figures also illustrate the influence of vibrational and electronic excited states on the rate coefficients. The electron power balance in pure N_2 plasma is illustrated in Fig. 8.6.

The influence of vibrationally excited molecules on the rate coefficients may be due to factors other than the deformation of the EEDF. Indeed, transitions from vibrationally excited levels, for which the thresholds are smaller, can also contribute to the total excitation or ionization rates. This effect has been numerically investigated for nitrogen plasmas in [69,77–79,83,84]. It has also been analytically estimated [37] by assuming a Maxwellian EEDF tail with some effective electron temperature T_t [47] (as E/N increases from 6 up to 12×10^{-16} Volt cm^2, T_t varies from ~ 0.9 to ~ 2 eV [47]). In this case, the sum of the transition rates from all vibrational levels can be estimated using the harmonic oscillator approximation for the Frank-Condon factors (see Chap. 5), and assuming a Boltzmann distribution with vibrational temperature T_V for the vibrational level populations of the ground electronic state. The calculation can be additionally simplified, if we assume that the elec-

8.2 Electron Rate Coefficients in Plasmas Maintained by an Electric Field 127

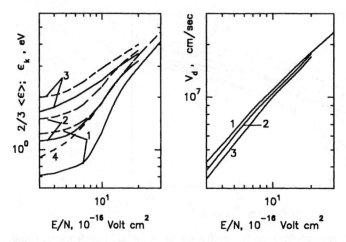

Fig. 8.1. Electron mean energy $\langle\epsilon\rangle$ (full curves), characteristic energy ϵ_k (broken curves 1–3) and electron drift velocity v_d as a function of the reduced field E/N in pure nitrogen plasmas for the N_2 vibrational "temperatures" $T_1 = 300$ K (curves 1); 4000 K (2); 6000 K (3). The curve 4 is a calculation of $\langle\epsilon\rangle$ assuming Boltzmann-like vibrational distribution function with vibrational temperature $T_V = 6000$ K [77]

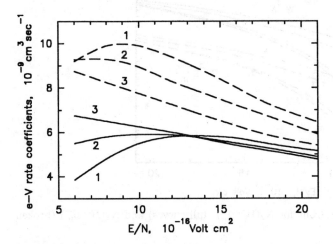

Fig. 8.2. Electron rate coefficients for vibrational excitation from $v = 0$ to $v = 1$ (full curves) and for the reverse process (broken curves) as a function of the reduced field E/N in pure nitrogen plasmas for the N_2 vibrational "temperatures" $T_1 = 2000$ K (curves 1); 4000 K (2); 6000 K (3) [77]

tronic terms involved in the transition have approximately equal frequencies $\overline{\omega} = (\omega' + \omega'')/2$, but have different equilibrium internuclear distances r'_e and r''_e. In this case, the total rate coefficient k' is given by [37]

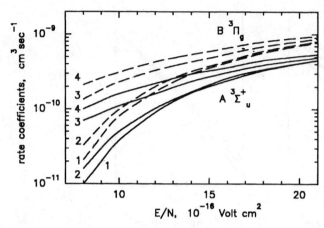

Fig. 8.3. Electron rate coefficients for excitation of $N_2(A^3\Sigma_u^+)$ (full curves) and $N_2(B^3\Pi_g)$ (broken curves) states as a function of the reduced field E/N in pure nitrogen plasmas for the N_2 vibrational "temperatures" $T_1 = 300$ K (curves 1); 2000 K (2); 4000 K (3); 6000 K (4) [77]

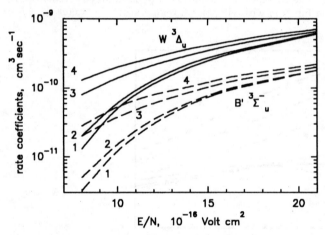

Fig. 8.4. As in Fig. 8.3 but for $N_2(W^3\Delta_u)$ (full curves) and $N_2(B'^3\Sigma_u^-)$ (broken curves) [77]

$$k' = k \times \exp\left\{ \Lambda \frac{X_V(1-Z_e)^2}{X_e(1-Z_V)} \right\}, \qquad (8.15)$$

where k is the rate coefficient without considering transitions from vibrationally excited molecules; $\Lambda = \mu\overline{\omega}(r'_e - r''_e)^2/(2\hbar)$; $X_V = \exp[-\hbar\overline{\omega}/kT_V]$; $X_e = \exp[-\hbar\overline{\omega}/kT_t]$; μ is the reduced mass of the oscillator.

The rate coefficients and average electron energy in pure nitrogen postdischarges have been calculated in [70–73,75,76]. In this case, the average electron energy, as well as the rate coefficients, can be controlled by vibrationally

8.2 Electron Rate Coefficients in Plasmas Maintained by an Electric Field   129

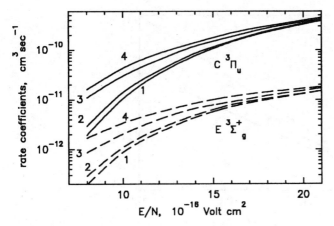

Fig. 8.5. As in Fig. 8.3 but for $N_2(C^3\Pi_u)$ (full curves) and $N_2(E^3\Sigma_g^+)$ (broken curves) [77]

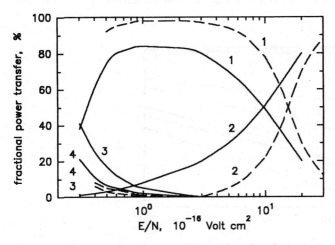

Fig. 8.6. Fractional power transferred by the electrons to the N_2 molecules in the vibrational excitation (1); electronic levels excitation and ionization (2); elastic collisions (3) and rotational excitation (4) as a function of the reduced field E/N in pure nitrogen plasmas for the N_2 vibrational "temperatures" $T_1 = 4000$ K (full curves) and 300 K (broken curves) [77]

or electronically excited molecules. Figure 8.7 illustrates the correlation between the electron and the vibrational temperatures for post-discharges in some molecular gases. Figure 8.8 shows the rate coefficients for some processes in pure nitrogen post discharge as a function of the relative concentrations φ_A and $\varphi_B = 0.1 \times \varphi_A$ of the states $N_2(A^3\Sigma_u^+)$ and $N_2(B^3\Pi_g)$. The curves 1, 2, 3, 4, 5, 6 in this figure correspond to the following transitions respectively: $N_2^+(X^2\Sigma_g^+) \to N_2^+(B^2\Sigma_u^-)$; $N_2(a'^1\Sigma_u^-) \to N_2^+(A^2\Pi_u)$; $N_2(A^3\Sigma_u^+)$

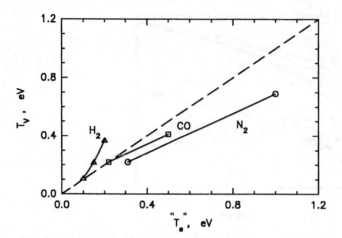

Fig. 8.7. Correlation between molecular vibrational temperature T_V and electron kinetic temperature T_e in pure N_2, CO, and H_2 post-discharges. The dashed line corresponds to $T_V = T_e$ [70].

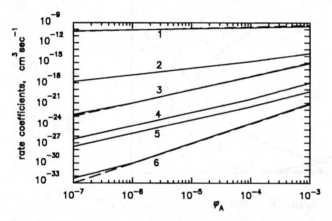

Fig. 8.8. Electron rate coefficients for N_2 ionization from different initial to different final states (curves 2-6, see text) and for N_2^+ excitation (curve 1, see text) after the electric field is switched off *versus* relative populations φ_A and $\varphi_B = 0.1 \times \varphi_A$, for the N_2 vibrational temperatures $T_V = 300$ K (dashed lines) and $T_V = 3400$ K (full lines) [73].

$\to N_2^+(A^2\Pi_u)$; $N_2(X^1\Sigma_g^+) \to N_2^+(X^2\Sigma_g^+)$; $N_2(X^1\Sigma_g^+) \to N_2^+(A^2\Pi_u)$; $N_2(X^1\Sigma_g^+) \to N_2^+(B^2\Sigma_u^+)$.

The time evolution of the electron kinetic coefficients and power balance when the applied electric field is suddenly changed were calculated in [66,67,84,85]. The typical time τ_ϵ for the power balance to reach a quasi-stationary regime in N_2 or CO plasmas is $\sim 10^9/N$ s cm^3 (N denoting the gas density in cm^{-3}), for moderate fields $E/N \sim (3\text{--}10) \times 10^{-16}$ Volt cm^2. Electron kinetic coefficients in nitrogen plasmas sustained by microwave fields

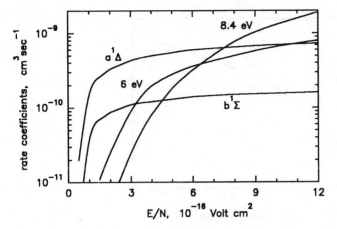

Fig. 8.9. Electron rate coefficients for excitation of different O_2 states in pure oxygen plasma as a function of E/N. The designations of electronic states or their energy thresholds are given near the curves [80]

with frequency $\omega \gg 1/\tau_\epsilon$ were calculated in [53,78] using the effective field approximation.

Maxwellization of the EEDF due to electron–electron collisions has been shown to start affecting the excitation and ionization rate coefficients for degrees of ionization $\geq 10^{-4}$ [47,66,68]. The effects are not the same for processes with different excitation thresholds. For example, the rate coefficients for vibrational excitation increase slightly with the degree of ionization, but those for excitation of $N_2(A^3\Sigma_u^+)$ decrease.

Calculations of the electron kinetic coefficients and power balance in oxygen plasmas have also been carried out [80,86–90]. Results of such calculations are shown in Figs. 8.9–8.11 as a function of E/N.

The ionization rate coefficients in nitrogen-oxygen mixtures were measured in [91], and numerically calculated in [82,92]. The most detailed calculations for the electron transport parameters and the rate coefficients for excitation, dissociation and ionization in N_2–O_2 and N_2–H_2 DC discharges have been carried out in [92]. These calculations were carried out by solving the stationary homogeneous electron Boltzmann equation for reduced electric fields $E/N = (4\text{–}20) \times 10^{-16}$ Volt cm^2, over the whole range of relative concentrations $\delta_M = 0 - 100$ %, where $\delta_M = [M]/([N_2] + [M])$ (M denoting O_2 or H_2), and for vibrational temperatures of $N_2(X,v)$ molecules in the range $T_V = 300 - 8000$ K. For example, Figs. 8.12 and 8.13 show the calculated rate coefficients for dissociation of N_2 and O_2 molecules and the ionization rate coefficients of N_2 and O_2, respectively. It is seem from these figures that the vibrational temperature of $N_2(X,v)$ molecules and the relative mixture composition, in addition to E/N, can strongly influence the rate coefficients. Fitting formulae for these numerical data have also been derived. It was found

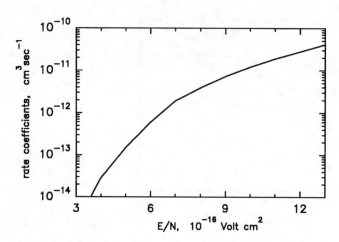

Fig. 8.10. Electron ionization rate coefficient of O_2 in pure oxygen plasma as a function of E/N [80]

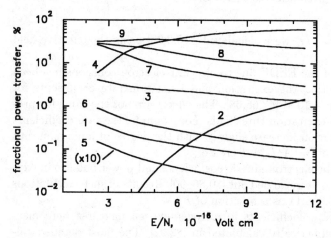

Fig. 8.11. Electron power balance in an oxygen plasma as a function of E/N. 1 – elastic collisions; 2 – ionization; 3 – vibrational excitation; 4 – excitation of molecular levels with thresholds ≥ 8.4 eV; 5 – rotational excitation; 6 – excitation of $O_2(b^1\Sigma)$; 7 – excitation of $O_2(^1\Delta_g)$ state; 8 – excitation of the states with thresholds 4.45 eV, 9.97 eV and 14.7 eV; 9 – excitation of state with threshold 10.6 eV [80]

that the logarithms of the rate coefficients can be approximated with sufficient accuracy by a polynomial in $(E/N)^{-1}$, δ_M and $X_V = \exp\left[-(E_{10}/kT_V)\right]$ ($E_{10} = 0.28889$ eV denoting the energy of the vibrational transition $0-1$ in N_2), that is:

$$\log_{10} k = \sum_{i=0}^{2} \sum_{j=0}^{2} \sum_{n=0}^{n_m} a_{ijn}(E/N)^{-i}(\delta_M)^j(X_V)^n; \quad M = O_2 \text{ or } H_2 \, . \quad (8.16)$$

8.2 Electron Rate Coefficients in Plasmas Maintained by an Electric Field

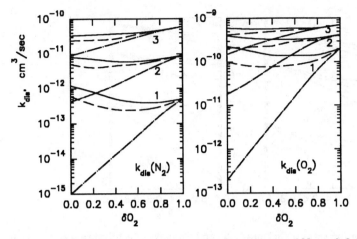

Fig. 8.12. Electron rate coefficients for dissociation of N_2 and O_2 in N_2-O_2 mixtures, as a function of $\delta_{O_2} = [O_2]/([N_2]+[O_2])$, for the following values of E/N in 10^{-16} Volt cm^2: (1) 4; (2) 6.67; (3) 10. The dash-dotted, broken and full lines are for $T_V(N_2)$=300, 4000 and 8000 K, respectively [92]

The mean square error of the fitting is smaller than 10% if $n_m = 3$, that is, for a quadratic polynomial in (N/E) and δ_M, and cubic in X_V.

It is worth noting that equation (8.16) is a generalization of the less accurate formulae obtained in pure N_2 [54] and in an $N_2:O_2 = 4:1$ mixture [82], where three fitting parameters have been used. In the present case, the calculation of each rate coefficient requires the knowledge of 36 fitting coefficients, a_{ijn}, due to the additional variable δ_M (M=O_2 or H_2) and to the higher accuracy desired. Fortunately this number may be reduced, as the coefficients a_{ij1}, a_{ij2} and a_{ij3} are the same for all inelastic processes with energy thresholds ≥ 4 eV. This is a consequence of the negligible electron cross-sections for excitation and de-excitation of $N_2(X,v)$ levels in this energy range, which leads to an identical influence of T_V on the high electron-energy tail of the EEDF and, therefore, on the rate coefficients of inelastic processes with energy thresholds larger than ~ 4 eV. The inelastic processes with energy thresholds ≥ 4 eV are dissociation, ionization and excitation of electronic states (with the exception of the states $O_2(a\ ^1\Delta_g)$ and $O_2(b\ ^1\Sigma_g^+)$).

The values of the fitting coefficients a_{ijn} are presented in Tables 8.4 to 8.10 for all the relevant inelastic processes of N_2 and O_2, in N_2-O_2 mixtures, with E/N expressed in 10^{-16} Volt cm^2 and T_V in K in (8.16). The rate coefficients so obtained are given in cm^{3}s^{-1}. In Table 8.6 the coefficients a_{ij1}, a_{ij2} and a_{ij3} expressing the dependence on T_V are presented. As noted above these coefficients are the same for all inelastic processes with energy thresholds ≥ 4 eV because they reflect the influence of T_V on the EEDF tail. For excitation of the states $O_2(a\ ^1\Delta_g)$ and $O_2(b\ ^1\Sigma_g^+)$, the coefficients a_{ij1}, a_{ij2} and a_{ij3}

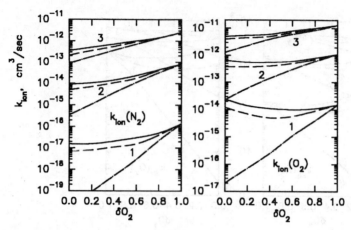

Fig. 8.13. Ionization rate coefficients of N_2 and O_2 in N_2-O_2 mixture, as a function of $\delta_{O_2} = [O_2]/([N_2]+[O_2])$ with the same notations as in Fig. 8.12 [92]

are given separately in Table 8.7, as these states have energy thresholds lower than 4 eV.

The rate coefficients for vibrational excitation and de-excitation by electron impact of $N_2(X, v)$ and $O_2(X, v)$, as well as those for dissociative attachment of $O_2(X, v)$, also have energy thresholds lower than 4 eV. Therefore, the values of a_{ij1}, a_{ij2} and a_{ij3} given in Table 8.6 cannot be used. Nevertheless, it is sufficient in this case to consider $n_m = 1$ in (8.16) in order to obtain the desired accuracy (better than 90%). The coefficients a_{ij0} and a_{ij1} obtained in (8.16) for these processes are presented in Tables 8.8, 8.9, excepting those for dissociative attachment of $O_2(X, v)$, which are given in Table 8.5.

The electron drift velocity v_d, characteristic energy $\epsilon_k = eD_e/\mu_e$ and kinetic temperature $T_e = \frac{2}{3}\langle\epsilon\rangle$ (D_e and μ_e denote the electron diffusion coefficient and mobility, respectively) can be fitted by the following expression [92]:

$$y = \sum_{i=0}^{i_m}\sum_{j=0}^{j_m}\sum_{n=0}^{n_m} a_{ijn}(E/N)^i(\delta_M)^j(X_V)^n, \qquad M = O_2 \text{ or } H_2. \qquad (8.17)$$

The accuracy of this approximation is always better than 90% in the following cases: for T_e and ϵ_k, with $i_m = 1$ and $j_m = n_m = 2$; for v_d, with $i_m = j_m = n_m = 1$. The coefficients a_{ijn} to be inserted in (8.17) to calculate T_e and ϵ_k (in eV) and v_d — (in 10^6 cm s^{-1}) are presented in Table 8.10, when E/N is expressed in 10^{-16} Volt cm^2.

The ionization rate coefficient for an HF discharge in air was measured in [91] and numerically calculated in [93]. Expression (8.16) can adequately describe the excitation and ionization rate constants in HF discharges when the effective field approximation applies. In this case, equations (8.16), (8.17), and the values a_{ijn} from Tables 8.4–8.10 can in principle be used for rough

8.2 Electron Rate Coefficients in Plasmas Maintained by an Electric Field

Table 8.4. Coefficients a_{ij0} in (8.16) for excitation of electronic states, dissociation and ionization of N_2 in N_2-O_2 mixtures [92]

	$N_2(A\ ^3\Sigma_u^+)$	$N_2(B\ ^3\Pi_g)$	$N_2(W\ ^3\Delta_u)$	$N_2(B'\ ^3\Sigma_u^-)$	$N_2(a'\ ^1\Sigma_u^-)$	$N_2(a\ ^1\Pi_g)$
a_{000}	-8.28	-8.10	-8.14	-8.61	-8.69	-8.26
a_{010}	-0.783	-0.832	-0.763	-0.749	-0.751	-0.696
a_{020}	0.578	0.556	0.585	0.565	0.560	0.595
a_{100}	-21.7	-20.4	-22.5	-23.6	-23.8	-24.9
a_{110}	20.2	20.8	20.0	19.9	19.9	19.4
a_{120}	-12.8	-12.7	-12.9	-12.5	-12.4	-12.7
a_{200}	1.08	-2.07	3.01	3.19	2.86	5.27
a_{210}	-19.0	-20.8	-18.8	-20.7	-21.7	-20.8
a_{220}	39.6	39.4	39.7	39.4	39.5	40.3

	$N_2(w\ ^1\Delta_u)$	$N_2(C\ ^3\Pi_u)$	$N_2(E\ ^3\Sigma_g^+)$	$N_2(a''\ ^1\Sigma_g^+)$	N_2(diss)	N_2(ion)
a_{000}	-8.58	-7.92	-9.62	-8.99	-7.68	-8.11
a_{010}	-0.770	-0.718	-0.805	-0.585	-0.513	-0.521
a_{020}	0.547	0.530	0.562	0.547	0.581	0.596
a_{100}	-23.6	-31.3	-24.0	-36.2	-41.0	-51.4
a_{110}	20.3	21.8	21.2	21.3	21.7	25.4
a_{120}	-12.3	-11.7	-12.5	-11.5	-11.7	-12.0
a_{200}	0.357	4.50	-9.66	11.4	16.7	17.1
a_{210}	-24.2	-32.6	-29.4	-31.0	-31.0	-31.6
a_{220}	39.8	40.5	42.7	40.0	40.9	41.9

Table 8.5. Coefficients a_{ij0} in (8.16) for excitation of electronic states and ionization of O_2, ionization of $O(^3P)$ and NO, and dissociative attachment of $O_2(X, v=0)$ in N_2–O_2 mixtures [92]

	O_2(4.5 eV)	O_2(6.0 eV)	O_2(8.4 eV)	O_2(ion)	O(ion)	NO(ion)	O_2(att)
a_{000}	-8.63	-8.19	-7.48	-8.24	-7.89	-7.96	-9.79
a_{010}	-0.980	-0.942	-0.812	-0.473	-0.595	-0.540	-0.997
a_{020}	0.510	0.532	0.579	0.611	0.581	0.614	0.502
a_{100}	-14.0	-16.0	-20.9	-38.2	-39.5	-31.1	-12.2
a_{110}	23.6	22.8	20.7	20.0	22.7	18.4	24.4
a_{120}	-12.9	-12.9	-12.9	-11.9	-12.0	-12.3	-13.1
a_{200}	-12.3	-8.61	-2.51	17.6	9.74	15.5	-15.6
a_{210}	-22.1	-22.3	-22.9	-28.1	-31.6	-21.6	-24.1
a_{220}	37.7	38.7	40.9	41.0	41.8	39.7	38.1

Table 8.6. Common coefficients a_{ijn} in (8.16) for all the processes listed in Tables 8.4 and 8.5. In this table and the following ones 0.22(4) means 0.22×10^4 [92]

n	1	2	3
a_{00n}	1.77	-9.53	8.83
a_{01n}	-3.83	22.9	-22.3
a_{02n}	2.06	-13.4	13.6
a_{10n}	-46.6	224	-200
a_{11n}	106	-579	550
a_{12n}	-59.8	359	-355
a_{20n}	227	-558	392
a_{21n}	-598	0.22(4)	-0.18(4)
a_{22n}	375	-0.16(4)	0.15(4)

Table 8.7. Coefficients a_{ijn} in (8.16) for excitation of the states $O_2(a\ ^1\Delta_g)$ and $O_2(b\ ^1\Sigma_g^+)$ in N_2-O_2 mixtures [92]

	$O_2(a\ ^1\Delta_g)$				$O_2(b\ ^1\Sigma_g^+)$			
n	0	1	2	3	0	1	2	3
a_{00n}	-8.34	0.214	-5.75	6.09	-8.90	-0.50(-1)	-5.45	5.99
a_{01n}	-1.08	-0.372	11.6	-12.9	-1.46	0.757	11.2	-13.6
a_{02n}	0.437	0.115	-5.77	6.74	0.762	-0.749	-5.66	7.54
a_{10n}	-18.8	-5.05	138	-143	-19.6	-0.715	128	-135
a_{11n}	31.3	9.52	-308	335	38.2	-11.7	-295	340
a_{12n}	-14.4	-4.37	169	-191	-20.8	13.4	166	-205
a_{20n}	39.4	-14.0	-217	245	32.5	-19.2	-125	142
a_{21n}	-112	1.08	920	-0.10(4)	-119	55.8	780	-907
a_{22n}	75.2	12.8	-709	784	91.1	-39.9	-662	779

Table 8.8. Coefficients a_{ijn} for excitation and de-excitation of $N_2(X,v)$ in N_2-O_2 mixtures [92]

	$N_2(0-1)$	$N_2(0-2)$	$N_2(0-3)$	$N_2(0-4)$	$N_2(0-5)$	$N_2(0-6)$	$N_2(0-7)$	$N_2(0-8)$
a_{000}	-8.24	-8.52	-8.70	-8.76	-8.85	-8.86	-9.04	-9.30
a_{010}	-0.451	-0.474	-0.513	-0.624	-0.600	-0.698	-0.788	-0.935
a_{020}	0.169	0.199	0.232	0.264	0.257	0.300	0.337	0.420
a_{100}	-0.59(-1)	1.30	1.53	-1.50	-0.910	-2.77	-4.86	-7.36
a_{110}	9.32	8.49	8.91	13.6	12.7	16.0	19.4	23.9
a_{120}	-5.68	-6.02	-6.65	-8.38	-8.01	-9.57	-10.9	-13.1
a_{200}	-6.24	-9.23	-11.3	-11.0	-10.7	-12.0	-10.6	-14.7
a_{210}	-24.2	-20.1	-18.4	-25.2	-24.4	-27.3	-34.0	-32.8
a_{220}	22.8	21.4	21.8	28.5	27.3	31.6	37.2	40.5
a_{001}	-0.292	-0.263	-0.269	-0.399	-0.372	-0.465	-0.558	-0.671
a_{011}	0.690	0.698	0.751	0.993	0.941	1.14	1.33	1.61
a_{021}	-0.395	-0.435	-0.484	-0.594	-0.569	-0.680	-0.771	-0.935
a_{101}	4.30	2.77	2.50	6.42	5.65	8.14	10.9	13.7
a_{111}	-13.2	-11.6	-12.1	-20.0	-18.4	-24.2	-30.0	-37.3
a_{121}	8.90	8.95	9.76	13.7	12.9	16.2	19.2	23.7
a_{201}	0.173	3.40	6.55	9.31	8.26	12.4	12.1	22.5
a_{211}	33.4	26.7	23.5	32.4	31.5	34.7	45.1	39.9
a_{221}	-34.0	-30.8	-30.9	-42.8	-40.7	-48.3	-58.6	-64.3
	$N_2(1-0)$	$N_2(2-0)$	$N_2(3-0)$	$N_2(4-0)$	$N_2(5-0)$	$N_2(6-0)$	$N_2(7-0)$	$N_2(8-0)$
a_{000}	-8.30	-8.56	-8.72	-8.85	-8.92	-8.96	-9.19	-9.50
a_{010}	-0.448	-0.581	-0.702	-0.743	-0.773	-0.801	-0.809	-0.812
a_{020}	0.238	0.362	0.465	0.499	0.521	0.539	0.542	0.543
a_{100}	3.79	6.24	7.81	8.37	8.67	8.88	8.89	8.90
a_{110}	5.70	5.84	6.69	7.02	7.48	7.95	8.14	8.20
a_{120}	-5.83	-8.13	-10.3	-11.1	-11.7	-12.2	-12.4	-12.4
a_{200}	-12.4	-17.3	-20.5	-21.6	-22.2	-22.7	-22.7	-22.8
a_{210}	-11.5	-8.54	-8.47	-8.60	-9.37	-10.3	-10.7	-10.8
a_{220}	16.4	18.1	21.2	22.4	23.6	24.8	25.2	25.4
a_{001}	-0.174	-0.164	-0.177	-0.184	-0.197	-0.212	-0.218	-0.220
a_{011}	0.607	0.741	0.862	0.903	0.937	0.971	0.981	0.984
a_{021}	-0.439	-0.587	-0.695	-0.728	-0.748	-0.766	-0.768	-0.770
a_{101}	-0.434	-2.80	-4.09	-4.50	-4.54	-4.49	-4.40	-4.37
a_{111}	-6.99	-6.71	-7.57	-7.93	-8.56	-9.25	-9.51	-9.61
a_{121}	7.62	9.84	12.0	12.8	13.5	14.1	14.2	14.3
a_{201}	6.32	10.6	13.0	13.7	14.0	14.1	14.1	14.2
a_{211}	14.3	9.99	9.99	10.3	11.5	13.0	13.5	13.6
a_{221}	-21.6	-22.3	-25.2	-26.4	-28.0	-29.6	-30.1	-30.3

8.2 Electron Rate Coefficients in Plasmas Maintained by an Electric Field

Table 8.9. Coefficients a_{ijn} in (8.16) for excitation and de-excitation of the $O_2(X,v)$ levels in N_2-O_2 mixtures [92]

	$O_2(0-1)$	$O_2(0-2)$	$O_2(0-3)$	$O_2(0-4)$	$O_2(1-0)$	$O_2(2-0)$	$O_2(3-0)$	$O_2(4-0)$
a_{000}	-8.57	-8.54	-8.81	-8.89	-8.53	-8.48	-8.72	-8.80
a_{010}	0.241	-0.296	-0.129	-0.483	0.215	-0.409	-0.316	-0.727
a_{020}	-0.310	-0.162	-0.271	-0.84(-1)	-0.315	-0.96(-1)	-0.167	0.80-01
a_{100}	-8.79	-17.9	-17.7	-21.1	-9.05	-18.0	-18.0	-20.8
a_{110}	-0.729	14.7	10.6	18.9	0.103	17.3	15.0	24.4
a_{120}	5.56	-0.709	2.25	-2.64	5.40	-2.45	-0.612	-6.72
a_{200}	23.4	44.6	44.5	51.1	24.0	44.5	44.5	49.6
a_{210}	-3.86	-64.6	-49.6	-78.9	-7.23	-73.1	-64.8	-96.6
a_{220}	-14.2	24.1	9.52	31.5	-12.3	31.6	22.8	48.7
a_{001}	-0.134	-0.708	-0.636	-0.896	-0.167	-0.766	-0.735	-0.990
a_{011}	-0.250	0.630	0.391	0.976	-0.217	0.801	0.666	1.34
a_{021}	0.399	0.128	0.292	-0.27(-1)	0.402	0.16(-1)	0.119	-0.301
a_{101}	6.94	20.4	19.0	24.4	7.69	21.5	20.9	26.0
a_{111}	-2.16	-26.9	-21.2	-34.9	-3.22	-30.8	-27.6	-42.8
a_{121}	-5.13	5.52	1.23	9.40	-4.89	8.27	5.64	15.8
a_{201}	-23.6	-57.5	-55.9	-67.4	-25.3	-59.0	-58.5	-68.4
a_{211}	11.0	105	81.0	127	15.4	118	104	155
a_{221}	14.0	-43.9	-21.9	-56.4	11.5	-55.9	-42.3	-83.9

Table 8.10. Coefficients a_{ijn} in (8.17) for the electron kinetic temperature T_e, the characteristic energy ϵ_k and drift velocity v_d in N_2-O_2 mixtures [92]

	T_e			ϵ_k			v_d	
n	0	1	2	0	1	2	0	1
a_{00n}	0.00747	0.191	3.41	0.658	-0.193	2.97	2.75	-2.26
a_{01n}	-1.21	2.02	-1.99	-1.80	1.94	-1.02	2.57	1.49
a_{02n}	2.86	-2.21	-1.38	2.93	-1.70	-1.79	-	-
a_{10n}	0.144	0.0303	-0.145	0.137	0.0434	-0.113	0.747	0.0497
a_{11n}	0.139	-0.187	0.0516	0.134	-0.157	0.00639	0.404	-0.0163
a_{12n}	-0.159	0.156	0.0896	-0.117	0.111	0.0972	-	-

estimates of the kinetic coefficients for N_2-O_2 HF discharges, by replacing the dc reduced field E/N with E_0/ω (E_0 denoting the amplitude of the HF field) according to the following scaling law

$$\frac{E}{N}(10^{-16}\text{Volt cm}^2) \simeq \frac{1.4}{\sqrt{2}}\frac{E_0}{\omega} \times 10^9 \ (\text{Volt s cm}^{-1}) \ . \tag{8.18}$$

Mixtures of nitrogen with other gases were investigated in [94–104]. The electron power balance, the average and characteristic energy of electrons, and the rate coefficients for vibrational excitation were calculated in these works for the laser mixtures $N_2-CO_2-He-CO$. Mixtures N_2-He were also investigated in [104]. Figure 8.14 shows the electron power balance in the laser mixture $N_2-CO_2-CO-He$ versus E/N. This figure also illustrates the role of superelastic electron collisions with vibrationally excited N_2, CO_2 and CO molecules in the electron power balance.

If some simplifying assumptions are made, it is possible to obtain analytical expressions for some electron energy losses. For example, assuming that

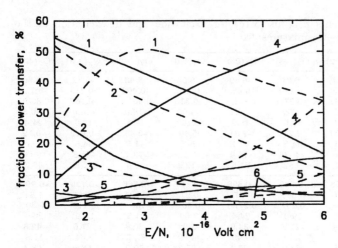

Fig. 8.14. Electron power balance in $N_2:CO_2:CO:He = 6:1:1:9$ mixture. The dashed curves are calculations neglecting vibrationally excited molecules; full curves - with vibrationally excited molecules for a vibrational temperature of 3000 K for N_2 and $CO_2(00v)$, and of 5000 K for CO. Curves 1, 2, 3 are the losses due to vibrational excitation of N_2, CO_2 and CO, respectively; curves 4, 5, 6 are the losses by excitation of N_2, CO_2 and CO electronic states, respectively [103]

$\nu_m \simeq \nu_{el}$ and that ν_m depends weakly on the electron energy, one can obtain from (8.6), (8.8), (8.12) and (8.14 a,b) [36, 105]:

$$\left(\frac{d\bar{\epsilon}}{dt}\right)_F \simeq \frac{E^2}{m_e \nu_m} ; \qquad (8.19)$$

$$\nu_m \simeq \frac{E}{m_e v_d} ; \qquad (8.20)$$

$$\left(\frac{d\bar{\epsilon}}{dt}\right)_{el} \simeq \delta_{el} \nu_m \left(\bar{\epsilon} - \frac{3}{2}kT\right) . \qquad (8.21)$$

To describe the energy exchange between electrons and rotational levels, one can use the continuous approximation and the Born approximation for the cross-sections of long-range electron–molecule "charge-quadrupole" interactions [42,106]. In this case, the value $\Delta\epsilon_r \nu_r$ in (4.7) and (8.14) for the gas mixture is

$$\Delta\epsilon_r \nu_r \equiv \Delta\epsilon_r \sigma_{rot} \sqrt{\frac{2\epsilon}{m_e}} N; \quad \Delta\epsilon_r \sigma_{rot} = \frac{32}{15} \pi a_0^2 \sum_s \alpha_s B_e^{(s)} q_s^2 N . \qquad (8.22)$$

Here ϵ and m_e are the electron kinetic energy and mass; a_0 is the Bohr radius; $B_e^{(s)}$ and q_s are the rotational constant (in eV) and the quadrupole moment (in units of ea_0^2) of "s" molecules; α_s is the relative concentration of these molecules in the gas mixture; N is the total gas density. The value of q_s is

1.01 for N_2 and 0.29 for O_2. By inserting (8.22) into (8.14c), integration of the latter yields

$$\left(\frac{d\bar{\epsilon}}{dt}\right)_{\text{rot}} \simeq \Delta\epsilon_r \sigma_{\text{rot}} \frac{4N}{\sqrt{3\pi m_e \bar{\epsilon}}} \left(\bar{\epsilon} - \frac{3}{2}kT_r\right). \tag{8.23}$$

Note that equation (8.22) does not take into account resonant excitation of rotational levels, which can play an important role [50,107,108]. This is why equations (8.22) and (8.23) must be used with care. They are correct only for small E/N (in pure nitrogen for $E/N < 10^{-16}$ V cm^2), when resonant rotational excitation is inefficient.

Using (8.13), (8.19)–(8.23) and experimental values of the electron drift velocity $v_d(E/N)$ and electric field E, the electron energy losses due to ionization and excitation of vibrational and electronic levels can be estimated by the expression [36,105]:

$$\left(\frac{d\bar{\epsilon}}{dt}\right)_{\text{in}} + \left(\frac{d\bar{\epsilon}}{dt}\right)_{\text{ion}}$$
$$\simeq \frac{E^2}{m_e \nu_m} - \delta_{\text{el}} \nu_m \left(\bar{\epsilon} - \frac{3}{2}kT\right) - \Delta\epsilon_r \sigma_{\text{rot}} \frac{4N}{\sqrt{3\pi m_e \bar{\epsilon}}} \left(\bar{\epsilon} - \frac{3}{2}kT_r\right) \tag{8.24}$$

8.3 Electron Loss Processes in the Plasma Bulk

The main electron loss processes in low-temperature plasmas include electron-ion recombination, attachment to neutral atoms and molecules, and diffusion losses. Diffusion losses are important at low pressures and are usually controlled by ambipolar diffusion. Ion mobilities, which determine the ambipolar diffusion coefficients, are given in Chap. 6. Here, the volume electron losses will be discussed and their rate coefficients will be given.

Recombination and attachment processes play a major part in the balance of charged particles and determine electron concentration in plasmas of molecular gases at relatively high densities and dimensions of the medium, when electron diffusion from the volume is ineffective.

When an electron and an ion recombine with each other, the electron-ion system goes from an initial charged state to a neutralized state of lower energy. To stabilize the recombination process, a means of removing energy from the electron + ion system must be provided at each recombination encounter; this energy may appear in the form of electromagnetic radiation, translational kinetic energy, or internal energy.

In the case of electron–atomic ion recombination, the most likely processes are the following

$$A^+ + e \rightarrow A^* + h\nu \tag{8.25a}$$

$$A^+ + e + e \rightarrow A^* + e \tag{8.25b}$$

$$A^+ + e + M \rightarrow A^* + M \qquad (8.25c)$$

Case (a) corresponds to the so-called radiative recombination, while cases (b) and (c) are three–body recombination processes in which the third body may be either an electron (b) or a heavy particle (c).

The recombination rate coefficient for these processes is related to the probability of a transition from a free electron state (positive-energy continuum state) into a bound (negative energy) state A^* of the resulting atom. This problem of multi-level collisional–radiative kinetics has been investigated in the context of both atomic physics and laser physics, due to the possibility of using plasmas decaying by recombination as laser media [32, 109–117]. It was found that the recombination rate coefficients for processes (8.25) depend rather weakly on the ion type. For example, the rate coefficient of radiative recombination (8.25a) is $\sim 10^{-12}$ cm^3/s [117]. The rate coefficients for three-body processes (8.25b) and (8.25c) are $\simeq 7 \times 10^{-20}(300/T_e)^{9/2}$ [32] and $\simeq 6 \times 10^{-27}(300/T_e)^{3/2}$ [32,118] cm^6/s, respectively (with T_e in K), for $T_e \leq 2000$ K. Detailed calculations of the rate coefficient for process (8.25b) with oxygen atoms atoms were carried out in [112,113] and with nitrogen atoms in [116,119]. Process (8.25c) was investigated in [32].

In the case of molecular ions, dissociative recombination

$$AB^+ + e \rightarrow A^* + B^* \qquad (8.26)$$

is usually the dominant electron-ion recombination process in low temperature plasmas. The energy surplus of this exothermic reaction is transformed into kinetic energy of the products A and B, or partly into internal energy of A and/or B. The cross-sections and the rate coefficients of dissociative recombination were calculated and measured for the dominant molecular ions in plasmas of atmospheric gases (see, for example, [37, 120–125]). Table 8.11 provides data on the corresponding rate coefficients, as well as the branching ratios for the formation of various reaction products.

In the lower ionosphere or in laboratory molecular plasmas in electronegative gases, electron attachment is an important electron loss channel. In particular, dissociative attachment processes, such as

$$O_2(v) + e \rightarrow O + O^- ; \qquad (8.27)$$

$$H_2(v) + e \rightarrow H + H^- ; \qquad (8.28)$$

$$H_2O + e \rightarrow H_2 + O^- (\text{or } OH + H^-) + e, \qquad (8.29)$$

are usually very effective. The processes (8.27)–(8.29) take place in two steps. First, there is an electron-impact transition (with some finite energy threshold) to a repulsive electronic state of the ion, which is followed and stabilized by a second transition into the electronic ground state of the molecule, which has a lower energy than the ion state. Note also that dissociative attachment strongly depends on the presence of Rydberg states [142,143]. Process (8.27) has been discussed in detail in [7,144,145]. The rate coefficients for processes

8.3 Electron Loss Processes in the Plasma Bulk

Table 8.11. Rate coefficients for dissociative recombination of molecular ions and branching ratios for the formation of some reaction products

Ion	rate coefficient (cm^3/s)	products	relative production	references
N_2^+	$1.8 \times 10^{-7} \left(\frac{300}{T_e}\right)^{0.39}$	$N(^4S)$	~ 1	[126]
		$N(^2D)$	~ 0.9	[127]
		$N(^3P)$	≤ 0.1	[128]
O_2^+	$2.7 \times 10^{-7} \left(\frac{300}{T_e}\right)^{0.7}$ ($T_e \leq 1200$ K)	$O(^3P)$	~ 1.15	[126,129]
		$O(^1D)$	~ 0.85	[130,131]
	$1.6 \times 10^{-7} \left(\frac{300}{T_e}\right)^{0.55}$ ($T_e > 1200$ K)	$O(^1S)$	≤ 0.1	[132]
NO^+	$4.2 \times 10^{-7} \left(\frac{300}{T_e}\right)^{0.85}$	$N(^4S)$	~ 0.2	[129,133]
		$N(^2D)$	~ 0.8	[134]
N_3^+	$2 \times 10^{-7} \left(\frac{300}{T_e}\right)^{0.5}$	N_2+N	-	[135]
N_4^+	$2.3 \times 10^{-6} \left(\frac{300}{T_e}\right)^{0.53}$	N_2+N_2	-	[136]
N_2O^+	$2 \times 10^{-7} \left(\frac{300}{T_e}\right)^{0.5}$	N_2+O	-	[135]
NO_2^+	$2 \times 10^{-7} \left(\frac{300}{T_e}\right)^{0.5}$	$NO+O$	-	[135]
O_4^+	$1.4 \times 10^{-6} \left(\frac{300}{T_e}\right)^{0.5}$	O_2+O_2	-	[137]
$O_2^+N_2$	$1.3 \times 10^{-6} \left(\frac{300}{T_e}\right)^{0.5}$	O_2+N_2	-	[135,137]
NO^+N_2	$1.3 \times 10^{-6} \left(\frac{300}{T_e}\right)^{0.5}$	$NO+N_2$	-	[135,137]
NO^+O_2	$1.3 \times 10^{-6} \left(\frac{300}{T_e}\right)^{0.5}$	$NO+O_2$	-	[135,137]
NO^+NO	$1.7 \times 10^{-6} \left(\frac{300}{T_e}\right)^{0.5}$	$NO+NO$	-	[138]
H_2^+	$3 \times 10^{-8} \left(\frac{300}{T_e}\right)^{0.5}$	$H(n=1)+$ $H(n \geq 2)$	1 for $n=1$; 0.1, 0.45, 0.22, 0.12, 0.11 for $n=2, 3, 4, 5, \geq 6$	[121,139]
H_3^+	$1.55 \times 10^{-7} \left(\frac{300}{T_e}\right)^{0.97}$	$3H$	0.71	[139,140]
		H_2+H	0.29	
H_5^+	$2.8 \times 10^{-8} \left(\frac{300}{T_e}\right)^{0.5}$	$2H_2+H$	-	[141]
HN_2^+	$7.1 \times 10^{-7} \left(\frac{300}{T_e}\right)^{0.72}$	N_2+H	-	[140]
$H^+(H_2O)_n$	$1.2 \times 10^{-6}(1.3n - 0.3)$	$H+nH_2O$	-	[142]

(8.27) and (8.28) in N_2-O_2 and N_2-H_2 mixtures have been calculated as a function of the reduced electric field E/N, relative concentration of N_2, and N_2 vibrational temperature in [92,146]. For the rate coefficient of process (8.27), the fitting formula (8.16) can be used with the coefficients a_{ijn} given in Table 8.5. The cross-section for process (8.29) is given in [147]. It should be noted that the cross-sections of processes (8.27), (8.28) depend strongly on the vibrational level involved in the transition. This is why the rate co-

Table 8.12. Electron attachment processes

No	Dissociative attachment	k_{att} (cm^3/s)	Ref.
R1	$e+O_3 \to O_2^- + O$	10^{-9}	[135]
R2	$e+O_3 \to O^- + O_2$	10^{-11}	[135]
R3	$e+NO_2 \to O^- + NO$	10^{-11}	[135]

	Three-body attachment	k_{att} (cm^6/s)	Ref.
R4	$e+O+O_2 \to O^- + O_2$	10^{-31}	[149]
R5	$e+O+O_2 \to O_2^- + O$	10^{-31}	[135]
R6	$e+O_2+O_2 \to O_2^- + O_2$	$1.4 \times 10^{-29}(300/T)$ $\times \exp(-600/T)\exp[(700(T_e - T)/(T_e T)]$	[150]
R7	$e+O_2+N_2 \to O_2^- + N_2$	$1.1 \times 10^{-31}(300/T)^2$ $\times \exp(-70/T)\exp[(1500(T_e - T)/(T_e T)]$	[150]
R8	$e+O_3+O_2 \to O_3^- + O_2$	10^{-31}	[145]
R9	$e+NO+NO \to NO^- + NO$	8×10^{-31}	[151]
R10	$e+N_2O+N_2O \to N_2O^- + N_2O$	6×10^{-33}	[152]
R11	$e+O_2+H_2O \to O_2^- + H_2O$	1.4×10^{-29}	[152]

efficients for these processes are functions of the vibrational temperature of O_2 and H_2 molecules.

Dissociative attachment can also take place in collisions of electrons with O_3 and NO_2 molecules. Besides, three-body attachment processes can also play an important role at higher pressures. Relevant processes of the above types and the corresponding rate coefficients are given in Table 8.12.

The total rate coefficient k_{att} of processes (R6),(R7) in Table 8.12 has been determined for an $N_2:O_2 = 4:1$ air mixture as a function E/N [146,148]. It was found that

$$k_{att}(\text{cm}^6/\text{s}) = \left(4.7 - 0.25 \times 10^{16} E/N\right) \times 10^{-31} \tag{8.30}$$

for reduced electric fields in the range $E/N = (1-10) \times 10^{-16}$ V cm^2 [146].

Attachment rates in decaying air plasmas are affected by the presence of metastable nitrogen molecules which change the electron energy distribution function [153].

8.4 Electron Rate Coefficients and Energy Balance in the Ionosphere Plasma

The Earth's ionosphere is a natural non-equilibrium low temperature plasma in atmospheric gases. Due to O_2 dissociation by U-V solar radiation, the main neutral components of this plasma at altitudes ≥ 100 km are N_2, O_2 and O species. The kinetic parameters and the energy balance of electrons in the presence of an electric field for ionospheric plasmas in N_2-O_2-O gas mixtures have been calculated in [154].

Note that the electron "temperature" in the ionosphere differs from the temperature of translational degrees of freedom of heavy particles. The same

8.4 Electron Rate Coefficients and Energy Balance in the Ionosphere Plasma

is true, for example, in the first stage of post-discharges (see Fig. 8.6) or in the presence of a source of electron heating due to the absorption by the gas of ionizing U-V radiation or beams of energetic particles such as fast electrons. For such conditions it is useful to know the rate coefficients for the different processes and the electron energy loss rates as a function of the electron kinetic temperature. Using the analytical approximations (8.1), (8.3), (8.4) and data from Tables 8.1–8.3, we have numerically calculated the electron rate coefficients of some collision processes with N_2, O_2 and O from (8.10) assuming a Maxwellian EEDF. These calculations were then fitted by the simple three-parameter formula

$$k_{nk}\left(\frac{\text{cm}^3}{\text{s}}\right) = A\left(\frac{T_e}{1000}\right)^\gamma \exp\left\{-\frac{\epsilon^*_{th}}{kT_e}\right\}, \tag{8.31}$$

where T_e is in K and A is in units of $10^{-10}\text{cm}^3/\text{s}$. The values of A, γ and energy thresholds ϵ^*_{kn} are given in Table 8.13 for excitation of different N_2, O_2 and O states from the ground state.

Approximate calculations of electron energy loss rates through some collision channels in N_2, O_2 and O were also carried out in [155–159] as a function of the electron temperature. Here we present analytical approximations for the rate coefficients of electron mean energy relaxation $(d\bar{\epsilon}/dt)$ through different collision processes with N_2, O_2, O and ionospheric ions X^+. All the rate coefficients given below are in erg cm^3 s^{-1}.

1. Elastic collisions [155,156]:

with N_2: $\quad 2.88 \times 10^{-31} T_e (1 - 1.21 \times 10^{-4} T_e)(T_e - T)$;

with O_2: $\quad 1.92 \times 10^{-30} T_e^{0.5}(1 + 3.6 \times 10^{-2} T_e^{0.5})(T_e - T)$;

Table 8.13. Values of the parameters in (8.31) for the rate coefficients for excitation of different N_2, O_2, and O states from the ground electronic state

State	ϵ^*_{th} (ev)	A	γ	State	ϵ^*_{th} (ev)	A	γ
N_2 molecule				O_2 molecule			
Triplets							
$A^3\Sigma_u^+$	6.17	3.19	0.717	$a^1\Delta g$	0.98	0.246	1.028
$B^3\Pi_g$	7.35	19.5	0.393	$b^1\Sigma g^+$	1.63	0.404	0.54
$W^3\Delta_u$	7.36	3.5	0.857	$(c^1\Sigma_u^- +$	-	-	-
$B'^3\Sigma_u^-$	8.16	3.44	0.566	$C^3\Delta_u +$			
$C^3\Pi_u$	11.03	97.2	0.138	$A^3\Sigma_u^+)$	$\simeq 4.5$	0.121	1.00
$E^3\Sigma_g^+$	11.88	0.0916	0.878	$B^3\Sigma_u^-$	7.1	33.6	0.554
$D^3\Sigma_u^+$	12.85	8.77	0.193	-	-	-	-
Singlets				O atom			
$a'^1\Sigma_u^-$	8.4	1.00	0.94	$2p^1D^0$	1.96	12.0	0.35
$a^1\Pi_g$	8.55	8.54	0.543	$2p^1S^0$	4.19	1.75	0.318
$w^1\Delta_u$	8.89	6.18	0.458	$3s^5S^0$	9.14	29.5	-0.043
$a''^1\Sigma_g^+$	12.25	2.78	0.527	$3s^3S^0$	9.52	1.74	0.755
$b^1\Pi_u$	12.5	12.8	0.589	-	-	-	-
$b'^1\Sigma_u^+$	12.85	11	0.58	-	-	-	-
$c'^1_4\Sigma_u^+$	12.94	15.5	0.577	-	-	-	-

with O: $8.48 \times 10^{-31} T_e^{0.5}(1 + 5.7 \times 10^{-4} T_e)(T_e - T)$; (8.32)

with X^+: $5.63 \times 10^{-19} \ln \Lambda \, T_e^{-1.5} M_X^{-1}(T_e - T_i)$.

2. Rotational excitation and de-excitation of N_2 and O_2 [157]:

for N_2: $4.5 \times 10^{-26} T_e^{-0.5}(T_e - T)$;

for O_2: $1.1 \times 10^{-25} T_e^{-0.5}(T_e - T)$. (8.33)

3. Vibrational excitation and de-excitation of N_2 and O_2 [158]:

for N_2:

$$4.78 \times 10^{-24} \exp\left(h_{N_2} \frac{T_e - 2000}{2000 T_e}\right) \left[\exp\left(-g \frac{T_e - T_V^{N_2}}{T_e T_V^{N_2}}\right) - 1\right], \quad (8.34)$$

where

$h_{N_2} = 1.06 \times 10^4 + 7.51 \times 10^3 \tanh[10^{-3}(T_e - 1800)]$;

$g = 3300 + 1.233(T_e - 1000) - 2.056 \times 10^{-4}(T_e - 1000)(T_e - 4000)$;

for O_2:

$$1.2 \times 10^{-24} \exp\left(h_{O_2} \frac{T_e - 700}{700 T_e}\right) \left[\exp\left(-3000 \frac{T_e - T_V^{O_2}}{T_e T_V^{O_2}}\right) - 1\right], \quad (8.35)$$

where

$h_{O_2} = 3.9 \times 10^3 + 4.38 \times 10^2 \tanh[4.56 \times 10^{-4}(T_e - 2400)]$.

4. Excitation and de-excitation of electronic level $O(^1D)$ [158]:

$$2.5 \times 10^{-24} \exp\left(h_O \frac{T_e - 3000}{3000 T_e}\right) \left[\exp\left(-22713 \frac{T_e - T_O}{T_e T_O}\right) - 1\right], \quad (8.36)$$

where

$h_O = 2.4 \times 10^4 + 0.3(T_e - 1500) - 1.95 \times 10^{-5}(T_e - 1500)(T_e - 4000)$.

5. Excitation of electronic levels of fine structure $O(P_{2,1,0}^3)$ [159]:

$$1.38 \times 10^{-17} T_e^{-0.5} \frac{1}{Z} \sum_{k=1}^{3} A_k B_k T_e^{B_k}$$

$$\times \left\{\epsilon_k (D_k - G_k) + 5.91 \times 10^{-9}(T - T_e)\right.$$

$$\left. \times \left[(1 + B_k) D_k + \left(\frac{E_k}{T_e} + 1 + B_k\right) G_k\right]\right\} \quad (8.37)$$

Here, $Z = 5 + 3\exp(-228/T_1) + \exp(-326/T_2)$. The values of the other parameters for the transitions $^3P_2 - ^3P_1$; $^3P_2 - ^3P_0$: $^3P_1 - ^3P_0$ are presented in Table 8.14.

Table 8.14. Values of the parameters in (8.37) for electron excitation and de-excitation of the fine structure levels $O(^3P_{2,1,0})$

k	A_k	B_k	E_k	ϵ_k	D_k	G_k
1	8.56×10^{-6}	1.02	228	0.02	$\exp(-228/T_1)$	$\exp(-228/T_e)$
2	7.2×10^{-6}	0.9	326	0.028	$\exp(-326/T_2)$	$\exp(-326/T_e)$
3	2.46×10^{-7}	1.27	98	0.008	$\exp(-326/T_2)$	$\exp(-98/T_e - 228/T_1)$

To analyze the electron energy balance equation, it is useful to have an approximate expression for the elastic momentum transfer collision frequency of electrons with N_2, O_2, O and ions. Taking into account (8.32) such an expression is

$$\nu_{el}(s^{-1})$$
$$\approx 3.58 \times 10^{-11} T_e (1 - 1.21 \times 10^{-4} T_e)[N_2]$$
$$+ 2.72 \times 10^{-10} T_e^{0.5}(1 + 3.6 \times 10^{-2} T_e^{0.5})[O_2]$$
$$+ 6.01 \times 10^{-11} T_e^{0.5}(1 + 5.7 \times 10^{-4} T_e)[O] + 2.5 \ln \Lambda \, T_e^{-1.5}[X^+] \,. \quad (8.38)$$

In (8.32)–(8.38) T and T_i are the translational temperatures of respectively neutral atoms and ions; $T_V^{N_2}$ and $T_V^{O_2}$ are the vibrational temperatures of N_2 and O_2; T_O, T_1 and T_2 are the excitation temperatures of the electronic levels ^1D, $^3P_1$ and $^3P_0$ (which determine the relative populations in these states); $\ln \Lambda$ is the Coulomb logarithm, where $\Lambda = 1.22 \times 10^4 T_e^{3/2}/[X^+]$. All temperatures are expressed in K. The concentrations of heavy particles are expressed in cm^{-3} and the mass M_{X^+} of X^+ ions is given in atomic units.

The expressions (8.32), (8.33), (8.36) are a concrete application of (8.21)–(8.23) for atmospheric gases which can be useful for practical calculations. Moreover, equation (8.31) can also be used in order to account for electron energy loss channels other than (8.32)–(8.37). Note however that the rate coefficient (8.36) for excitation of $O(^1D)$ is based on different approximation from (8.31). Note also that the values of the quadrupole moments for N_2 and O_2 used to obtain (8.33) differ from those mentioned in Sect. 8.2 (1.01 and 0.29, respectively, in units ea_0^2). Moreover, the contribution of resonant excitation of rotational levels was not taken into account in (8.33). Finally, it should be recalled that equations (8.32)-(8.38) have been obtained for a Maxwellian EEDF and apply for low electron temperatures only ($T_e \leq 4000$ K).

8.5 Excitation, Dissociation and Ionization of N_2, O_2 and O Species by an Electron Beam

As in the case of plasmas sustained by an electric field, the electron rate coefficients for different processes and the electron power balance in electron beam sustained discharges are determined by the EEDF. In this case, it is convenient to express the excitation rates for the various processes in terms of the fractional power losses associated with each process. The fractional power loss P_{snk} for the inelastic process $n \to k$ involving the gas component s is

$$P_{snk} = \frac{\int_{\epsilon^*_{snk}}^{\infty} \epsilon^*_{snk} f(\epsilon) \sqrt{\epsilon} \nu_{snk} d\epsilon}{\sum_{s',n',k'} \int_{\epsilon^*_{s'n'k'}}^{\infty} \epsilon^*_{s'n'k'} f(\epsilon) \sqrt{\epsilon} \nu_{s'n'k'} d\epsilon} , \qquad (8.39)$$

where the same notations as in Chap. 4 have been used. The denominator in this equation is summed over all processes for all gas components. The fractional losses (8.39) are connected with other important parameters characterizing the dissipation of the beam energy in the gas, particularly the energy cost U_{snk} of the different inelastic processes:

$$U_{snk} = \epsilon^*_{snk}/P_{snk} . \qquad (8.40)$$

It should be emphasized that P_{snk} and U_{snk} are mostly determined by the cross-sections of the processes in the energy range of the corresponding thresholds (where, usually, these cross-sections are maximum), and do not in fact depend on the energy of the primary beam electrons ϵ_b if $\epsilon_b \geq 300$ eV. This is why the values of P_{snk} and U_{snk} can be considered as characteristics of the medium itself under interaction with high-energy electrons.

Calculations of P_{skn} and U_{skn} have been made for N_2 [160–168], for O_2 [160,162,166,167,169], for O [167,170], for N_2–O_2 [160,162,171,172], for N_2–O_2–O [160,173,174], for N_2–Ar [175], and for O_2–Ar [176]. Note that the energetic cost U_{snk} for any process in pure N_2, O_2 gases and in an N_2–O_2 air mixture can be obtained using the expression

$$U_{snk} = U_i \left[\alpha_s \int_0^{\infty} \Phi(\epsilon) \sigma_{snk}(\epsilon) d\epsilon \right]^{-1} . \qquad (8.41)$$

Here $\sigma(\epsilon)$ is the cross-section of the process; α_s is the relative concentration of component s in the mixture; $\Phi(\epsilon)$ is the electron flux (4.37), given for N_2, O_2 and air in Fig. 4.9; U_i is the energetic cost for the production of an electron-ion pair (33, 34.4 and 34 eV, respectively, for N_2, O_2 and the air mixture).

A simple approximation for the values P_{snk} in N_2 :O_2 :O mixtures for any relative concentrations α_s of the components, and for typical relative concentrations at different ionospheric altitudes, were obtained in [173,174]. This approximation can be expressed by the expression

$$P_{snk} = P^0 \frac{C_{snk}\alpha_s}{\sum_{s'=N_2,O_2,O} C_{s'nk}\alpha_s} ; \quad s, s' = N_2, O_2, O , \qquad (8.42)$$

where P^0 and C are parameters that depend on the particular component $s = N_2, O_2, O$ and elementary process $n \to k$. The parameter P^0 is the fractional power lost by the electrons in the excitation process $n \to k$ in the pure gas "s" (that is, when $\alpha_s = 1$, and thus $\alpha_{s'} = 0$ for $s' \neq s$).

8.5 Excitation, Dissociation and Ionization of N_2, O_2, O by an Electron Beam

Using for the coefficients C_{snk} in (8.42) the normalization condition $C_{N_2,n,k} + C_{O_2,n,k} + C_{O,n,k} = 1$ for any process $n \to k$, Tables 8.15–8.17 give values of the parameters P^0 and C_{N_2}, C_{O_2}, C_O for various electron processes with N_2, O_2 and O in N_2 :O_2 :O mixtures, which are typical for ionospheric altitudes (i.e., in mixtures where $\alpha_{O_2} < 0.22$ always) [174]. The notation of the excited states, the energy thresholds ϵ_{th}^*, and the mean square error δ of the approximation with respect to numerical calculations are also presented.

The mean square error δ does not exceed 1% for any inelastic processes with thresholds higher than 12 eV, and does not exceed 3% for thresholds higher than $\sim$7 ev. This error is somewhat larger for inelastic processes with lower thresholds and for elastic collisions. The sum of all P^0 values for each component N_2, O_2, O given in Tables 8.15–8.17 does not exactly equal 100% due to these errors (this is especially noticeable for the O atom). However, the accuracy of this approximation is sufficient for practical calculations taking into account possible inaccuracies in the cross-sections used. The fact that some coefficients C_{O_2} in Table 8.17 vanish is a choice motivated by minimization of the errors δ for typical ionosphere N_2 :O_2 :O gas mixtures, at altitudes where the relative concentration of O_2 is smaller than 0.2. The application of (8.42) over a wider range of relative concentrations α_{N_2}, α_{O_2}, α_O (including the case $\alpha_{O_2} = 1$) results in C_{N_2}, C_{O_2} and C_O values which differ

Table 8.15. Parameters of (8.42) for the interaction of an electron beam with N_2 molecules in the ground electronic state [174]

State or process	ϵ_{th}^* (ev)	P_0(%)	C_{N_2}	C_{O_2}	C_O	δ(%)
elastic	0	0.426	0.154	0.657	0.189	13
rotational excitation	-	1.03	0.114	0.681	0.205	16
vibrational excitation	-	6.3	0.407	0.520	0.073	9.0
$A^3\Sigma_u^+$	6.17	3.49	0.293	0.529	0.178	6.3
$B^3\Pi_g$	7.35	3.63	0.453	0.389	0.158	2.7
$W^3\Delta_u$	7.36	3.01	0.474	0.365	0.161	2.3
$B'^3\Sigma_u^-$	8.16	0.934	0.513	0.328	0.159	1.7
$a'^1\Sigma_u^-$	8.40	0.676	0.529	0.323	0.148	1.6
$a^1\Pi_g$	8.55	2.87	0.500	0.320	0.180	1.7
$w^1\Delta_u$	8.89	0.87	0.546	0.311	0.142	1.6
$C^3\Pi_u$	11.03	2.35	0.575	0.270	0.155	1.7
$a''^1\Sigma_g^+$	12.25	0.42	0.470	0.319	0.211	1.3
Rydberg states	13.75	18.52	0.407	0.407	0.186	< 1
all electronic states	6.17	36.9	0.422	0.405	0.173	1.8
dissociation	9.76	21.3	0.417	0.395	0.187	< 1
N_2 ionization to different N_2^+ states						
$X^2\Sigma_g^+$	15.6	8.30	0.405	0.413	0.182	< 1
$A^2\Pi_u$	10.8	9.20	0.412	0.403	0.185	< 1
$B^2\Sigma_u^+$	18.7	5.68	0.412	0.407	0.181	< 1
$D^2\Pi_g$	22.0	3.75	0.406	0.419	0.175	< 1
$C^2\Sigma_u^+$	23.8	3.08	0.413	0.413	0.174	< 1
dissociative	25.0	13.7	0.417	0.412	0.171	< 1
all	15.6	53.9	0.405	0.417	0.179	< 1

Table 8.16. As in Table 8.15 but for O_2 molecules in the ground electronic state [174]

State or process	ϵ_{th}^* (ev)	P_0(%)	C_{N_2}	C_{O_2}	C_O	δ(%)
elastic	0	0.222	0.426	0.350	0.224	4.4
rotational excitation	-	0.372	0.295	0.339	0.366	9.2
vibrational excitation	-	1.88	0.124	0.775	0.101	5.2
attachment	-	0.433	0.183	0.435	0.382	7.2
$a^1\Delta_g$	0.98	5.17	0.340	0.191	0.469	14
$b^1\Sigma_g^+$	1.64	1.41	0.237	0.196	0.567	11
$A^3\Sigma_u^+$	4.5	8.41	0.297	0.402	0.301	8.4
$B^3\Sigma_u^-$	8.4	15.7	0.456	0.351	0.193	< 1
-	9.9	0.945	0.427	0.372	0.201	< 1
Rydberg	13.5	19.3	0.410	0.399	0.191	< 1
all electronic	0.98	50.7	0.405	0.365	0.230	3.6
dissociation	5.12	21.7	0.456	0.368	0.176	1.3
O_2 ionization to different O_2^+ states						
$X^2\Pi_g$	12.1	5.84	0.422	0.388	0.190	< 1
$a^4\Pi_u$	16.1	14.0	0.410	0.407	0.183	< 1
$A^2\Pi_u$	16.8	3.38	0.410	0.410	0.180	< 1
$b^4\Sigma_g^-$	18.2	5.92	0.410	0.410	0.180	< 1
dissociative	20.0	17.8	0.410	0.418	0.172	< 1
all	12.1	47.1	0.412	0.408	0.180	< 1

Table 8.17. As in Tables 8.15, 8.16 but for O atoms in the ground electronic state [174]

State or process	ϵ_{th}^* (ev)	P_0(%)	C_{N_2}	C_{O_2}	C_O	δ(%)
elastic	0 -	0.783	0.721	0	0.279	4.2
$^3P_{1/2}, {}^3P_{3/2}$	0.02	3.86	0.573	0	0.427	8.6
$2p^1D^0$	1.96	9.73	0.375	0	0.625	16
$2p^1S^0$	4.18	1.16	0.500	0	0.500	12
$3s^5S^0, 3s^3S^0$	9.14	2.61	0.776	0	0.224	2.1
$3p^5P, 3p^3P$	10.7	1.06	0.784	0	0.216	2.1
$3d^3D^0$	12.03	0.544	0.712	0.016	0.272	1.2
$3s^3D^0$	12.48	1.73	0.659	0.077	0.264	1.1
Rydberg	15.5	16.3	0.586	0.162	0.252	1.2
all	-	35.4	0.573	0	0.427	11
$O(^3P)$ ionization in different O^+ states						
^4S	13.6	21.7	0.455	0.343	0.202	< 1
^2D	16.9	23.5	0.446	0.356	0.148	< 1
^2P	18.5	11.3	0.435	0.373	0.192	< 1
all	13.6	56.6	0.444	0.359	0.197	< 1

slightly from those given in Tables 8.15–8.17 (with no vanishing values) [173]. However, in this case the errors of this approximation are larger.

The fractional power losses P_{snk} and energy costs U_{snk} also permit the volume inelastic rates q_{snk} to be determined as follows:

$$q_{snk} = \frac{U_i}{U_{snk}} q_{ion} \equiv \frac{Q}{U_{snk}} \equiv \frac{P_{snk}}{\epsilon_{snk}^*} Q, \qquad (8.43)$$

where q_{ion} and Q are, respectively, the volume ionization rate (in cm^{-3} s^{-1}) and the energy dissipation rate of high-energy beam electrons in the gas mixture (in eV cm^{-3} s^{-1}). The rate Q is related to the energy dissipation function $\lambda(\overline{Z}/R)$ through the approximate expression [177]

$$Q\left(\frac{\text{eV}}{\text{cm}^3\text{s}}\right) \simeq \frac{10^3 \epsilon_b}{R} \mu [N_{\text{tot}}] \lambda\left(\frac{\overline{Z}}{R}\right), \qquad (8.44)$$

where

$$\lambda\left(\frac{\overline{Z}}{R}\right)$$

$$\simeq 4.2 \frac{\overline{Z}}{R} \exp\left\{-\left[\left(\frac{\overline{Z}}{R}\right)^2 + \frac{\overline{Z}}{R}\right]\right\} + 0.48 \exp\left\{-17.4\left(\frac{\overline{Z}}{R}\right)^{1.37}\right\}. \quad (8.45)$$

Herein, ϵ_b is expressed in keV; $R = 4.6 \times 10^{-6} \epsilon_b^{1.65}$ is the mean depth of penetration of an electron with initial energy ϵ_b into the medium (in g/cm^2); $\overline{Z} = \mu[N_{\text{tot}}]z$ is the current depth of penetration (in g/cm^2) expressed in terms of the linear depth z (cm); μ is the molecular weight of the gas mixture (in grams); and $[N_{\text{tot}}]$ is the total particle concentration of the gas mixture (in cm^{-3}).

In conclusion, equations (8.42)–(8.45) with the values of the parameters P^0 and C_{snk} in Tables 8.15–8.17 (or in [173]) enable the energy losses and the volume rates of different processes to be calculated when an electron beam with electron energy $\epsilon_b \geq 0.3$ keV interacts with an N$_2$–O$_2$–O mixture.

References

1. Burke P.G. and Berrington K.A (1993) *Atomic and Molecular Processes: R-Matrix Approach*, IOP Publishing, Bristol
2. Itikawa Y. (1994) Adv. Atom. Mol. and Opt. Phys. **33**, 253
3. Capitelli M,, Celiberto R. and Cacciatore M. (1994) Adv. Atom. Mol. and Opt. Phys. **33**, 322
4. Huo W. M and Gianturco F. A. (1995) *Computational Methods for Electron-Molecule Collisions*, Plenum Press, New York; Huo W. M. and Thummel H. T. (1996) in *Molecular Physics and Hypersonic Flows*, ed. M.Capitelli, Kluwer, Dordrecht, NATO ASI Series, Vol. 482, 115
5. Chutjian A., Garscadden A. and Wadehra J. M.(1996) Phys. Reports **264**, 393
6. Vanshtein L. A., Sobelman I. I. and Uykov E. A. (1973) *Cross Sections for Excitation of Atoms and Ions by Electrons*, Moscow Nauka (*in Russian*)
7. Christophorou L. G. (1984) *Electron-Molecule Interactions and their Applications*, Academic Press, New York
8. Uyrova I.U. and Ivanov B.E. (1989) *Cross Sections for Scattering of Electrons by Atmospheric Gases*, Nauka, Leningrad (*in Russian*)
9. Cartwright D.C., Trajmar S., Chutjian A. and Williams W. (1977) Phys. Rev. **A 16**, 1041

10. Trajmar S., Register D.F. and Chutjian A. (1983) Phys. Rev. **97**, 219
11. Haddad G.N. (1984) Aust. J. Phys. **37**, 487
12. Pitchford L.C. and Phelps A.V. (1982) it Bull. Am. Phys. Soc. **47**, 109
13. Yoshida S., Phelps A.V., Pitchford L.C. (1983) Phys. Rev. **A 27**, 2858
14. Phelps A.V., Pitchford L.C. (1985) Phys. Rev. **A 31**, 2932
15. Itikawa Y., Hayashi M., Ichimura A., Onda K., Sakomoto A., Takayanagi K., Nakamura M., Nishimura H. and Takayanagi T. (1986) J. Phys. Chem. Ref. Data. **15**, 985
16. Phelps A.V. (1985) *JILA Information Center Report*, **28**, University of Colorado, Boulder, Colorado, USA
17. Phelps A.V. (1987) in *Swarm Studies and Inelastic Electron-Molecular Collisions*, ed. L.C. Pitchford, B.V. McKoy, A. Chutjian and S. Trajmar, Springer, Berlin Heidelberg 127
18. Schulz G.J. (1973) Rev. Mod. Phys. **45**, 378; Schulz G.J. (1976) in *Principles of Laser Plasma*, ed. G. Bekefi, John Wiley and Sons, New York
19. Mihajlov A. A., Stojanovic V.D. and Petrovic Z. (1999) J. Phys. D: Appl. Phys. **32**, 2620
20. Opal C.B., Peterson W.K. and Beaty E.C. (1971) J. Chem. Phys. **55**, 4100
21. Gryzinski M.(1965) Phys. Rev. **138 A**, 322
22. Dravin H.W. (1969) Ztschr. Phys. **255**, 470
23. Jackman C.H., Garvey R.H. and Green A.E.S. (1977) J. Geophys. Res. **82**, 5081
24. Gorse C., Celiberto R., Cacciatore M., Lagana A. and Capitelli M. (1992) Chem. Phys. **161**, 217
25. Cacciatore M., Capitelli M. and Gorse C.(1992) Chem. Phys. **66**, 141
26. Capitelli M. and Celiberto R. (1998) in *Novel Aspects of Electron-Molecule Collisions*, ed. K.H. Becker, World Scientific, Singapore, p.283
27. Shkarovsky I., Jonston T., Bachinsky M. (1969) *Kinetics of Plasma Particles*, Atomizdat, Moscow (*in Russian*)
28. Granovsky V.L. (1971) *Electric Currents in Gases*, Nauka, Moscow (*in Russian*)
29. Gurevich A.V. and Shwartsburg A.B. (1973) *Nonlinear Theory of Propagation of Radio Waves in the Ionosphere*, Nauka, Moscow (*in Russian*)
30. Huxley L.G.H. and Crompton R.W. (1974) *The Diffusion and Drift of Electrons in Gases*, John Wiley and Sons, New York
31. Cercignani C. (1988) *The Boltzmann Equation and its Applications*, Springer, Berlin Heidelberg
32. Biberman L.M., Vorob'ev V.S. and Yakubov I.T. (1987) *Kinetics of Nonequilibrium Low Temperature Plasmas*, Consultants Bureau, London
33. Raizer Yu.P. (1986) *Gas Discharge Physics*, Springer, Berlin Heidelberg
34. Velihov E.P., Kozlov A.S. and Rahimov A.T. (1987) *Physical Phenomena in Discharge Plasmas*, Nauka, Moscow (*in Russian*)
35. Gurevich A.V. (1980) Adv. Phys. Sci **132**, 685 (*in Russian*)
36. Eletsky A.V., Palkina L.A. and Smirnov B.M. (1975) *Transport Phenomena in Weakly Ionized Plasmas*, Atomizdat, Moscow (*in Russian*)
37. Mnatsakanian A.H. and Naidis G.V. (1987) in *Plasma Chemistry*, ed. B.M. Smirnov, Energoizdat, Moscow, Vol. 14, p. 227 (*in Russian*)
38. Nighan W.L. (1970) Phys. Rev. A. **2**, 1989
39. Lukovnikov A.I., Fetisov E.E. and Trehov E.S. (1980) J. Techn. Phys. **40**, 1916 (*in Russian*)
40. Latypova R.A., Lukovnikov A.I. and Fetisov E.P. (1972) J. Techn. Phys. **42**, 115 (*in Russian*)
41. Osipov A.P. and Rahimov A.T. (1977) Physics of Plasmas **3**, 644 (*in Russian*)

42. Frost L.S. and Phelps A.V. (1962) Phys. Rev. **127**, 1621
43. Engelhardt A.G., Phelps A.V. and Risk C.G. (1964) Phys. Rev. A **135**, 1566
44. Winkler R. and Pfau S. (1973) Beitr. Plasmaphys. **13**, 273
45. Winkler R. and Pfau S. (1974) Beitr. Plasmaphys. **14**, 169
46. Polak L.S. and Slovetsky D.I. (1972) Chem. High Energy, **7**, 87 (*in Russian*)
47. Mnatsakanian A.H. and Naidis G.V. (1976) Physics of Plasmas **2**, 152 (*in Russian*)
48. Morison M.A. and Lane N.F. (1977) Phys. Rev. A **16**, 375
49. Oksyuk Yu.D. (1965) J. Exp. Theor. Phys. **49**, 1261 (*in Russian*)
50. Konev Yu.V., Kochetov I.V., Marchenko V.S. and Pevgov V.G. (1977) Quantum Electronics **4**, 1359 (*in Russian*)
51. Aleksandrov N.L., Konchakov A.M. and Son E.E. (1978) *Physics of Plasmas* **4**, 169; 1182 (*in Russian*)
52. Taniguchi T., Tagashira H. and Sakai Y. (1978) J. Phys. D: Appl. Phys. **11**, 1757
53. Nikonov S.V., Osipov A.P. and Rahimov A.T. (1979) Quantum Electronics **6**, 1258 (*in Russian*)
54. Aleksadrov N.L., Konchakov A.M. and Son E.E. (1979) High Temp. **17**, 179
55. Tachibana K. and Phelps A.V. (1979) J. Chem. Phys. **71**, 3544
56. Tagashira H., Taniguchi T. and Sakai Y. (1980) J. Phys. D: Appl. Phys. **13**, 235
57. Newman L.A. and De Temple T.A. (1976) J. Appl. Phys. **47**, 1912
58. Levron D. and Phelps A.V. (1979) Bull. Am. Phys. Soc. **24**, 129.
59. Slovetsky D.I. (1980) *Mechanisms of Chemical Reactions in Non-equilibrium Plasmas*, Nauka, Moscow (*in Russian*)
60. Dutton J. (1975) J. Phys. Chem. Ref. Data **4**, 577
61. Haydon S.C. and Williams O.M. (1976) J. Phys. D: Appl. Phys. **9**, 523
62. Buszevski W.W., Enright M.J. and Proud M. (1982) IEEE Trans. Plasma Sci. **PS-10**, 281
63. Gallagher J.W., Beaty E.C., Dutton J. and Pitchford L.C. (1983) J. Phys. Chem. Ref. Data **12**, 109
64. Bondarenko A.V., Vysikailo F.I. and Smakotin M.M. (1984) Thermophys. High Temp. **22**, 602 (*in Russian*)
65. Brunet H., Vincent P. and Rocca Serra J. (1983) J. Appl. Phys. **54**, 4951
66. Capitelli M., Gorse C., Wilhelm J. and Winkler J. (1981) Lettere Nuovo Cimento **32**, 225
67. Capitelli M., Celiberto R., Gorse C., Winkler R. and Wilhelm J. (1988) J. Phys. D: Appl. Phys. **21**, 691
68. Colonna G., Gorse C., Capitelli M,, Winkler R., and Wilhelm J. (1993) Chem. Phys. Lett. **213**, 5
69. Cacciatore M., Capitelli M. and Gorse C. (1982) Chem. Phys. **66**, 141
70. Gorse C., Capitelli M. and Ricard A. (1985) J. Chem. Phys. **82**, 1900
71. Paniccia F., Gorse C., Bretagne J. and Capitelli M. (1986) J. Appl. Phys. **59**, 4004
72. Capitelli M., Gorse C. and Ricard A. (1986) in *Non-equilibrium Vibrational Kinetics*. ed. M. Capitelli, Springer, Berlin Heidelberg, p. 315
73. Paniccia F., Gorse C., Cacciatore M. and Capitelli M. (1987) J. Appl. Phys. **61**, 1323
74. Aleksandrov N.L. and Kochetov I.V. (1987) Thermophys. High Temp. **25**, 1062 (*in Russian*)
75. Gorse C. and Capitelli M. (1987) J. Appl. Phys. **62**, 4072
76. Gorse C., Cacciatore M., Capitelli M., De Benedictis S. and Dilecci G. (1988) Chem. Phys. **119**, 63

77. Loureiro J. and Ferreira C.M. (1986) J. Phys. D: Appl. Phys. **19**, 17
78. Loureiro J. and Ferreira C.M.(1989) J. Phys. D: Appl. Phys. **22**, 67
79. Loureiro J., Ferreira C.M., Capitelli M., Gorse C. and Cacciatore M. (1990) J. Phys. D: Appl. Phys. **23**, 1371
80. Gousset G., Ferreira C.M., Pinheiro M., Sa P.A., Touzeau M., Vialle M. and Loureiro J. (1991) J. Phys. D: Appl. Phys. **24**, 290
81. Diatko N.A., Kochetov I.V., Napartovich A.P. and Taran M.D. (1983) Preprint of Kurchatov IAE **3842/12** (*in Russian*)
82. Aleksandrov N.L, Vysikailo F.I., Islamov R.Sh., Kochetov I.V., Napartovich A.P. and Pevgov V.G. (1981) High Temp. **19**, 342
83. Cacciatore M., Capitelli M., Gorse C., Massabieaux B. and Ricard A. (1982) Lettere Nuovo Cimento **34**, 417
84. Capitelli M., Gorse C., Wilhelm J. and Winkler R. (1984) Ann. der Phys. **41**, 119
85. Diatko N.A., Kochetov I.V. and Napartovich A.P. (1985) *Physics of Plasmas* **11**, 739 (*in Russian*)
86. Lucas J., Price D.A. and Moruzzi J.L. (1973) J. Phys. D: Appl. Phys. **6**, 1503
87. Masek K., Ruzicka T. and Laska L. (1977) Czechosl. J. Phys. **27**, 888
88. Lawton S.A. and Phelps A.V. (1978) J. Chem. Phys. **69**, 1055
89. Capitelli M., Dilonardo M. and Gorse C. (1980) Beitr. Plasmaphys. **20**, 83
90. Masek K. and Rohlena K. (1984) Czechosl. J. Phys. **B34**, 1227
91. Mayhan J.T. and Fante R.D. (1969) J. Appl. Phys. **40**, 5207
92. Guerra V., Pinheiro M.J., Gordiets B.F., Loureiro J. and Ferreira C.M. (1997) Plasma Sources Sci. Technol. **6**, 220; Gordiets B., Ferreira C. M., Pinheiro M., and Ricard A. (1998) Plasma Sources Sci. Technol. **7**, 363; Pintassilgo C. D., Loureiro J., Cernogora G., and Touzeau M. (1999) Plasma Sources Sci. Technol. **8**, 463
93. Diatko N.A., Kochetov I.V.and Napartovich A.P. (1987) Engineer. Phys. J. **52**, 95 (*in Russian*)
94. Nighan W.L. and Bennet J.H. (1969) Appl. Phys. Lett. **14**, 240
95. Lowke J.J., Phelps A.V. and Irwin B.W. (1973) J. Appl. Phys. **44**, 4664
96. Rockwood S.D. (1974) J. Appl. Phys. **45**, 5229
97. Lobanov A.N. and Suchkov A.F. (1974) Quantum Electronics **1**, 1527 (*in Russian*)
98. Judd O.P. (1974) J. Appl. Phys. **45**, 4572
99. Klagge S. (1977) Beitr. Plasmaphys. **17**, 237
100. Winkler R. and Pfau S. (1977) Beitr. Plasmaphys. **17**, 317
101. Pfau S. and Winkler R. (1977) Beitr. Plasmaphys. **17**, 397
102. Michel R., Pfau S. and Winkler R. (1978) Beitr. Plasmaphys. **18**, 131
103. Capitelli M., Gorse C., Berardini M. and Braglia G.L. (1981) Lettere Nuovo Cimento **31**, 231
104. Gorse C., De Benedictis S., Dilecce G. and Capitelli M. (1990) Spectrochemica Acta **41 B**, 521
105. Eletsky A.V. (1977) Plasma Physics **3**, 657 (*in Russian*)
106. Smirnov B.M. (1972) *Physics of Weakly Ionized Gas*, Nauka, Moscow (*in Russian*)
107. Oksyuk Yu.D. (1965) J. Exp. Theor. Phys. **49**, 1261 (*in Russian*)
108. Morrison M.A. and Lane N.F. (1977) Phys. Rev. A **16**, 975
109. Bates D.R., Kingston A.E. and McWhriter R.W.P.(1963) Proc. Roy. Soc. **A 267**, 297
110. Gurevich A.V. and Pitaevski L.P. (1964) J. Exp. Theor. Phys. **46**, 1281, (*in Russian*)
111. Gordiets B.F., Gudzenko A.I. and Shlepin L.A. (1968) JQSRT **8**, 791.

112. Cacciatore M. and Capitelli M. (1976) Z. Naturforsch. **31A**, 362
113. Cacciatore M. and Capitelli M. (1976) JQSRT **16**, 325
114. Cacciatore M., Capitelli M. and Dravin H.W. (1976) Physica **84C**, 267
115. Gudzenko A.I. and Yakovlenko S.I. (1978) *Plasma Lasers*, Atomizdat, Moscow (*in Russian*)
116. Potapov A.V., Tsvetkova L.I., Antropov V.I. and Volgova G.I. (1977) Optics and Spectroscopy **43**, 112 (*in Russian*)
117. Bates D. (1962) *Atomic and Molecular Processes*, Academic Press, New York
118. Aleksandrov N.L., Konchakov A.M., Shachkin L.V. and Shashkov V.M. (1986) Plasma Physics **12**, 1218 (*in Russian*)
119. Gorse C., Cacciatore M. and Capitelli M. (1978) Z. Naturforsch **33A**, 895
120. Smirnov B.M. (1974) *Ions and Excited Atoms in Plasma*, Atomizdat, Moscow (*in Russian*)
121. Eletsky A.V. and Smirnov B.M. (1982) Advan. Phys. Sci. **136**, 25 (*in Russian*)
122. Bardsley J.N. and Biondi M.A. (1970) in *Advances in Atomic and Molecular Physics*, eds. D.R. Bates. and I.Esterman, Academic Press, New York, Vol. 6, p. 1
123. Mitchell J.B.A. and McGowan J.W. (1983) *Physics of Ion-Ion and Electron-Ion Collisions*, Plenum Press, New York, p. 279
124. McGowan J.W. and Mitchell J.B.A. (1984) in *Electron-Molecule Interactions and Their Applications*, ed. L.G. Christophorou, Academic Press, New York
125. Biondi M.A. (1976) in *Principles of Laser Plasmas*, ed. G. Bekefi, John Wiley and Sons, New York
126. Mehr F.I. and Biondi M.A. (1969) Phys. Rev. **181**, 264
127. Frederick J.E. and Rusch D.W. (1977) J. Geophys. Res. **82**, 3509
128. Zipf E.C., Espy P.J. and Boyle C.F. (1980) J. Geophys. Res. **85**, 687
129. Walls F.L. and Dunn F.H. (1974) J. Geophys. Res. **79**, 1911
130. Torr D.G. et al. (1979) J. Geophys. Res. **84**, 3360
131. Hays P.B., Rusch D.W., Roble R.G. and Walker J.C.G. (1978) Rev. Geophys. Space Phys. **16**, 225
132. Zipf E.C. (1980) J. Geophys. Res. **85**, 4232
133. Torr M.R. and Torr D.G. (1979) Planet. Space Sci. **27**, 1233
134. Kley D., Lawrence G.W. and Stone E.J. (1977) J. Chem. Phys. **66**, 4157
135. Kossyi I.A., Kostinsky Yu.A., Matveyev A.A. and Silakov V. (1992) Plasma Sources Sci.Technol. **1**, 207
136. Fitaire M., Pointu A.M. and Stathopoulos D. (1984) J. Chem. Phys. **81**, 1753
137. Smirnov B.M. (1983) *Complex Ions*, Nauka, Moscow (*in Russian*)
138. Biondi M.A. (1969) Can. J. Chem. **47**, 1711
139. Janev R.K., Langer W.D., Evans K. and Post D.E. (1987) *Elementary Processes in Hydrogen-Helium Plasmas*, Springer, Berlin Heidelberg
140. Amano T. (1990) J. Chem. Phys. **92**, 6492
141. Michel P., Pfau S., Rutscher A. and Winkler R. (1980) Beitr. Plasmaphys. **20**, 25
142. Ding W.X., Pinnaduwage L.A., Tav C., and McCorle D.L. (1999) Plasma Sources Sci. Technol. **8**, 384; Datskos P.G., Pinnaduwage L.A. and Kielkpof J.F. (1997) Phys. Rev. **A55**, 4131
143. Hassouni K., Gicquel A. and Capitelli M. (1998) Chem. Phys. Lett. **290**, 502; Hassouni K., Gicquel A., Capitelli M. and Loureiro J.(1999) Plasma Sources Sci. Tecnhol. **8**, 494
144. Bottcher C. (1978) J. Phys. B: At. Mol. Phys. **11**, 3887
145. Massey H.S.W. (1976) *Negative Ions*, Cambridge University Press, Cambridge; Smirnov B.M. (1978) *Negative Ions*, Atomizdat, Moscow (*in Russian*)

146. Aleksandrov N.L., Vysikailo F.I., Islamov. R.Sh., Kochetov I.V., Napartovich A.P. and Pevgov V.G. (1981) Thermphys. High Temp. **19**, 22 (*in Russian*)
147. Yousfi M., Azzi N., Segur P. et al. (1987) *Rapport Interne du Centre de Physique Atomique de Toulouse*, University of Toulouse, Toulouse (*in French*)
148. Fournier G., Bonnet G. and Pigache D. (1980) J. Phys. Lett. **41**, L143
149. Bastien F., Haug R. and Lecuiller M. (1975) J. Chem. Phys. **72**, 105
150. Aleksandrov N.L. (1988) Adv. Phys. Sci. **154**, 177 (*in Russian*)
151. Parkes D.A. and Sugden T.M. (1972) J. Chem. Soc. Faraday Trans. **11, 68**, 600
152. Phelps A.V. (1969) Can. J. Chem. **47**, 1783
153. Dyatko N.A., Capitelli M., Celiberto R. and Napartovich A.P. (1996) Chem. Phys. Lett. **263**, 305
154. Megill L.R. and Cahn J.H. (1967) J. Geophys. Res. **69**, 5041; Diatko N.A., Kochetov I.V., Mishin E.V., and Telegin V.A. (1989) Geomagnetism and Aeronomy **29**, 275 (*in Russian*)
155. Banks P. (1966) Planet Space Sci. **14**, 1085
156. Henry R.J.W. and McElroy M.B. (1968) *The Atmospheres of Mars and Venus*, Gordon and Bridge, New York
157. Dalgarno A., McElroy M.B., Rece M.N. and Walker J.C.G. (1968) Planet Space Sci. **16**, 1371
158. Stuba P. and Varnum W.S. (1982) Planet. Space Sci. **20**, 1126
159. Holgy W.R. (1976) Geophys. Res. Lett. **3**, 541.
160. Konovalov V.P. and Son E.E. (1987) in *Plasma Chemistry*, ed. B.M.Smirnov, Energoizdat, Moscow **17**, 194 (*in Russian*)
161. Ryjov V.V. and Yastremsky A.G. (1978) Physics of Plasmas **4**, 1262 (*in Russian*)
162. Konovalov V.P. and Son E.E. (1980) J. Techn. Phys. **50**, 300 (*in Russian*)
163. Suhre D.R. and Verdeen J.T. (1976) J. Appl. Phys. **47**, 4484
164. Oks E.A., Rusanov V.D. and Sholin G.V. (1979) Physics of Plasmas **5**, 211 (*in Russian*)
165. Khare S.P. and Kumar A.J. (1977) J. Phys. B: At. Mol. Phys. **10**, 2239
166. Khare S.P. and Kumar A. (1977) Planet. Space Sci. **25**, 555
167. Sergienko T.I., Ivanov G.A., Ivanov B.E. and Kirilov A.S. (1987) Geomagnetism and Aeronomy **27**, 948 (*in Russian*)
168. Slinkler S.P., Ali A.W. and Taylor R.D. (1989) J. Appl. Phys. **66**, 5216
169. Khare S.P. and Kumar A. (1978) J. Phys. B: At. Mol. Phys. **11**, 2403
170. Slinkler S.P., Taylor R.D. and Ali A.W. (1988) J. Appl. Phys. **63**, 1
171. Medvedev Yu.A. and Hohlov V.D. (1979) J. Techn. Phys. **49**, 309 (*in Russian*)
172. Lappo G.B., Prudnikov M.M. and Checherin V.G. (1980) Thermophys. High Temp. **18**, 677 (*in Russian*)
173. Gordiets B.F. and Konovalov V.P. (1990) in *Plasma Jets in the Development of New Materials Technology*, ed. O.P. Solonenko and A.I. Fedorehenko, VSP, Utrecht Tokyo, p.617
174. Gordiets B.F. and Konovalov V.P. (1991) Geomagnetism and Aeronomy **31**, 649 (*in Russian*)
175. Ryjov V.V. and Yastremsky A.G. (1979) J. Techn. Phys. **49**, 2141 (*in Russian*)
176. Keto J.W. (1981) J. Chem. Phys. **74**, 4445
177. Lazarev V.I. (1967) Geomagnetism and Aeronomy **7**, 278 (*in Russian*)

9. Electronic State Relaxation Rates

9.1 Radiative Lifetimes

The kinetics of excited electronic states of atoms and molecules in gases and low-temperature plasmas is determined not only by the processes involving free electrons considered in Chap. 8, but also by spontaneous radiative transitions and collisions with heavy particles (atoms and molecules).

The probabilities of radiative transitions between electronic states also determine the intensities of different optical emissions. Spectral measurements of the intensities of electronic-vibrational–rotational lines and bands give various kinds of information on the properties of gases and plasmas and their kinetic characteristics. For this reason, information on the probabilities $A_{v'v''}^{nm}$ of optical transitions $(n, v') \to (m, v'')$ is very useful. Here n, m and v', v'' are the initial and final electronic states and vibrational levels respectively. The radiative lifetime $\tau_{v'}^n$ of electronic-vibrational state (n, v') is given by

$$\tau_{v'}^n = 1/A_{v'}^n \equiv 1/\sum_m A_{v'}^{nm} \equiv 1/\sum_m \sum_{v''} A_{v'v''}^{nm}, \qquad (9.1)$$

where $A_{v'}^n$ and $A_{v'}^{n,m}$ are, respectively, the probabilities of optical transitions from (n, v') state to all lower (m, v'') states and to all vibrational levels v'' of electronic state m.

The optical characteristics of N_2 molecules have been widely investigated. This is due to the importance of nitrogen emissions in observations of air luminescence in various non-equilibrium conditions, including the upper terrestrial atmosphere (aurora, air glow). Detailed information on the optical characteristics of N_2 is presented in [1], where, in particular, values of wavelengths $\lambda_{v'v''}$, Franck–Condon factors $q_{v'v''}$ and Einstein coefficients (optical probabilities $A_{v'v''}^{nm}$) are given for 18 systems of electronic transitions between states $X^1\Sigma_g^+$, $a^1\Pi_g$, $a''\Sigma_u^-$, $w'\Delta_u$, $b^1\Pi_u$, $b'^1\Sigma_u^+$, $A^3\Sigma_u^+$, $B'^3\Sigma_u^-$, $W^3\Delta_u$, $C^3\Pi_u$, $D^3\Sigma_u^+$, $E^3\Sigma_g^+$. The probabilities $A_{v'}^{nm}$ of different optical transitions $n \to m$ are presented in Fig. 9.1 as a function of the vibrational level v'. It can be seen that the dependence on v' for many cases is rather weak. However, for some transitions (such as $W^3\Delta_u \to B^3\Pi_g$, $a^1\Pi_g \to a'^1\Sigma_u^-$, and $a'^1\Sigma_u^- \to a^1\Pi_g$) this dependence is strong, and for the transition $A^3\Sigma_u^+ \to X^1\Sigma_g^+$ (the Vegard-Kaplan system) it is non-monotonic. This results from taking into consider-

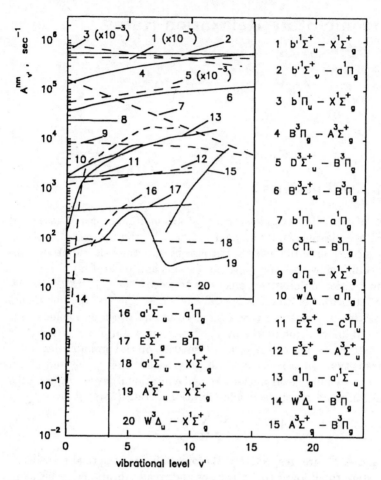

Fig. 9.1. Dependence on vibrational level v' of the probabilities $A_{v'}^{nm}$ for optical transitions $(n, v' \to m, \sum v'')$ in N_2 [1]

ation the dependence of the transition strength on the transition wavelength [1]. Data ignoring such dependence [2,3] exhibit a monotonic behaviour.

As the probabilities of transitions between molecular electronic states depend on the vibrational levels involved, the population kinetics for such states should be investigated together with a detailed vibrational kinetic analysis. However, in practice, this problem is often simplified and only the kinetics of total electronic state populations (summed over all vibrational levels) is investigated. In this approach, however, the transition probabilities used (for both optical and collision processes) are approximate averages over the vibrational distribution of the initial electronic state. Since this vibrational distribution depends on the conditions of the medium, the same probability as for the ground vibrational level is often assumed for all vibrational levels of the initial electronic state. This is justified by the fact that the $v'=0$ level usually has maximum population, and the probabilities are in many cases weakly dependent on level v' (see Fig. 9.1).

Tables 9.1 and 9.2 present the main optical transitions in N_2 and O_2, together with the probabilities $A_{v'=0}^{nm}$ of transitions from level $v'=0$ to all levels v'' of the lower electronic state and the wavelengths of some transitions from ground vibrational levels $v' = 0$.

It can be seen from Fig. 9.1 and Tables 9.1 and 9.2 that the radiative transition probabilities for some electronic states are rather small. For this reason, the populations of such states can be high in low pressure plasmas of atmospheric gases, and these populations can play an important role in plasma chemistry. This is particularly true for the metastable states $N_2(A^3\Sigma_u^+, a'^1\Sigma_u^-)$, $O_2(a^1\Delta_g, b^1\Sigma_g^+)$. Note that some electronic states with energy higher than the energy of dissociation have a probability of decay considerably exceeding the optical transition probability. Some channels for predissociation of N_2 and O_2 molecules and the corresponding probabilities D are also given in Tables 9.1 and 9.2.

Table 9.1. Optical transitions and predissociation of N_2

No	Transition $n \to m$	$A_{v'=0}^{nm}$ (s^{-1})	Wavelength for $v' \to v''$
1	$A^3\Sigma_u^+ \to X^1\Sigma_g^+$	0.5	2930 Å; 0 → 7
2	$B^3\Pi_g \to A^3\Sigma_u^+$	1.34×10^5	10450 Å; 0 → 0
3	$W^3\Delta_u \to X^1\Sigma_g^+$	0.154	2080 Å; 0 → 5
4	$B'^3\Sigma_u^- \to B^3\Pi_g$	3.4×10^4	15240 Å; 0 → 0
5	$C^3\Pi_u \to B^3\Pi_g$	2.45×10^7	3365 Å; 0 → 0
6	$E^3\Sigma_g^+ \to A^3\Sigma_u^+$	1.2×10^3	2553 Å; 0 → 5
7	$E^3\Sigma_g^+ \to B^3\Pi_g$	3.46×10^2	2874 Å; 0 → 1
8	$E^3\Sigma_g^+ \to C^3\Pi_u$	1.73×10^3	15000 Å; 0 → 0
9	$D^3\Sigma_u^+ \to B^3\Pi_g$	7.15×10^7	2347 Å; 0 → 1
10	$a'^1\Sigma_u^- \to X^1\Sigma_g^+$	$1. \times 10^2$	1771 Å; 0 → 5
11	$a^1\Pi_g \to X^1\Sigma_g^+$	8.55×10^3	1552 Å; 0 → 2
12	$a^1\Pi_g \to a'^1\Sigma_u^-$	1.3×10^2	82520 Å; 0 → 3
13	$w^1\Delta_u \to a^1\Pi_g$	1.51×10^3	35780 Å; 0 → 0
14	$b^1\Pi_u \to X^1\Sigma_g^+$	7.7×10^8	1116 Å; 0 → 5
15	$b^1\Pi_u \to a^1\Pi_g$	1.59×10^8	3133 Å; 0 → 0
16	$b'^1\Sigma_u^+ \to X^1\Sigma_g^+$	4.8×10^8	1354 Å; 0 → 14
17	$b'^1\Sigma_u^+ \to a^1\Pi_g$	5.56×10^5	3531 Å; 0 → 4

predissociation of N_2

No	Process	D (s^{-1})
18	$N_2(b^1\Pi_u) \to N(^4S)+N(^2D)$	8×10^{10}
19	$N_2(b'^1\Sigma_u^+) \to N(^4S)+N(^2D)$	8×10^{10}
20	$N_2(c'^1\Sigma_u^+) \to N(^4S)+N(^2D)$	8×10^{10}

Table 9.2. Optical transitions and predissociation of O_2

No	Transition	A_{ij} (s^{-1})	Spectrum region
1	$a^1\Delta_g \to X^3\Sigma_g^-$	2.6×10^{-4}	~ 1.27 mkm
2	$b^1\Sigma_g^+ \to a^1\Delta_g$	1.5×10^{-3}	~ 1.9 mkm
3	$b^1\Sigma_g^+ \to X^3\Sigma_g^-$	8.5×10^{-2}	~ 7620 Å
4	$(A^3\Sigma_u^+ + C^3\Delta_u + c^1\Sigma_u^-)$ $\to X^3\Sigma_g^-$	11	2430–3060 Å
5	$B^3\Sigma_u^- \to X^3\Sigma_g^-$	$\sim 2 \times 10^6$	~ 2000 Å

predissociation of O_2

No	Process	D_{ij} (s^{-1})
6	$O_2(B^3\Sigma_u^- \to O(^3P)+O(^1D)$	$\sim 2 \times 10^{10}$

9.2 Relaxation in Collisions with Atoms and Molecules

Collisions of electronically excited particles can cause a transfer of energy into translational and vibrational degrees of freedom (E-T and E-V processes) and also into other excited electronic states (E-E processes) of the collision partners. Changes in vibrational excitation often occur in E-E exchange processes, that is, an E-EV process does in fact take place if the partner is a molecule. In this case the total change in internal energy of the colliding particles may be minimal, that is, the process can be resonant or quasi-resonant, which generally leads to a large transition cross-section.

Most E-E processes are non-adiabatic. Due to intermolecular interaction, the position of the molecular electronic states changes during the collision and a pseudo-intersection between them is possible. The probability P of a transition between two electronic states $U_1(r)$ and $U_2(r)$ at their approach or intersection is determined by the Landau–Zener formula [4-6]:

$$P = 2 \times \exp\left\{-\frac{2\pi V^2}{\hbar c \Delta U'}\right\} \left(1 - \exp\left\{-\frac{2\pi V^2}{\hbar c \Delta U'}\right\}\right), \quad (9.2)$$

where $\Delta U' = \left|\frac{dU_2(r)}{dr} - \frac{dU_1(r)}{dr}\right|_{r=r_0}$. Here r_0 is the point of intersection, V is the perturbation energy at point r_0, and c is the relative velocity of the colliding particles.

Theoretical methods to analyze non-adiabatic electronic transitions are available in the literature [7–11]. However, it is practically impossible to carry out reliable theoretical calculations of the corresponding transition probabilities owing to the lack of accurate information on the structure and intersections of the colliding particles' electronic terms. Experiment is thus the main source of information on the rate coefficients for such processes.

The rate coefficients for relaxation of various electronic states - $N_2(A^3\Sigma_u^+, B^3\Pi_g, a'^1\Sigma_u^-, a^1\Pi_g, C^3\Pi_u, a'''^1\Sigma_g^+)$, $O_2(a^1\Delta_g, b^1\Sigma_g^+)$, $N(^2D, ^2P)$, $O(^1D, ^1S)$ - have been experimentally determined by different authors. Relaxation

Table 9.3. Rate coefficients for quenching and excitation of N_2 electronic states by collisions with atoms and molecules

No	Process	k(cm^3/s)	Ref.
R1	$N_2(A)+O(^3P) \rightarrow NO+N(^2D)$	7×10^{-12}	[12–14]
R2	$N_2(A)+O(^3P) \rightarrow N_2(X)+O(^1S)$	2.1×10^{-11}	[12–14]
R3	$N_2(A)+N(^4S) \rightarrow N_2(X)+N(^4S)$	2×10^{-12}	[15]
R4	$N_2(A)+N(^4S) \rightarrow N_2(X)+N(^2P)$	$4 \times 10^{-11}(300/T)^{2/3}$	[16–18]
R5	$N_2(A)+O_2(X) \rightarrow N_2(X)+O_2(B)$	$2.1 \times 10^{-12}(T/300)^{0.55}$	[19–22]
R6	$N_2(A)+O_2(X) \rightarrow N_2(X)+O_2(a)$	$2 \times 10^{-13}(T/300)^{0.55}$	[20–22]
R7	$N_2(A)+O_2(X) \rightarrow N_2(X)+O_2(b)$	$2 \times 10^{-13}(T/300)^{0.55}$	[20–22]
R8	$N_2(A)+O_2(X) \rightarrow N_2O+O(^3P)$	$2 \times 10^{-14}(T/300)^{0.55}$	[20–22]
R9	$N_2(A)+N_2(X) \rightarrow 2N_2(X)$	3×10^{-16}	[23]
R10	$N_2(A)+NO \rightarrow N_2(X)+NO(A)$	6.9×10^{-11}	[24]
R11	$N_2(A)+N_2O \rightarrow N_2(X)+N(^4S)+NO$	10^{-11}	[25,26]
R12	$N_2(A)+NO_2 \rightarrow N_2(X)+O(^3P)+NO$	10^{-12}	[27]
R13	$N_2(A)+H_2O \rightarrow N_2(X)+H+OH$	5×10^{-14}	[28]
R14	$N_2(A)+OH \rightarrow N_2(X)+OH(A)$	10^{-10}	[29]
R15	$N_2(A)+OH \rightarrow N_2(X)+O+H$	10^{-11}	[29]
R16	$N_2(A)+NH_3 \rightarrow N_2(X)+H+NH_2$	8.5×10^{-11}	[24]
R17	$N_2(A)+H \rightarrow N_2(X)+H$	2.1×10^{-10}	[29]
R18	$N_2(A)+H_2 \rightarrow N_2(X)+2H$	$2 \times 10^{-10}\exp(-3500/T)$	[30,31]
R19	$N_2(A)+N_2(A) \rightarrow N_2(X)+N_2(B)$	3×10^{-10}	[32,33]
R20	$N_2(A)+N_2(A) \rightarrow N_2(X)+N_2(C)$	1.5×10^{-10}	[34]
R21	$N_2(A)+N_2(X,v \geq 6) \rightarrow N_2(B)+N_2(X,v-6)$	3×10^{-11}	[35]
R22	$N_2(B)+N_2(X,v-6) \rightarrow N_2(A)+N_2(X,v \geq 6)$	3×10^{-11}	[35]
R23	$N_2(B)+N_2(X) \rightarrow 2N_2(X)$	2×10^{-12}	[36]
R24	$N_2(B)+O_2(X) \rightarrow N_2(X)+O+O$	3×10^{-10}	[28]
R25	$N_2(B)+NO \rightarrow N_2(A)+NO$	2.4×10^{-10}	[25]
R26	$N_2(B)+H_2 \rightarrow N_2(A)+H_2$	2.5×10^{-11}	[30,37]
R27	$N_2(C)+N_2(X) \rightarrow N_2(a')+N_2(X)$	10^{-11}	[28,38]
R28	$N_2(C)+O_2(X) \rightarrow N_2(X)+O(^3P)+O(^1S)$	3×10^{-10}	[28,39]
R29	$N_2(a')+N_2(X) \rightarrow N_2(B)+N_2(X)$	1.9×10^{-13}	[40,41]
R30	$N_2(a')+O_2(X) \rightarrow N_2(X)+O+O$	2.8×10^{-11}	[41]
R31	$N_2(a')+NO \rightarrow N_2(X)+N+O$	3.6×10^{-10}	[41]
R32	$N_2(a')+H \rightarrow N_2(X)+H$	1.5×10^{-10}	[42]
R33	$N_2(a')+H_2 \rightarrow N_2(X)+2H$	2.6×10^{-11}	[41]
R34	$N_2(a')+M \rightarrow N_2(a)+M$	$1.8 \times 10^{-11}\exp(-1700/T)$	[43]
R35	$N_2(a)+M \rightarrow N_2(a')+M$	9×10^{-12}	[43]
R36	$N_2(a')+N_2(A) \rightarrow N_4^+ + e$	4×10^{-12}	[42]
R37	$2N_2(a') \rightarrow N_4^+ + e$	10^{-11}	[42,44]
R38	$N_2(v>16)+N_2(v \geq 16) \rightarrow N_2(a')+N_2(X)$	$2 \times 10^{-14}\exp(-1700/T)$	[42]
R39	$N_2(a'')+N_2(X) \rightarrow N_2(X)+N_2(X)$	2.3×10^{-10}	[45]

No	Three-body collisions	k (cm^6/s)	Ref.
R40	$N+N+M \rightarrow N_2(A)+M$	1.7×10^{-33} (M=N_2,O_2,NO) 10^{-32} (M=N,O)	[46–48]
R41	$N+N+M \rightarrow N_2(B)+M$	2.4×10^{-33} (M=N_2,O_2,NO) 1.4×10^{-32} (M=N,O)	[47–50]

through different channels upon collisions with various atoms and molecules (including atmospheric components) has been investigated. Experimental rate coefficients for deactivation and excitation of the above-mentioned states are presented in Tables 9.3-9.6. Unfortunately, the temperature dependence

Table 9.4. Rate coefficients for quenching and excitation of O_2 electronic states by collisions with atoms and molecules

No	Process	k(cm³/s)	Ref.
R1	$O_2(a)+O(^3P) \to O_2(X)+O$	7×10^{-16}	[39,51]
R2	$O_2(a)+N(^4S) \to NO+O$	$2 \times 10^{-14} \exp(-600/T)$	[39,51]
R3	$O_2(a)+O_2(X) \to O_2(X)+O_2(X)$	$3.8 \times 10^{-18} \exp(-205/T)$	[52]
R4	$O_2(a)+N_2(X) \to O_2(X)+N_2(X)$	3×10^{-21}	[39]
R5	$O_2(a)+NO \to O_2(X)+NO$	2.5×10^{-11}	[53]
R6	$O_2(a)+O_3 \to 2O_2(X)+O(^1D)$	$5.2 \times 10^{-11} \exp(-2840/T)$	[54]
R7	$O_2(a)+O_2(a) \to O_2(X)+O_2(b)$	$7 \times 10^{-28} T^{3.8} \exp(700/T)$	[55,56]
R8	$O(^3P)+O_3 \to O_2(X)+O_2(a)$	$10^{-11} \exp(-2300/T)$	[57]
R9	$O_2(b)+O(^3P) \to O_2(a)+O(^3P)$	8.1×10^{-14}	[58]
R10	$O_2(b)+O(^3P) \to O_2(X)+O(^1D)$	$3.4 \times 10^{-11}(300/T)^{0.1} \exp(-\frac{4200}{T})$	[39,59]
R11	$O_2(b)+O_2(X) \to O_2(a)+O_2(X)$	$4.3 \times 10^{-22} T^{2.4} \exp(-281/T)$	[60,61]
R12	$O_2(b)+N_2(X) \to O_2(a)+N_2(X)$	$1.7 \times 10^{-15}(T/300)$	[62,63]
R13	$O_2(b)+NO \to O_2(a)+NO$	6×10^{-14}	[64,65]
R14	$O_2(b)+O_3 \to 2O_2+O$	2.2×10^{-11}	[58,63]
R15	$O_2(c)+O \to$ products	6×10^{-12}	[66]
R16	$O_2(c)+O_2 \to$ products	3.1×10^{-14}	[66]
R17	$O_2(c)+O_2(a) \to$ products	6×10^{-12}	[66]
R18	$O_2(A)+O \to$ products	9×10^{-12}	[66,67]
R19	$O_2(A)+O_2 \to$ products	$3. \times 10^{-13}$	[66,67]
R20	$O_2(A)+N_2 \to$ products	$9. \times 10^{-15}$	[66,67]
R21	$O_2(^5\Pi_g)+O_2 \to$ products	$k^{(21)} > 10^{-12}$	[68]
R22	$O_2(^5\Pi_g)+O_2 \to O_2(a)+O_2$	$0.5 \times k^{(21)}$	[69]
R23	$O_2(^5\Pi_g)+O \to O_2+O$	$3 \times k^{(21)}$	[69]
R24	$O_2(^5\Pi_g)+O \to O_2(b)+O_2$	$0.31 \times k^{(21)}$	[69]
	Three-body collisions	k (cm⁶/s)	Ref.
R25	$O+O+M \to O_2(a)+M$	$0.07 \times k_{tot}$ (M)	[69]
R25	$\to O_2(b)+M$	$< 0.01 \times k_{tot}$ (M)	[68]

In Table 9.4 k_{tot} is the total rate coefficient for three-body recombination (see reaction (2) in Table 10.2).

of many rate coefficients is still unknown, these constants usually being measured at room temperature.

Note that the rate coefficients for production of electronically excited O_2 molecules due to three–body recombination are much smaller then the total rate coefficient k_{tot} for such recombination, according to current knowledge [68,69,92–94] (see Table 9.4). These states are mainly formed in these processes via a precursor, the electronic state $^5\Pi_g$, which has a high probability of being formed as a result of recombination. The coefficients for processes (21)–(24) presented in Table 9.4 allow a quantitative explanation of different emission intensities in the upper terrestrial atmosphere at an altitude of ~100 km. Such emissions result from excited O_2 electronic states formed by $O+O+M \to O_2+M$ recombination.

The rate coefficients for deactivation of N_2 and O_2 electronically excited states are presented in Tables 9.3 and 9.4 mainly for transitions from the

Table 9.5. Rate coefficients for deactivation of N metastable levels

No	Process	k(cm^3/s)	Ref.
R1	$N(^2D)+O(^3P) \to N(^4S)+O(^1D)$	4×10^{-13}	[70,71]
R2	$N(^2D)+O_2 \to NO+O$	5.2×10^{-12}	[72,73]
R3	$N(^2D)+N_2 \to N(^4S)+N_2$	6×10^{-15}	[39,74]
R4	$N(^2D)+NO \to N_2+O$	1.8×10^{-10}	[74,75]
R5	$N(^2D)+N_2O \to NO+N_2$	3.5×10^{-12}	[74,76]
R6	$N(^2D)+H_2 \to NH+H$	2.3×10^{-12}	[77]
R7	$N(^2D)+NH_3 \to NH+NH_2$	1.1×10^{-10}	[78]
R8	$N(^2P)+N(^4S) \to N(^2D)+N(^4S)$	1.8×10^{-12}	[28]
R9	$N(^2P)+O_2(X) \to NO+O$	2.6×10^{-15}	[73,79]
R10	$N(^2P)+N_2(X) \to N(^4S)+N_2(X)$	2×10^{-18}	[75]
R11	$N(^2P)+NO \to N_2+O$	3×10^{-11}	[39]
R12	$N(^2P)+H_2 \to H+NH$	2.5×10^{-14}	[80]

Table 9.6. Rate coefficients for relaxation of O metastable levels

No	Process	k(cm^3/s)	Ref.
R1	$O(^1D)+O(^3P) \to O(^3P)+O(^3P)$	8×10^{-12}	[81]
R2	$O(^1D)+O_2(X) \to O(^3P)+O_2(X)$	$6.4 \times 10^{-12} \exp(67/T)$	[54]
R3	$O(^1D)+O_2(X) \to O(^3P)+O_2(a)$	10^{-12}	[57]
R4	$O(^1D)+O_2(X) \to O(^3P)+O_2(b)$	$2.6 \times 10^{-11} \exp(67/T)$	[54]
R5	$O(^1D)+N_2(X) \to O(^3P)+N_2(X)$	2.3×10^{-11}	[54,82]
R6	$O(^1D)+O_3 \to O_2(X) + 2O(^3P)$	1.2×10^{-10}	[54]
R7	$O(^1D)+O_3 \to 2O_2$	1.2×10^{-10}	[54]
R8	$O(^1D)+NO \to O_2(X)+N(^4S)$	1.7×10^{-10}	[39]
R9	$O(^1D)+N_2O \to NO+NO$	7.2×10^{-11}	[54]
R10	$(^1D)+N_2O \to O_2(X)+N_2(X)$	4.4×10^{-11}	[54]
R11	$O(^1S)+O(^3P) \to O(^1D)+O(^1D)$	$5 \times 10^{-11} \exp(-300/T)$	[84]
R12	$O(^1S)+N(^4S) \to O(^3P)+N$	10^{-12}	[85]
R13	$O(^1S)+O_2(X) \to O(^3P)+O_2(X)$	$1.3 \times 10^{-12} \exp(-850/T)$	[86,87]
R14	$O(^1S)+O_2(X) \to O(^3P)+O_2(A,c,C)$	$3 \times 10^{-12} \exp(-850/T)$	[86,87]
R15	$O(^1S)+N_2(X) \to O(^3P)+N_2(X)$	10^{-17}	[82]
R16	$O(^1S)+O_2(a) \to O(^3P)+O_2(C+A)$	1.1×10^{-10}	[88,89]
R17	$O(^1S)+O_2(a) \to O(^1D)+O_2(b)$	2.9×10^{-11}	[88,89]
R18	$O(^1S)+O_2(a) \to 3O(^3P)$	3.2×10^{-11}	[88,89]
R19	$O(^1S)+NO \to O(^3P)+NO$	2.9×10^{-10}	[90,91]
R20	$O(^1S)+NO \to O(^1D)+NO$	5.1×10^{-10}	[90,91]
R21	$O(^1S)+O_3 \to 2O_2(X)$	2.9×10^{-10}	[39]
R22	$O(^1S)+O_3 \to O_2(X) + O(^3P)+O(^1D)$	2.9×10^{-10}	[39]
R23	$O(^1S)+N_2O \to O(^3P)+N_2O$	6.3×10^{-12}	[39,91]
R24	$O(^1S)+N_2O \to O(^1D)+N_2O$	3.1×10^{-12}	[39,91]
R25	$O(^1S)+H_2O \to$ products	10^{-10}	[90]

ground vibrational level $v'=0$. The rate coefficients are, of course, dependent on v'. Unfortunately, this dependence has so far been obtained for some transitions only. The dependence on v' of the rate coefficients for deactivation $N_2(A^3\Sigma_u^+, v')$, $N_2(B^3\Pi_g(v'))$ and $N_2(a^1\Sigma_u^-(v'))$ states are shown in Fig. 9.2. It can be seen that this dependence is rather marked, and for deactivation by N_2 molecules increases sharply with v'. This sharp increase can be explained as an E-E deactivation process through the following channels [47]:

$$N_2(A, v' \geq 2) + N_2(X, v=0) \to N_2(A, v''=v'-2) + N_2(X, v=1) \quad (9.3)$$

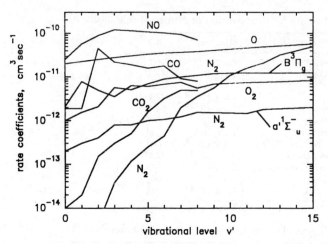

Fig. 9.2. Rate coefficients for deactivation of $N_2(A, v')$ by collisions with N_2, O_2, O, NO, CO, CO_2 and of $N_2(B, v')$, $N_2(a', v')$ by collisions with N_2 [1,3,28]

$$N_2(A, v' \geq 7) + N_2(X) \rightarrow N_2(B, v = 0) + N_2(X) \ . \tag{9.4}$$

Resonance is increasingly approached in process (9.3) as v' increases from 2 to $\sim$10, which explains why the rate coefficients also increase. Mechanism (9.4) switches on at $v' \geq 7$ and yields an additional increase in the deactivation rate coefficient (note that process (9.4) is endothermic for $v' < 7$ and, consequently, is unimportant at room temperature).

The difference between the rate coefficient values presented in Fig. 9.2 and Tables 9.3 for $v'=0$ shows that the vibrational kinetics of $N_2(A^3 \Sigma_u^+ (v'))$, $N_2(B^3 \Pi_g(v'))$ and $N_2(a^1 \Sigma_u^- (v'))$ states can be sufficiently significant to influence their deactivation and, as a result, their total populations.

Most of the processes presented in Tables 9.3–9.6 are exothermic. The electronic energy difference ΔE is distributed between translational, rotational and vibrational degrees of freedom after collision. In some cases, reverse processes can play an important role in relaxation. Under conditions of equilibrium between these degrees of freedom, the rate constants for reverse processes can easily be obtained from the principle of detailed balance by multiplying the coefficients of direct processes by the ratio of the statistical weights of the final and initial states and by the Boltzmann factor $\exp(-\Delta E/kT)$ (T denoting the temperature of the translational, rotational and vibrational degrees of freedom). However, the vibrational energy stored in molecular gases is in most cases out of equilibrium. Consider, for example, process (4) in Table 9.3 yielding $N(^2P)$ atoms and vibrationally excited $N_2(X, v)$ molecules. The degree of N_2 vibrational excitation can considerably exceed the equilibrium value (that is, $T_V > T$) and the reverse process of (4) can be essential in low-temperature plasmas at low gas temperature T. To obtain the total rate coefficient for the reverse process in such a non-equilibrium case, detailed information is required on the rates of direct E-EV processes for each vibrational level.

The influence of N_2 vibrational excitation on the total rate coefficients for the direct and reverse processes

$$N_2(A) + N(^4S) \longleftrightarrow N_2(X) + N(^2P) \tag{9.5}$$

was studied in [11]. It was found that

$$k^-(T, T_V) = \frac{1}{n} k^+ \exp\left(-\frac{\Delta E_1}{kT} - \frac{\Delta E_2}{kT_V}\right), \tag{9.6}$$

where $k^+ = 4 \times 10^{-11}(300/T)^{2/3} \text{cm}^3\text{s}^{-1}$ is the rate coefficient of the direct process, $n=10$ is the number of energetically possible decaying channels of the complex N_2-N, and $\Delta E_1 \simeq 0.11$ eV and $\Delta E_2 \simeq 2.48$ eV are the parts of the reaction energy barrier which are overcome due to translational energy and to $N_2(X)$ vibrational energy, respectively

References

1. Gurvich A.V., Kuznetsova L.A., Kuzmenko N.E., Kuzyakov Y.Y., Mikirov A.E. and Smirnov D.Y. (1985) Proceedings of Institute of Applied Geophysics, Gidrometeoizdat, Moscow, Vol. 64 (*in Russian*)
2. (1976) *Spectroscopic Data: Homonuclear Diatomic Molecules*, ed. S.N. Suchard, Plenum Press, New York
3. Cartwright G.C. (1978) J. Geophys. Res. A **83**, 517
4. Landau L.D. (1932) Sow. Phys. **2**, 46
5. Zener C. (1932) Proc. Roy. Soc. **A137**, 696
6. Bates D. (1962), in *Atomic and Molecular Processes*, ed. D.R. Bates, Academic Press, New York
7. Nikitin E.E. (1974) *Theory of Elementary Atomic and Molecular Processes in Gases*, Clarendon Press, Oxford
8. Smirnov B.M. (1973) *Asymptotic Methods in the Theory of Atomic Collisions*, Atomizdat, Moscow (*in Russian*)
9. Hirst D.M. (1985) *Potential Energy Surfaces, Molecular Structure and Reaction Dynamics*, Academic Press, New York
10. Child M.S. (1979), in *Atom-Molecule Collision Theory*, ed. R.B. Bernstein, Plenum Press, New York, p. 427
11. Dvoryankin A.N., Ibragimov L.B., Kulagin Yu.A. and Shelepin L.A. (1987), in *Plasma Chemistry*, ed. B.M. Smirnov, Energoiatomizdat, Moscow, Vol. 14, 102 (*in Russian*)
12. Piper L.G. (1982) J. Chem. Phys. **77**, 2373
13. De Souza A.R. and Touzeau M. (1083) Proceed. XVI ICPIG, Duesseldorf, Vol. 4, p. 355
14. De Souza A.R., Gousset G., Touzeau M. and Tu Khiet (1985) J. Phys.B: At. Mol. Phys. **18**, L661
15. Gilmore F.R., Baner E. and McGowan J.W. (1969) JQSRT **9**, 157
16. Flagan R.C. and Appleton J.P. (1972) J. Chem. Phys. **56**, 1163
17. Golde M.F. and Moyle A.M. (1985) Chem. Phys. Lett. **117**, 375
18. Piper L.G. (1989) J. Chem. Phys. **90**, 7087

19. Callear A.B. and Wood P.M. (1971) Trans. Faraday Soc. **67**, 272.
20. Jannuzzi M.P., Jeffries J.B. and Kaufman F. (1982) Chem. Phys. Lett. **87**, 570
21. De Sousa A.R. and Touzeau M. (1985) Chem. Phys. Lett. **121**, 423
22. Fraizer M.E. and Piper L.G. (1989) J. Phys. Chem. **93**, 1107
23. Clarc W.G. and Setser D.W. (1980) J. Phys. Chem. **84**, 2225
24. Cao D.Z. and Setser D.W. (1988) J. Phys. Chem. **92**, 1169
25. Young R.A., Black G. and Slanger T.G. (1969) J. Chem. Phys. **50**, 303
26. Taylor G.W. and Setser D.W. (1971) J. Amer. Chem. Soc. **93**, 4930
27. Maetzing H (1991), in *Chemical Kinetics of Flue Gas Cleaning by Irradiation with Electrons*, Adv. in Chem. Phys., ed. I. Prigogine and S.A. Rice, Vol. LXXX, John Wiley and Sons, New York
28. Slovetsky D.I. (1980) *Mechanisms of Chemical Reactions in Non-equilibrium Plasmas*, Nauka, Moscow (*in Russian*)
29. Grace H.Ho and Golde M.F. (1991) J. Chem. Phys. **95**, 8866
30. Slanger T. G., Wood B. J. and Black G., J. (1973/74) Photochem. **2**, 63
31. Sperlein A., (1988) J. Chem. Phys. **89**, 3113
32. Hays G.N. and Oskam H.J. (1973) J. Chem. Phys. **59**, 6088
33. Piper L.G. (1988) J. Chem. Phys. **88**, 6911
34. Piper l.G. (1988) J. Chem. Phys. **88**, 231
35. Piper l.G. (1988) J. Chem. Phys. **88**, 864
36. Heidner R.F., Sutton D.G. and Suchard S.N. (1976) Chem. Phys. Lett. **37**, 243
37. Lee L.C. and Suto M. (1984) J. Chem. Phys. **80**, 4718
38. Calo J.M. and Axtmann R.C. (1971) J. Chem. Phys. **54**, 1332
39. Kossiy I.A., Kostinsky Yu.A., Matveyev A.A. and Silakov V. (1992) Plasma Sources Sci.Technol. **1**, 207
40. Dreyer J.W. and Perner D. (1972) Chem. Phys. Lett. **16**, 169
41. Piper L.G. (1987) J. Chem. Phys. **87**, 1625
42. Gordiets B.F., Pinheiro M., Ferreira C.M. and Ricard A. (1998) Plasma Sources Sci. Technol. **7**, 363; 379.
43. Magne L., Cernogora G. and Veis P. (1992) J. Phys. D: Appl. Phys. **25**, 472
44. Gordiets B., Ferreira C.M., Nahorny J., Pagnon D., Touzeau M., and Vialle M. (1996) J. Phys. D: Appl.Phys. **29**, 1021
45. Wedding A.B., Borysow J. and Phelps A.V. (1993) J. Chem. Phys. **98**, 6227
46. Shemansky D.E. (1976) J. Chem. Phys. **64**, 565
47. Delcroix J.L., Ferreira C.M. and Ricard A. (1976), in *Principles of Laser Plasmas*, ed. G. Bekefi, John Wiley and Sons, New York
48. Krivonosova O.E., Losev S.A., Nalivaiko V.P., Mukosev Yu.K. and Shatalov O.P. (1987), in *Plasma Chemistry*, ed. B.M. Smirnov, Energoatomizdat, Moscow, Vol. 14, p. 3 (*in Russian*)
49. Becker K.H., Fink E.H., Groth W., Jud W. and Kley D. (1972) Farad. Discuss. Chem. Soc. **53**, 35
50. Partridge H., Langhoff S.R., Bauschlicher C.W. and Schwenke D.W. (1988) J. Chem. Phys. **88**, 3174
51. O'Keefe A., Mauclaire G., Parent D. and Bowers M.T. (1986) J. Chem. Phys. **84**, 215
52. Billington A.P. and Borell P. (1986) J. Chem. Soc. Faraday. Trans. **82**, 963
53. Yaron M., von Engel A. and Vidaut P.H. (1976) Chem. Phys. Lett. **37**, 159
54. Baulch D., Cox R:, Crutzen P., Hampson R., Kerr J., Troe J. and Watson R. (1982) J. Phys. Chem. Ref. Data **11**, 327
55. Heidner R.F., Gardner C.E., El-Sayed T.M., Segal G.I. and Kasper V.V. (1981) J. Chem. Phys. **74**, 5618
56. Cohen N. and Westerberg K. (1983) J. Phys. Chem. Ref. Data **12**, 531

57. Eliasson B. and Kogelschatz U. (1986) *Brown Boveri Report* **KLR 86-11C**, Baden
58. Slanger T.G. and Black G. (1979) J. Chem. Phys. **70**, 3434
59. Zinn J., Sutherland C.D., Stone S.N. and Duncan L.M. (1982) J. Atm. Terr. Phys. **44**, 1143
60. Borrell P.M., Pedley M.D., Borell P. and Grant K.R. (1979) Proc. R. Soc. **A367**, 395
61. Borrell P.M., Borell P., Grant K.R. and Pedley M.D. (1982) J. Phys. Chem. **86**, 700
62. Kohse-Hoinghaus K. and Stuhl F. (1980) J. Chem. Phys. **72**, 3720
63. Choo K.Y. and Leu M.T. (1985) Int. J. Chem. Kinet. **17**, 1155
64. O'Brien R.J. and Myers G.H. (1970) J. Chem. Phys. **53**, 3832
65. Becker K.H., Groth W. and Schurath U. (1971) Chem. Phys. Lett. **8**, 259
66. Kenner R.D. and Ogryzlo E.A. (1983) Can. J. Chem. **61**, 921
67. Kenner R.D. and Ogryzlo E.A. (1980) Int. J. Chem. Kinet. **12**, 501
68. Lopez-Gonzalez M.J., Lopez-Moreno J.J. and Rodrigo R. (1992) Planet. Space Sci. **40**, 783
69. Lopez-Gonzalez M.J. and Rodrigo R. (1992) Proceed. 19th Annual European Meeting on Atmospheric Studies by Optical Methods, Kiruna, Sweden, 76
70. Frederick J.E. and Rusch D.W. (1977) J. Geophys. Res. **82**, 3509
71. Davenport J.E., Slanger T.G. and Black G. (1976) J. Geophys. Res. **81**, 12
72. Lin C.L. and Kaufman F. (1971) J. Chem. Phys. **55**, 3760
73. Husain D., Mitra S.K. and Young A.N. (1974) J. Chem. Soc. Farad. Trans. **70**, 1721
74. Black G., Slanger T.G., John G.A.St. and Young B.A. (1969) J. Chem. Phys. **51**, 116
75. Donovan R. and Husain D. (1970) Chem. Rev. **70**, 489
76. Black G., Sharpless R.L., Slanger T.G. and Lorents D.C. (1975) J. Chem. Phys. **62**, 4266
77. Piper L.G., Donahue M.E. and Rawlins W.T. (1987) J. Phys. Chem. **91**, 3833
78. Black G. (1969) J. Chem. Phys. **51**, 116
79. Husain D. (1977) Ber. Bunsen Ges. Phys. Chem. **81**, 168
80. Young R.A. and Dunn O.J. (1975) J. Chem. Phys. **63**, 1150
81. Abren V.J., Yee J.H., Solomon S.C. and Dalgarno A. (1986) Planet. Space Sci. **34**, 1143
82. Levine J.S., ed. (1985) *The Photochemistry of the Atmosphere*, Academic Press, New York
83. Wine P.H. and Ravishankara A.R. (1981) Chem. Phys. Lett. **77**, 103
84. Slanger T.G. and Black G. (1976) J. Chem. Phys. **64**, 3763
85. Felder W. and Young R.A. (1972) J. Chem. Phys. **56**, 6028
86. Slanger T.G., Wood B.G. and Black G. (1972) Chem. Phys. Lett. **17**, 401
87. Slanger T.G. and Black G. (1978) J. Chem. Phys. **68**, 998
88. Slanger T.G. and Black G. (1981) J. Chem. Phys. **75**, 2247
89. Kenner R.D. and Ogryzlo E.A. (1982) J. Photochemistry **18**, 379
90. Filseth S.V., Stuhl F. and Welge K.H. (1972) J. Chem. Phys. **57**, 4064
91. Slanger T.G. and Black G. (1978) J. Chem. Phys. **68**, 989
92. Wright P.S. (1982) Planet. Space Sci. **30**, 251
93. Smith I.W.M. (1984) Int. J. Chem. Kinet. **16**, 423
94. Bates D. (1988) Planet. Space Sci. **36**, 875

10. Rate Coefficients of Chemical Reactions

Non-equilibrium gases and low-temperature plasmas are media in which the internal (rotational, vibrational and electronic) degrees of freedom of the particles and the chemical composition are in general far from thermodynamic equilibrium with the translational degrees of freedom. For example, ionization and dissociation of atoms and molecules in plasmas by electron impact lead to the creation of chemically active atoms, radicals and ions. The collisions of these heavy particles with the main gas species initiate various ion–molecule reactions, resulting in a non-equilibrium chemical composition of the medium. To control such reactions in order to obtain valuable chemical products is the main aim of plasma-chemical technology. The kinetics of non-equilibrium plasma-chemical reactions plays an important role in the design of laboratory plasma devices, in high-temperature gas dynamics, and in such natural objects as planetary atmospheres and ionospheres.

In order to analyze the kinetics of chemical and plasma-chemical processes, it is of course necessary to know the rate coefficients of different reactions involving atoms, molecules and ions. To accurately describe these phenomena, it is often necessary to account for a considerable quantity of such reactions. A large number of reactions for atmospheric gases are presented in [1]. More up-to-date information for reactions with nitrogen–oxygen and nitrogen–oxygen–hydrogen species has been used in preparing the tables in this chapter.

10.1 Reactions of Neutral Species

The processes and rate coefficients for reactions involving nitrogen-oxygen neutral species are given in Tables 10.1–10.3. Most of these data are taken from [2] and from a database developed at the Institute of Mechanics, Moscow State University [3].

Reactions involving hydrogen, oxygen, and nitrogen and the corresponding rate coefficients are given in Table 10.4. All data are taken from [14,15]. Note that the list of reactions in Table 10.4 was used [14,15] to model the combustion of ammonia in air, that is, for gas temperatures $T > 1000$ K. Therefore, the accuracy of the rate coefficients presented in Table 10.4 may be

Table 10.1. Rate coefficients for bimolecular nitrogen-oxygen reactions

No	Process	$k(\text{cm}^3/\text{s})$	T (K)	Ref.
R1	$N+NO \rightarrow O+N_2$	$1.8 \times 10^{-11}(T/300)^{0.5}$	200–4000	[2,4]
R2	$N+O_2 \rightarrow O+NO$	$3.2 \times 10^{-12}(T/300)$ $\times \exp(-3150/T)$	300–3000	[2,5]
R3	$N+NO_2 \rightarrow 2O+N_2$	9.1×10^{-13}	298	[2,6]
R4	$N+NO_2 \rightarrow O+N_2O$	3×10^{-12}	298	[2,7]
R5	$N+NO_2 \rightarrow N_2+O_2$	7×10^{-13}	298	[2,6]
R6	$N+NO_2 \rightarrow NO+NO$	2.3×10^{-12}	298	[2,6]
R7	$N+O_3 \rightarrow NO+O_2$	$\leq 2 \times 10^{-16}$	300	[2,8]
R8	$O+N_2 \rightarrow N+NO$	$3 \times 10^{-10} \exp(-38370/T)$	$(2-5) \times 10^3$	[3]
R9	$O+NO \rightarrow N+O_2$	$7.5 \times 10^{-12}(T/300)$ $\times \exp(-19500/T)$	$(1-3) \times 10^3$	[2,5]
R10	$O+NO \rightarrow NO_2 + h\nu$	4.2×10^{-18}	300	[1]
R11	$O+N_2O \rightarrow N_2+O_2$	$8.3 \times 10^{-12} \exp(-14000/T)$	1000–2500	[3]
R12	$O+N_2O \rightarrow NO+NO$	$1.5 \times 10^{-10} \exp(-14090/T)$	300–2500	[3]
R13	$O+NO_2 \rightarrow NO+O_2$	$9.1 \times 10^{-12}(T/300)^{0.18}$	230–2500	[2]
R14	$O+O_3 \rightarrow O_2+O_2(a)$	$2 \times 10^{-11} \exp(-2280/T)$	220–1000	[2,9]
R15	$O+NO_3 \rightarrow O_2+NO_2$	10^{-11}	298	[2,10]
R16	$O+N_2O_5 \rightarrow$ products	$\leq 3 \times 10^{-16}$	220–360	[2,11]
R17	$N_2+O_2 \rightarrow O+N_2O$	$2.5 \times 10^{-10} \exp(-50390/T)$	$300-10^4$	[2]
R18	$NO+NO \rightarrow N+NO_2$	$3.3 \times 10^{-16}(300/T)^{0.5}$ $\times \exp(-39200/T)$	300–2500	[3]
R19	$NO+NO \rightarrow O+N_2O$	$2.2 \times 10^{-12} \exp(-32100/T)$	1200–2500	[3]
R20	$NO+NO \rightarrow N_2+O_2$	$5.1 \times 10^{-13} \exp(-33660/T)$	10^3-10^4	[2]
R21	$NO+O_2 \rightarrow O+NO_2$	$2.8 \times 10^{-12} \exp(-23400/T)$	300–550	[2,5]
R22	$NO+O_3 \rightarrow O_2+NO_2$	$2.5 \times 10^{-13} \exp(-765/T)$	250–350	[3]
R23	$NO+N_2O \rightarrow N_2+NO_2$	$4.6 \times 10^{-10} \exp(-25170/T)$	900–2500	[3]
R24	$NO+NO_3 \rightarrow NO_2+NO_2$	1.7×10^{-11}	300–5000	[2]
R25	$O_2+O_2 \rightarrow O+O_3$	$2 \times 10^{-11} \exp(-49800/T)$	220–2500	[3]
R26	$O_2+NO_2 \rightarrow NO+O_3$	$2.8 \times 10^{-12} \exp(-25400/T)$	200–500	[2]
R27	$NO_2+NO_2 \rightarrow 2NO+O_2$	$3.3 \times 10^{-12} \exp(-13500/T)$	300–3000	[2,5]
R28	$NO_2+NO_2 \rightarrow NO+NO_3$	$4.5 \times 10^{-10} \exp(-18500/T)$	1000–2600	[3]
R29	$NO_2+O_3 \rightarrow O_2+NO_3$	$1.2 \times 10^{-13} \exp(-2450/T)$	230–360	[2,12]
R30	$NO_2+NO_3 \rightarrow$ $NO+NO_2+O_2$	$2.3 \times 10^{-13} \exp(-1600/T)$	300–850	[2,5]
R31	$NO_3+O_2 \rightarrow NO_2+O_3$	$1.5 \times 10^{-12} \exp(-15020/T)$	230–360	[3]
R32	$NO_3+NO_3 \rightarrow O_2 + 2NO_2$	$4.3 \times 10^{-12} \exp(-3850/T)$	600–1100	[2]
	associative ionization			
R33	$N+N \rightarrow N_2^+ + e$	$2.7 \times 10^{-11} \exp(-6.7410^4/T)$	≥ 4000	[3]
R34	$N+O \rightarrow NO^+ + e$	$1.6 \times 10^{-12}(T/300)^{0.5}(0.19 + 8.6 \times T)$ $\exp(-32000/T)$	≥ 4000	[3]

smaller at lower temperatures. Lists of different reactions and corresponding rate coefficients can also be found in [13,16–25].

In general, the rate coefficients for chemical processes depend not only on the translational temperature T, but also on the level of excitation of the internal degrees of freedom of the reactants. For example, molecular dissociation in collisions with heavy particles usually takes place through multi-step excitation of vibrational molecular levels followed by a transition from a level close to the dissociation boundary, into the continuous (see Chap. 3). A sim-

Table 10.2. Rate coefficients for dissociation of nitrogen-oxygen molecules

No	Process	$k(\mathrm{cm}^3/\mathrm{s})$	T (K)	Ref.
R1	$N_2+M \to N+N+M$	$k_{N_2} = 5.4 \times 10^{-8}[1 - \exp(-3354/T)]$ $\times \exp(-113200/T)$ $k_{O_2} = k_{NO} = k_{N_2},$ $k_O = k_N = 6.6 k_{N_2}$	$300–1.5 \cdot 10^4$	[2]
R2	$O_2+M \to O+O+M$	$k_{N_2} = 6.1 \times 10^{-9}[1 - \exp(-2240/T)]$ $\times \exp(-59380/T)$ $k_{O_2} = 5.9 k_{N_2}, k_O = 21 k_{N_2}$ $k_N = k_{NO} = k_{N_2},$ $k_{Ar} = 0.3 k_{N_2}, k_{CO_2} = 2 k_{N_2}$	$300–10^4$	[2]
R3	$NO+M \to N+O+M$	$k_{N_2} = 8.7 \times 10^{-9} \exp(-75994/T)$ $k_{O_2} = k_{N_2},$ $k_O = k_N = k_{NO} = 20 k_{N_2}$	$(4–7.5) \cdot 10^3$	[2]
R4	$O_3+M \to O_2+O+M$	$k_{N_2} = 6.6 \times 10^{-10} \exp(-11600/T)$ $k_{O_2} = 0.38 k_{N_2}$ $k_O = k_N = 6.3 \times \exp(170/T) k_{N_2}$ $k_{Ar} = 0.63 \times \exp(170/T) k_{N_2}$	290–4000	[3]
R5	$N_2O+M \to N_2+O+M$	$k_{N_2} = 1.2 \times 10^{-8}(300/T)$ $\times \exp(-29000/T)$ $k_{O_2} = k_{N_2}, k_{NO} = 2 k_{N_2}$ $k_{N_2O} = 4 k_{N_2}, k_{Ar} = 0.74 k_{N_2}$	$(1.5–2.5) \cdot 10^3$	[3]
R6	$NO_2+M \to NO+O+M$	$k_{N_2} = 6.8 \times 10^{-6}(300/T)^2$ $\times \exp(-36180/T)$ $k_{O_2} = 0.78 k_{N_2}, k_{NO} = 7.8 k_{N_2}$ $k_{NO_2} = 5.9 k_{N_2}, k_{Ar} = 0.59 k_{N_2}$	200–4000	[3]
R7	$NO_3+M \to NO_2+O+M$	$k_{N_2} = 3.1 \times 10^{-5}(300/T)^2$ $\times \exp(-25000/T)$ $k_{O_2} = k_{NO} = k_{N_2},$ $k_N = k_O = 10 k_{N_2}, k_{Ar} = 1.2 k_{N_2}$	200–1500	[3]
R8	$NO_3+M \to NO+O_2+M$	$k_{N_2} = 6.2 \times 10^{-5}(300/T)^2$ $\times \exp(-25000/T)$ $k_{O_2} = k_{NO} = k_{N_2},$ $k_N = k_O = 12 k_{N_2}, k_{Ar} = 1.2 k_{N_2}$	200–1500	[3]
R9	$N_2O_5+M \to NO_2+NO_3+M$	$k_M = 2.1 \times 10^{-11}(300/T)^{4.4}$ $\times \exp(-11080/T)$		[13]

ilar mechanism may operate for other reactions with an activation energy. The vibrational kinetics of diatomic molecules when dissociation involving a boundary vibrational level is present was discussed in Chap. 3, where the equations determining the total dissociation rate coefficient were also given. This rate coefficient depends on both the translational and vibrational temperatures T and T_V of the dissociating molecule; it is the product of the equilibrium rate coefficient $K(T)$ by a non-equilibrium factor $\Phi(T, T_V)$ which is a function of T and T_V (see (3.127),(3.128)).

For dissociation of polyatomic molecules under moderate deviations from equilibrium, this non-equilibrium factor can be written as [26,27]

Table 10.3. Recombination rate coefficients of nitrogen-oxygen species

No	Process	k (cm^6/s)	T (K)	Ref.
R1	N+N+M→ N$_2$+M	$k_{N_2} = 8.3 \times 10^{-34} \exp(500/T)$ $k_{N_2} = 1.91 \times 10^{-33}$ $k_{O_2} = 1.8 \times 10^{-33} \exp(435/T)$ $k_{NO} = k_{O_2}, k_O = k_N = 3k_{O_2}$ $k_{CO_2} = 3.5k_{O_2}, k_{Ar} = 0.5k_{O_2}$	100–600 600–6300 290–400	[3] [3] [3] [3] [3]
R2	O+O+M→ O$_2$+M	$k_{N_2} = 2.8 \times 10^{-34} \exp(720/T)$ $k_{N_2} = 10^{-33}(300/T)^{0.41}$ $k_{O_2} = 4 \times 10^{-33}(300/T)^{0.41}$ $k_N = 0.8 k_{O_2}, k_O = 3.6 k_{O_2},$ $k_{NO} = 0.17 k_{O_2}$ $k_{CO_2} = 2.5 \times 10^{-33}(300/T)^{0.41}$ $k_{CO_2} = 3.6 \times 10^{-34} \exp(900/T)$	190–500 500–4000 290–4000 290–500 500–4000	[3] [3] [3] [3] [3] [3] [3]
R3	N+O+M→ NO+M	$k_{N_2} = 10^{-32}(300/T)^{0.5}$ $k_N = 1.8 \times 10^{-31}(300/T)$ $k_{O_2} = k_{N_2}, k_O = k_{NO} = k_N$ $k_{CO_2} = 2.9 k_N, k_{Ar} = 0.05 k_N$	200–4000	[3] [3] [3] [3]
R4	O+O$_2$+M→ O$_3$+M	$k_{N_2} = 5.8 \times 10^{-34}(300/T)^{2.8}$ $k_{N_2} = 5.4 \times 10^{-34}(300/T)^{1.9}$ $k_{O_2} = 7.6 \times 10^{-34}(300/T)^{1.9} (M = O_2)$ $k_N = 3.9^{-33}(300/T)^{1.9}$ $k_N = 1.1^{-34} \exp(1060/T)$ $k_O = k_N, k_{NO} = k_{O_2}, k_{Ar} = 0.1 k_N$	200–1000 1000–4000 200–4000 200–1000 1000–4000 1000–4000	[3] [3] [3] [3] [3] [3]
R4	O+N$_2$+M→ N$_2$O+M	$k_M = 3.9 \times 10^{-35} \exp(-10400/T)$ for any M		[14]
R4	O+NO+M→ NO$_2$+M	$k_{N_2} = 1.2 \times 10^{-31}(300/T)^{1.8}$ $k_{O_2} = k_{NO} = 0.78 k_{N_2}$ $k_{Ar} = 0.6 k_{N_2}$	200–2500	[3] [3] [3]
R5	O+NO$_2$+M→ NO$_3$+M	$k_{N_2} = 8.9 \times 10^{-32}(300/T)^2$ $k_{O_2} = k_{N_2}, k_N = k_O = 13 k_{N_2}$ $k_{NO} = 2.4 k_{N_2}, k_{Ar} = 1.2 k_{N_2}$	200–400 1000–4000 1000–4000	[3] [3] [3]
R6	NO$_2$+NO$_3$+M→ N$_2$O$_5$+M	$k_M = 3.7 \times 10^{-30}(300/T)^{4.1}$ for any M	200–400	[13]

$$\Phi(T, T_{Vj}) = \frac{\prod_{j=1}^{m} Q_j(T)}{\prod_{j=1}^{m} Q_j(T_{Vj})} \exp\left[-E_v^* \left(\frac{m}{\sum_{j=1}^{m} kT_{Vj}} - \frac{1}{kT}\right)\right], \quad (10.1)$$

where m is the number of vibrational modes involved and Q_j is the vibrational partition function of mode j. To calculate Q_j it is often possible to use the harmonic oscillator model, which yields $Q_j(T) \simeq G_j[1 - \exp(-E_{10}^{(j)}/kT)]$, $Q_j(T_V) \simeq G_j[1 - \exp(-E_{10}^{(j)}/kT_V)]$, where $E_{10}^{(j)}$ is the energy of the lower vibrational quantum of mode j and G_j is the degeneracy of mode j. The parameter E_v^* in (10.1) is an effective vibrational energy threshold. It divides the vibrational energy space of the molecules into a region of fast V-V energy exchange and a region (for vibrational energies $E_v > E_v^*$) of fast V-T energy exchanges. For diatomic molecules (see Chap. 3)

10.1 Reactions of Neutral Species

$$E_v^* = n^{**} E_{10}, \qquad (10.2)$$

where n^{**} is determined by (3.85). For polyatomic molecules in accordance with estimations [28]

$$E_v^* = \overline{E_{10}} \left\{ \left[\frac{D\overline{E_{10}}}{B_e^2} \right]^{1/(3+m)} \times \exp\left(\frac{1-m}{3+m}\right) - \frac{1}{2} \right\}, \qquad (10.3)$$

where $\overline{E_{10}}$ is the averaged (over all vibrational modes) number of vibrational quanta of the molecule, D is the dissociation energy and B_e is the rotational

Table 10.4. Rate coefficients for hydrogen-oxygen-nitrogen reactions

		$k_{\pm} = A(T/300)^n \exp(-E_a/T)$					
No	Reaction	k_+ (cm^3/s)			k_- (cm^6/s) or (cm^3/s)		
		A	n	E_a (K)	A	n	E_a (K)
	Reactions with H$_2$O, OH, O, H, H$_2$, O$_2$						
1	H$_2$O+M=OH+H+M	5.9×10^{-7}	-2.2	59000	6.7×10^{-31}	-2	0
2	H$_2$+M=2H+M	3.7×10^{-10}	0	48300	8.3×10^{-33}	-1	0
3	OH+M=H+O+M	4.7×10^{-8}	-1	50830	6.5×10^{-32}	$-1.$	0
4	O$_2$+H=OH+O	3.7×10^{-10}	0	8455	2.2×10^{-11}	0	350
5	H$_2$+O=OH+H	9×10^{-12}	1	4480	4.1×10^{-12}	1	3500
6	H$_2$O+H=OH+H$_2$	1.4×10^{-10}	0	10116	3.2×10^{-11}	0	2600
7	H$_2$O+O= 2OH	9.6×10^{-11}	0	9059	8.8×10^{-12}	0	503
8	H$_2$+O$_2$ = 2OH	2.8×10^{-9}	0	24200	2.8×10^{-11}	0	24100
	Reactions with HO$_2$						
9	HO$_2$+M=H+O$_2$+M	3.5×10^{-9}	0	23000	4.1×10^{-33}	0	-500
10	H$_2$+O$_2$ =H+HO$_2$	3.2×10^{-11}	0	24100	2.2×10^{-11}	0	0
11	H$_2$O+O=H+HO$_2$	2.8×10^{-12}	0.37	28743	1.7×10^{-11}	0	540
12	H$_2$O+O$_2$ =OH+HO$_2$	4.3×10^{-9}	0.5	36600	5×10^{-10}	0	0
13	H$_2$O+OH=H$_2$+HO$_2$	1.4×10^{-13}	0	36100	1.1×10^{-12}	0	9400
14	OH+OH=H+HO$_2$	2×10^{-11}	0	20200	4.2×10^{-10}	0	950
15	OH+O$_2$ =O+HO$_2$	2.2×10^{-11}	0	28200	8.3×10^{-11}	0	500
	Reactions with H$_2$O$_2$						
16	H$_2$O$_2$+M=OH+OH+M	2×10^{-7}	0	22900	2.5×10^{-32}	0	-2650
17	H+H$_2$O$_2$ =HO$_2$+H$_2$	2.8×10^{-12}	0	1900	1×10^{-12}	0	9300
18	H+H$_2$O$_2$ =H$_2$O+OH	8.3×10^{-10}	0	5000	4×10^{-10}	0	40500
19	HO$_2$+HO$_2$ =H$_2$O$_2$+O$_2$	3×10^{-11}	0	500	5×10^{-11}	0	21600
20	HO$_2$+H$_2$O=H$_2$O$_2$+OH	3×10^{-11}	0	15100	1.7×10^{-11}	0	910
21	OH+HO$_2$ =H$_2$O$_2$+O	1.5×10^{-12}	0.5	10600	3.3×10^{-11}	0	2950
22	H$_2$O+O$_2$ =H$_2$O$_2$+O	9.8×10^{-8}	0.5	44800	1.4×10^{-12}	0	2130
	Reactions with O$_3$						
23	O$_3$+H=OH+O$_2$	2.8×10^{-11}	0.75	0	2.7×10^{-13}	1.44	38600
24	O$_3$+OH=HO$_2$+O$_2$	1.6×10^{-12}	0	1000	1.5×10^{-15}	0	0
25	O$_3$+HO$_2$ =OH+2O$_2$	3.3×10^{-14}	0	1000	-	-	-
26	O$_3$+H$_2$ =OH+HO$_2$	10^{-13}	0	10000	-	-	-
	Reactions with N, NO, NO$_2$						
27	N+HO$_2$ =NO+OH	1.7×10^{-11}	0	1000	4.5×10^{-12}	0	41630
28	NO+H=N+OH	2.8×10^{-10}	0	24500	7.5×10^{-11}	0	0
29	NO+OH=H+NO$_2$	5.8×10^{-12}	0.5	15500	5.8×10^{-10}	0	740
30	NO$_2$+OH=NO+HO$_2$	2.9×10^{-12}	0.5	6000	8.6×10^{-11}	0.5	1200
	Reactions with HNO$_2$						
31	HNO$_2$+M=NO+OH+M	2.8×10^{-9}	-1	25000	2.2×10^{-32}	0	-1000
32	H$_2$+HO$_2$ =H+HNO$_2$	4×10^{-11}	0	14500	1.4×10^{-11}	0.5	1500
33	O+HNO$_2$ =OH+NO$_2$	10^{-12}	0	2000	-	-	-
34	OH+HNO$_2$ =H$_2$O+NO$_2$	2.5×10^{-12}	0	28	1.4×10^{-12}	0	21136

Table 10.4. (Continued)

No	Reaction	k_+ (cm³/s)			k_- (cm⁶/s) or (cm³/s)		
		A	n	E_a (K)	A	n	E_a (K)
	Reactions with HNO, HNO₃, HNO₄						
35	O+HNO₂ =O₂+HNO	5×10^{-12}	0	8000	-	-	-
36	HNO+O=NO+OH	1.4×10^{-11}	0.5	1000	3.4×10^{-12}	0.5	27468
37	HNO+OH=NO+H₂O	3.6×10^{-11}	0.5	1000	7.4×10^{-11}	0.5	36338
38	NO+HO₂ =O₂+HNO	3.3×10^{-13}	0	1000	-	-	-
39	HNO+HO₂ =NO+H₂O₂	9.1×10^{-12}	0.5	1000	1.1×10^{-12}	0.5	18468
40	HNO+H=NO+H₂	2.1×10^{-11}	0	2000	9.1×10^{-12}	0	29686
41	HNO+M=H+NO+M	2.3×10^{-8}	0	24494	1.5×10^{-32}	0	-300
42	HNO₃+O=O₂+HNO₂	10^{-11}	0	8000	-	-	-
43	HNO₃+M=OH+NO₂+M	2.7×10^{-9}	0	15400	6.6×10^{-32}	0	-1016
44	HNO₄+OH=H₂O+NO₂+O₂	1.3×10^{-12}	0	380	-	-	-
	Reactions with N₂O, NO₃						
45	N₂O+OH=N₂+HO₂	10^{-12}	0	5000	1.2×10^{-12}	0	20160
46	N₂O+H=N₂+OH	1.3×10^{-10}	0	7600	4.2×10^{-12}	0	39967
47	HNO₃+M=H+NO₃+M	2.7×10^{-9}	0	15400	-	-	-
48	HNO₃+O=OH+NO₃	10^{-12}	0	4000	-	-	-
49	NO₃+H=NO₂+OH	5.8×10^{-10}	0	750	-	-	-
	Reactions with N_nH_m						
50	N₂H₄+M= 2NH₂+M	6.6×10^{-9}	0	20600	-	-	-
51	N₂H₄+M=N₂H₃+H+M	1.7×10^{-9}	0	32000	-	-	-
52	N₂H₃+M=N₂H₂+H+M	1.7×10^{-8}	0	25000	-	-	-
53	N₂H₃+M=NH₂+NH+M	1.7×10^{-8}	0	21000	-	-	-
54	N₂H₂+M= 2NH+M	5.2×10^{-8}	0	50000	-	-	-
55	N₂H+M=N₂+H+M	3.3×10^{-10}	0	10000	5.5×10^{-34}	1	6189
56	N₂H+M=NH+N+M	1.7×10^{-9}	0	35000	-	-	-
57	NH₃+M=NH₂+H+M	4.2×10^{-8}	0	47200	2.2×10^{-32}	1	-7056
58	NH₃+M=NH+H₂+M	10^{-9}	0	47000	-	-	-
59	NH₂+M=NH+H+M	5.8×10^{-6}	-2	46000	-	-	-
60	NH+M=N+H+M	5.8×10^{-8}	-2	42000	3.2×10^{-32}	-0.5	0
61	N₂H₄+H=N₂H₃+H₂	2.1×10^{-11}	0	1260	-	-	-
62	N₂H₄+H=NH₂+NH₃	1.9×10^{-15}	0	1560	-	-	-
63	N₂H₃+H=N₂H₂+H₂	1.7×10^{-12}	0	1000	-	-	-
64	N₂H₃+H= 2NH₂	2.6×10^{-12}	0	0	-	-	-
65	N₂H₃+H=NH+NH₃	1.7×10^{-13}	0	0	-	-	-
66	N₂H₂+H=N₂H+H₂	1.7×10^{-11}	0	500	1.9×10^{-13}	0.5	23184
67	N₂H+H=N₂+H₂	6.6×10^{-11}	0	1500	7.1×10^{-10}	0	56950
68	NH₃+H=NH₂+H₂	2.1×10^{-10}	0	10820	2.1×10^{-11}	0.5	9274
69	NH₂+H=NH+H₂	3.2×10^{-11}	0	0	2.2×10^{-11}	0	3880
70	NH+H=N+H₂	8.3×10^{-11}	0	1000	3.9×10^{-10}	0	15775
71	N₂H₄+NH=NH₂+N₂H₃	2.9×10^{-11}	0.5	1000	-	-	-
72	N₂H₂+NH=N₂H+NH₂	1.7×10^{-11}	0	500	-	-	-
73	N₂H+NH=N₂+NH₂	5.8×10^{-12}	0.5	1000	8.5×10^{-11}	0.5	52920
74	NH+NH=NH₂+N	5.8×10^{-12}	0.5	1000	4×10^{-11}	0.5	11890
75	NH+NH=N₂H+H	2.3×10^{-11}	0.5	500	2.9×10^{-10}	0.5	43240
76	N₂H₄+NH₂ =N₂H₃+NH₃	1.1×10^{-11}	0.5	1000	-	-	-
77	N₂H₃+NH₂ =N₂H₂+NH₃	2.9×10^{-12}	0.5	0	-	-	-
78	N₂H₂+NH₂ =N₂H+NH₃	1.7×10^{-11}	0	2000	-	-	-
79	N₂H₂+NH₂ =NH+N₂H₃	2.9×10^{-12}	0.5	17000	-	-	-
80	N₂H+NH=N₂+NH₃	1.7×10^{-11}	0	0	1.7×10^{-9}	0	57960

10.1 Reactions of Neutral Species   173

Table 10.4. (Continued)

No	Reaction	k_+ (cm³/s)			k_- (cm⁶/s) or (cm³/s)		
		A	n	E_a (K)	A	n	E_a (K)
81	$NH_3+NH_2 = N_2H_3+N_2$	2.3×10^{-11}	0.5	10850	-	-	-
82	$NH_2+NH_2 = NH+NH_3$	10^{-11}	0	5000	7.4×10^{-11}	0	15470
83	$NH_2+NH_2 = N_2H_2+H_2$	6.6×10^{-11}	0	6000	6.2×10^{-9}	0	25910
84	$NH_2+NH = N_2H_2+H$	5.2×10^{-11}	0	500	6.8×10^{-9}	0	11995
85	$N_2H_4+N_2H_2 = 2N_2H_3$	7.2×10^{-13}	0.5	15000	-	-	-
86	$N_2H_3+N_2H_2 = N_2H_4+N_2H$	1.7×10^{-11}	0	5000	-	-	-
87	$N_2H_2+N_2H_2 = N_2H+N_2H_3$	1.7×10^{-11}	0	5000	-	-	-
88	$N_2H+N_2H = N_2H_2+N_2$	1.7×10^{-11}	0	5000	8.3×10^{-11}	0	33260
89	$NH+N = N_2+H$	1.8×10^{-11}	0.5	0	4.7×10^{-10}	0.5	75900
90	$N_2H+N = NH+N_2$	5.2×10^{-11}	0	1000	1.2×10^{-10}	0	41830
91	$N_2H_4+O = N_2H_2+H_2O$	10^{-10}	0	600	-	-	-
92	$N_2H_4+O = N_2H_3+OH$	4.2×10^{-12}	0	600	-	-	-
93	$N_2H_3+O = N_2H_2+OH$	9.1×10^{-12}	0.5	0	-	-	-
94	$N_2H_3+O = N_2H+H_2O$	9.1×10^{-12}	0.5	0	-	-	-
95	$N_2H_2+O = N_2H+OH$	2.9×10^{-12}	0.5	0	1.6×10^{-13}	0	22680
96	$N_2H+O = N_2+OH$	1.7×10^{-11}	0	2500	-	-	-
97	$N_2H+O = N_2O+H$	1.7×10^{-11}	0	1500	7.1×10^{-8}	0	23380
98	$NH_3+O = NH_2+OH$	3.3×10^{-11}	0	4470	1.4×10^{-12}	0	1360
99	$NH_2+O = NH+OH$	9.6×10^{-12}	-0.5	0	3×10^{-12}	-0.5	7360
100	$NH_2+O = HNO+H$	6.1×10^{-11}	-0.5	0	3.6×10^{-10}	-0.5	14160
101	$NH+O = N+OH$	1.8×10^{-11}	0.5	4000	3.7×10^{-11}	0.5	17840
102	$NH+O = NO+H$	1.8×10^{-11}	0.5	0	10^{-10}	0.5	38010
103	$N_2H_4+OH = N_2H_3+H_2O$	6.6×10^{-11}	0	0	-	-	-
104	$N_2H_3+OH = N_2H_2+H_2O$	1.7×10^{-11}	0	1000	-	-	-
105	$N_2H_2+OH = N_2H+H_2O$	1.7×10^{-11}	0	1000	5.1×10^{-12}	0	30240
106	$N_2H+OH = N_2+H_2O$	5.2×10^{-11}	0	0	7.6×10^{-11}	0	64510
107	$NH_3+OH = NH_2+H_2O$	9.5×10^{-11}	0	4055	5×10^{-11}	0	9680
108	$NH_2+OH = NH+H_2O$	1.4×10^{-11}	0.5	1000	4.9×10^{-11}	0.5	17090
109	$NH+OH = N+H_2O$	1.4×10^{-11}	0.5	1000	3×10^{-9}	0.5	23370
110	$HNO+H = NH+OH$	5.8×10^{-10}	0.5	12720	2.9×10^{-11}	0.5	1010
111	$N_2H_4+HO_2 = N_2H_3+H_2O_2$	6.6×10^{-11}	0	1000	-	-	-
112	$N_2H_3+HO_2 = N_2H_2+H_2O_2$	1.7×10^{-11}	0	1000	-	-	-
113	$N_2H_2+HO_2 = N_2H+H_2O_2$	1.7×10^{-11}	0	1000	2.1×10^{-12}	0	13100
114	$N_2H+HO_2 = N_2+H_2O_2$	1.7×10^{-11}	0	1000	3.5×10^{-10}	0	46870
115	$NH_3+HO_2 = NH_2+H_2O_2$	4.2×10^{-12}	0	12000	8.3×10^{-13}	0	100
116	$NH_2+HO_2 = NH_3+O_2$	1.7×10^{-11}	0	1000	-	-	-
117	$NH_2+HO_2 = NH+H_2O_2$	1.7×10^{-11}	0	1000	2×10^{-11}	0	330
118	$NH+HO_2 = NHO+OH$	1.7×10^{-11}	0	1000	3.6×10^{-11}	0	29970
119	$N+HO_2 = NH+O_2$	1.7×10^{-11}	0	1000	7.1×10^{-12}	0	11490
120	$NHO+N = NO+NH$	1.7×10^{-11}	0	1000	2×10^{-12}	0	18500
121	$NO+NH = N_2O+H$	1.7×10^{-12}	0	250	4.5×10^{-10}	0	15020
122	$NH_2+NO = N_2+H_2O$	6.7×10^{-11}	-2.5	960	2×10^{-9}	-2.5	64110
123	$NH_2+NO = N_2+H+OH$	6.7×10^{-11}	-2.5	950	-	-	-
124	$NH_2+NO = N_2O+H_2$	8.3×10^{-11}	0	12450	6.9×10^{-9}	0	34520
125	$NO_2+NH = NHO+NO$	2.9×10^{-12}	0.5	2020	1.6×10^{-12}	0.5	24040
126	$NO_2+N_2H = N_2O+H_2O$	1.2×10^{-11}	-3	0	-	-	-
127	$N_2O+NH = N_2+HNO$	3.3×10^{-12}	0	3000	-	-	-
128	$NH_2+O_2 = HNO+OH$	3.3×10^{-12}	0	7560	1.8×10^{-12}	0	13360
129	$NH_2+O_2 = NH+HO_2$	1.7×10^{-10}	0	25200	5.5×10^{-11}	0	2750
130	$NH+O_2 = HNO+O$	1.7×10^{-12}	0	1610	4.1×10^{-12}	0	5280
131	$N_2H+O = N_2+HO_2$	1.7×10^{-11}	0	2020	5.8×10^{-11}	0	29230

Table 10.5. Values of E_v^* for some triatomic dissociating molecules

No	Molecule	Decay products	E_v^* (K)
1	CO_2	$CO-O$	15200
2	H_2O	H_2-O	21900
3	N_2O	N_2-O	11500
4	NO_2	$NO-O$	22600

constant of the molecule. Estimated values of E_v^* using (10.3) for some triatomic dissociating molecules [27] are given in Table 10.5.

Note that (10.3) and the data in Table (10.5) can be used only for rough estimations. For example, for dissociation of CO_2, higher values of $E_v^*(\simeq 35000$ K) have been obtained [29].

The corrective factors $\Phi(T, T_1)$ can also be markedly different from those given by (3.128) and (10.1) when $T_V > T$. A simultaneous analysis of chemical and vibrational kinetics is in general necessary to solve this problem. This was numerically investigated for N_2 [30] (see Chap. 12). An analytical approximation has also been developed [31] (see (3.129)).

The rate coefficients for bimolecular reactions presented in Tables 10.1–10.4 apply only to the equilibrium regime, when $T = T_V$. Vibrational excitation of a reacting molecule, when $T_V > T$, increases the rate of a reaction with activation energy E_a, that is, the rate of reaction through excited vibrational levels v

$$A + BC(v) \to AB + C \tag{10.4}$$

increases with v.

In order to quantitatively describe the influence of vibrational excitation on such reactions, a parameter $\beta \leq 1$, called the coefficient of utilization of vibrational energy [32–34], is often used. In particular, this parameter is included in the Le Roy formula [35] for the probability P_v of a reaction from vibrational level v, which is written

$$P_v = \begin{cases} 0 & \text{for } E_T < E_a - \beta E_v; \quad E_a \geq \beta E_v \\ 1 - \frac{E_a - \beta E_v}{E_T} & \text{for } E_T > E_a - \beta E_v; \quad E_a \geq \beta E_v \\ 1 & \text{for any } E_T, \text{ if } E_a < \beta E_v \end{cases} \tag{10.5}$$

E_T and E_v denoting the translational and vibrational energies respectively. By averaging (10.5) over kinetic energy E_T using a Maxwellian velocity distribution, it is easy to obtain the following expression for the rate coefficient k_v for a reaction from vibrational level v:

$$k_v = \begin{cases} k_T^0 \exp\left(-\frac{E_a}{kT}\right) \exp\left(\frac{\beta E_v}{kT}\right) & \text{if } \beta E_v < E_a \\ k_T^0 & \text{if } \beta E_v \geq E_a \end{cases} \tag{10.6}$$

Equation (10.6) means that the actual activation energy decreases by the amount βE_v for a reacting molecule in vibrational level v, with energy E_v

10.1 Reactions of Neutral Species   175

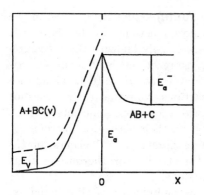
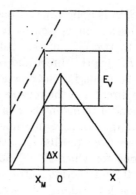

Fig. 10.1. Reaction path A+BC(v) →AB+C involving excited vibrational states of BC molecules with energy E_v

A simple method for estimating β has been proposed [34,36,37] which is based on the analysis of reaction paths x through vibrationally excited molecules with energy E_v (see Fig. 10.1). The potential energies of these terms are approximated by exponential functions. The potential energies of reactants ($U_1(x)$) and products ($U_2(x)$) along the reaction path x then change according to the following law:

$$U_1(x) \simeq E_a \exp(\gamma_1 x) + E_v \quad \text{for } x \leq x_m$$
$$U_2(x) \simeq E_a^- \exp(-\gamma_2 x) + E_a - E_a^- \quad \text{for } x > x_m . \quad (10.7)$$

Here E_v is the vibrational energy of the reacting molecule, E_a and E_a^- are the activation energies of the direct and reverse reactions involving a reacting molecule in the ground vibrational state (i.e., for $E_v=0$), x is the reaction coordinate measured from the point where the top of the reaction barrier is reached (see Fig. 10.1) with a non-excited reacting molecule (i.e., $x_m=0$ for $E_v=0$). If $E_v > 0$, the top of the barrier shifts along the axis x by a distance Δx and reaches the point $x_m \equiv \Delta x < 0$. The energy of this barrier is increased by the value $\Delta E = U_1(x_m) - U_2(x=0)$. The activation energy also changes by the value ΔE_a. If the inequalities

$$\gamma_1 |\Delta x| \ll 1; \qquad \gamma_2 |\Delta x| \ll 1 \quad (10.8)$$

apply, one easily obtains from (10.7) and condition $U_1(x_m) = U_2(x_m)$

$$\Delta x \simeq \frac{-E_v}{\gamma_1 E_a + \gamma_2 E_a^-}, \quad \Delta E_a = E_a - E_v + \Delta E \simeq \frac{-\gamma_1 E_a}{\gamma_1 E_a + \gamma_2 E_a^-} E_v \quad (10.9)$$

by expanding the exponential functions in (10.7) in the vicinity of $x=0$. This means that the coefficient β is given by

$$\beta \simeq \frac{\gamma_1 E_a}{\gamma_1 E_a + \gamma_2 E_a^-} . \quad (10.10)$$

To estimate β, it is appropriate to use (10.10) assuming $\gamma_1 \simeq \gamma_2$, in the absence of information on γ_1 and γ_2 for most reactions. In this case, β is determined only by two parameters that are usually known - the activation energies E_a and E_a^- for the direct and reverse reactions. Such a simple estimation of β can be made for the reactions in Tables 10.1, 10.4, taking into account both the direct and the reverse reactions. Note that the presence or absence of exponential terms in the expressions for rate coefficients of these processes provides quantitative information on their activation energies E_a and E_a^-. It follows from (10.10) that the maximum $\beta = 1$ occurs for endothermic reactions, for which $E_a^- = 0$. For thresholdless exothermic reactions $\beta \simeq 0$, while $\beta \simeq 0.5$ for thermoneutral reactions.

Another approach to taking into account the influence of vibrational excitation on reaction rates has been proposed [38]. Using a classical impulsive model, it was found that, instead of (10.6), the rate coefficients k_v are determined by the following expression

$$k_v = k_T^0 \exp\left\{-\frac{F(E_v)}{kT}\right\}, \tag{10.11}$$

where

$$F(E_v) = \begin{cases} W + \frac{1}{1-\delta_m}\left[(E_a - W)^{1/2} - (\delta_m E_v)^{1/2}\right]^2, & \text{if } E_v \leq \frac{E_a - W}{\delta_m} \\ W, & \text{if } E_v > \frac{E_a - W}{\delta_m} \end{cases} \tag{10.12}$$

$$\delta_m = \frac{m_C(m_A + m_B + m_C)}{(m_A + m_C)(m_B + m_C)}, \tag{10.13}$$

$$W = E_a(1 - \xi/\delta_m). \tag{10.14}$$

Here $F(E_v)$ is a threshold function, m_A, m_B, m_C are the masses of atoms A, B, C in reaction (10.4), and ξ is the energy fraction that goes into vibrations in the reverse exothermic reaction, which is usually determined from experiments. It is interesting to note that, for a Boltzmann vibrational distribution function with temperature T_V (harmonic oscillator model), (10.11) yields the following total two-temperature rate coefficient $k(T, T_V)$ for reaction (10.4) [39]:

$$k(T, T_V) = k_T^0 \exp\left\{-\frac{E_a - W}{k[\delta_m T_V + (1-\delta_m)T]} - \frac{W}{kT}\right\}. \tag{10.15}$$

Considerable attention is currently being focused on the following reaction involving vibrationally excited nitrogen:

$$N_2(v) + O \to N + NO. \tag{10.16}$$

Such attention is motivated by research on non-traditional methods for isotope separation (see Chap. 13), the synthesis of nitrogen oxides, and the role of the latter in the chemistry of the upper atmosphere (see Chap. 14).

Table 10.6. Parameters a_i for the rate coefficient (10.17) of reaction (10.16)

v	a_1	a_2	a_3	a_4	a_5
$0 \leq v \leq 8$	−0.419312	−0.37836	−23.04468	−0.992436	0.989385
$9 \leq v \leq 12$	−3.42306	−1.4234	1.423118	−0.919692	0.917323
$13 \leq v \leq 23$	6.4805404	-0.279371	-96.75885	−0.037869	0.019647

Theoretical calculations of rate coefficients for reaction (10.16) have been performed [34,36,37,40–44]. Some calculated results at different gas temperatures are shown in Fig. 10.2.

The calculations [44] have been approximated by means of the expression

$$k_v(T)\left(\frac{cm^3}{s}\right) = \frac{(e_v + 3000)^{a_1}}{T^{a_2}} \exp\left(a_3 + \frac{38370}{T}a_4 + \frac{E_v}{T}a_5\right), \quad (10.17)$$

where $E_v = 3395v[1 - 6.217 \times 10^{-2}(v+1)]$ in Kelvin. The values of the parameters a_i are given in Table 10.6.

The rate coefficients for reaction (10.16) involving N_2 vibrational levels $v \geq 12$ have been estimated from experiment [45]. It was found that $k_v \geq 10^{-11}$ cm^3/s, which is in agreement with calculations [44] and (10.17).

Chemical reactions may be accelerated not only by vibrational excitation of the reactants, but also by excitation of their electronic states. Reactions involving electronically excited species are often very effective quenching channels for such excited electronic states. Examples of such reactions have been presented in Tables 9.3–9.6, for collisions with neutrals, and are presented in Tables 10.7 and 10.8, for collisions with ions.

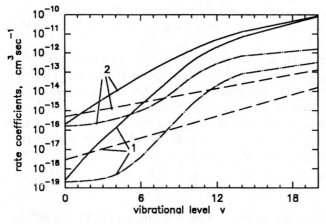

Fig. 10.2. Rate coefficients for reaction $N_2(v)+O \rightarrow N+NO$, calculated by different authors. The solid curves are from [44], the broken curves from [43], and the dash-dotted curves from [42]. Curves 1 are for T=2000 K, and curves 2 for T=3000 K [44]

10.2 Ion–Molecule Reactions

Reactions involving neutral atoms or molecules and ions [46] play an important role in energy transfer and the formation of chemical components in low-temperature plasmas. The absence of activation energy for many exothermic processes and the considerable magnitudes ($\sim 10^{-10}$–10^{-9} cm^3/s) of their rate coefficients, without noticeable dependence on temperature, are typical properties of such ion–molecule reactions. This is explained by the effective polarization interaction between the colliding ion and neutral. The potential energy of such an interaction, for a neutral particle with polarizability α, is

$$U(r) = -\frac{\alpha e^2}{2r^4}, \tag{10.18}$$

where e is the electron charge and r is the distance between the colliding particles. According to the classical Langevin theory, there is a critical impact parameter r_0 for such a collision. The parameter r_0 determines the reaction cross-section

$$\sigma_{\mathrm{im}}(c) = \pi r_0^2 \equiv \frac{\pi}{c}\left(\frac{4\alpha e^2}{\mu}\right)^{1/2}, \tag{10.19}$$

where c is the relative velocity of the colliding particles and μ is their reduced mass. It follows from (10.19) that the rate coefficient for the ion–molecule reaction is

$$k_{\mathrm{im}} = 2\pi\left(\frac{\alpha e^2}{\mu}\right)^{1/2} \tag{10.20}$$

and therefore does not depend on temperature.

Ion–molecule reactions and the corresponding rate coefficients for oxygen–nitrogen–hydrogen low-temperature plasma are given in Tables 10.7 and 10.8 [1,47,48].

The rate coefficients are, in accordance with (10.20), independent of temperature for most of the ion–neutral reactions listed in the above tables. However, a dependence on temperature does occur for some processes. Note that in low pressure plasmas with a strong electric field the ion kinetic energy can be different from the kinetic energy of neutral particles. In this case, an effective temperature T_{eff} must be used for calculations of such rate coefficients. This effective temperature T_{eff} is given by

$$T_{\mathrm{eff}} = \frac{m_{\mathrm{i}} T_{\mathrm{g}} + m_{\mathrm{g}} T_{\mathrm{i}}}{m_{\mathrm{i}} + m_{\mathrm{g}}} + \frac{m_{\mathrm{i}} m_{\mathrm{g}}}{3k(m_{\mathrm{i}} + m_{\mathrm{g}})} v_{\mathrm{di}}^2, \tag{10.21}$$

where T_{g} and T_{i} are respectively the neutral and the ion translational temperature, m_{g} and m_{i} are their masses, and v_{di} is the ion drift velocity.

Table 10.7. Reactions of positive nitrogen and oxygen ions

No	Process	$k(\text{cm}^3/\text{s})$	Ref.
R1	$N^+ + O \to N + O^+$	10^{-12}	[47]
R2	$N^+ + O_2 \to O_2^+ + N$	2.8×10^{-10}	[47]
R3	$N^+ + O_2 \to NO^+ + O$	2.5×10^{-10}	[49]
R4	$N^+ + O_2 \to O^+ + NO$	2.8×10^{-11}	[49]
R5	$N^+ + O_3 \to NO^+ + O_2$	5×10^{-10}	[50]
R6	$N^+ + NO \to NO^+ + N$	8×10^{-10}	[47]
R7	$N^+ + NO \to N_2^+ + O$	3×10^{-12}	[47]
R8	$N^+ + NO \to O^+ + N_2$	10^{-12}	[47]
R9	$N^+ + N_2O \to NO^+ + N_2$	5.5×10^{-10}	[47]
R10	$O^+ + N_2 \to NO^+ + N$	$(1.5 - 2 \times 10^{-3} T + 9.6 \times 10^{-7} T^2) \times 10^{-12}$	[50,52]
R11	$O^+ + O_2 \to O_2^+ + O$	$2 \times 10^{-11}(300/T)^{0.5}$	[53]
R12	$O^+ + O_3 \to O_2^+ + O_2$	10^{-10}	[53]
R13	$O^+ + NO \to NO^+ + O$	2.4×10^{-11}	[1]
R14	$O^+ + NO \to O_2^+ + N$	3×10^{-12}	[1]
R15	$O^+ + N(^2D) \to N^+ + O$	1.3×10^{-10}	[50]
R16	$O^+ + N_2O \to NO^+ + NO$	2.3×10^{-10}	[1]
R17	$O^+ + N_2O \to N_2O^+ + O$	2.2×10^{-10}	[1]
R18	$O^+ + N_2O \to O_2^+ + N_2$	2×10^{-11}	[53]
R19	$O^+ + NO_2 \to NO_2^+ + O$	1.6×10^{-9}	[54]
R20	$N_2^+ + O_2 \to O_2^+ + N_2$	$6 \times 10^{-11}(300/T)^{0.5}$	[47,55]
R21	$N_2^+ + O \to NO^+ + N(^4S,^2D)$	$1.3 \times 10^{-10}(300/T)^{0.5}$	[47]
R22	$N_2^+ + O \to O^+ + N_2$	$10^{-11}(300/T)^{0.5}$	[47]
R23	$N_2^+ + O_3 \to O_2^+ + O + N_2$	10^{-10}	[55]
R24	$N_2^+ + N \to N^+ + N_2$	$7.2 \times 10^{-13}(T/300)$	[17]
R25	$N_2^+ + NO \to NO^+ + N_2$	3.3×10^{-10}	[47]
R26	$N_2^+ + N_2O \to N_2O^+ + N_2$	5×10^{-10}	[53]
R27	$N_2^+ + N_2O \to NO^+ + N + N_2$	4×10^{-10}	[53]
R28	$O_2^+ + N_2 \to NO^+ + NO$	10^{-17}	[56]
R29	$O_2^+ + N \to NO^+ + O$	1.2×10^{-10}	[47]
R30	$O_2^+ + NO \to NO^+ + O_2$	6.3×10^{-10}	[57]
R31	$O_2^+ + NO_2 \to NO^+ + O_3$	10^{-11}	[47,58]
R32	$O_2^+ + NO_2 \to NO_2^+ + O_2$	6.6×10^{-10}	[1]
R33	$N_3^+ + O_2 \to O_2^+ + N + N_2$	2.3×10^{-11}	[47]
R34	$N_3^+ + O_2 \to NO_2^+ + N_2$	4.4×10^{-11}	[47]
R35	$N_3^+ + N \to N_2^+ + N_2$	6.6×10^{-11}	[17]
R36	$N_3^+ + NO \to NO^+ + N + N_2$	7×10^{-11}	[47]
R37	$N_3^+ + NO \to N_2O^+ + N_2$	7×10^{-11}	[47]
R38	$NO_2^+ + NO \to NO^+ + NO_2$	2.9×10^{-10}	[47]
R39	$N_2O^+ + NO \to NO^+ + N_2O$	2.9×10^{-10}	[47]
R40	$N_4^+ + N_2 \to N_2^+ + 2N_2$	$2.1 \times 10^{-16} \exp(T/121)$	[17]
R41	$N_4^+ + O_2 \to O_2^+ + 2N_2$	2.5×10^{-10}	[47]
R42	$N_4^+ + O \to O^+ + 2N_2$	2.5×10^{-10}	[47]
R43	$N_4^+ + N \to N^+ + 2N_2$	10^{-11}	[29]
R44	$N_4^+ + NO \to NO^+ + 2N_2$	4×10^{-10}	[47]
R45	$O_4^+ + N_2 \to O_2^+ N_2 + O_2$	$4.6 \times 10^{-12}(T/300)^{2.5} \times \exp(-2650/T)$	[47]
R46	$O_4^+ + O_2 \to O_2^+ + 2O_2$	$3.3 \times 10^{-6}(300/T)^4 \times \exp(-5030/T)$	[47,56]
R47	$O_4^+ + O_2(a,b) \to O_2^+ + 2O_2$	10^{-10}	[47,56]
R48	$O_4^+ + O \to O_2^+ + O_3$	3×10^{-10}	[47,56]
R49	$O_4^+ + NO \to NO^+ + 2O_2$	10^{-10}	[47]
R50	$O_2^+ N_2 + N_2 \to O_2^+ + 2N_2$	$1.1 \times 10^{-6}(300/T)^{5.3} \times \exp(-2360/T)$	[47]
R51	$O_2^+ N_2 + O_2 \to O_4^+ + N_2$	10^{-9}	[47]
R52	$NO^+ N_2 + N_2 \to NO^+ + 2N_2$	$6.3 \times 10^{-8}(300/T)^{5.4} \times \exp(-2450/T)$	[47,56]

Table 10.7. (Continued)

	Three-body collisions	k (cm^6/s)	Ref.
R53	$N^+ + N_2 + N_2 \rightarrow N_3^+ + N_2$	$1.7 \times 10^{-29}(300/T)^{2.1}$	[59]
R54	$N^+ + O + M \rightarrow NO^+ + M$	10^{-29} (M=N$_2$,O$_2$)	[58]
R55	$N^+ + N + M \rightarrow N_2^+ + M$	10^{-29} (M=N$_2$,O$_2$)	[58]
R56	$O^+ + N_2 + M \rightarrow NO^+ + N + M$	$6 \times 10^{-29}(300/T)^2$ (M=N$_2$,O$_2$)	[47]
R57	$O^+ + O + M \rightarrow O_2^+ + M$	10^{-29} (M=N$_2$,O$_2$)	[58]
R58	$O^+ + N + M \rightarrow NO^+ + M$	10^{-29} (M=N$_2$,O$_2$)	[58]
R59	$N_2^+ + N_2 + N_2 \rightarrow N_4^+ + N_2$	$5.2 \times 10^{-29}(300/T)^{2.2}$	[59]
R60	$N_2^+ + N + N_2 \rightarrow N_3^+ + N_2$	$9 \times 10^{-30} \exp(400/T)$	[17]
R61	$O_2^+ + O_2 + O_2 \rightarrow O_4^+ + O_2$	$2.4 \times 10^{-30}(300/T)^{3.2}$	[47,56]
R62	$O_2^+ + N_2 + N_2 \rightarrow O_2^+ N_2 + N_2$	$9 \times 10^{-31}(300/T)^2$	[47]
R63	$NO^+ + N_2 + N_2 \rightarrow NO^+ N_2 + N_2$	$2 \times 10^{-31}(300/T)^{4.4}$	[47,56]
R64	$NO^+ + O_2 + N_2 \rightarrow NO^+ O_2 + N_2$	3×10^{-31}	[47,60]
R65	$NO^+ + O_2 + O_2 \rightarrow NO^+ O_2 + O_2$	9×10^{-32}	[47,60]

Table 10.8. Reactions of positive hydrogen, nitrogen, and oxygen ions

No.	Process	k(cm^3/s)	Ref.
R1	$H^+ + O \rightarrow O^+ + H$	3.8×10^{-10}	[1]
R2	$H^+ + NO \rightarrow NO^+ + H$	1.9×10^{-9}	[1]
R3	$N^+ + H_2 \rightarrow NH^+ + H$	$6 \times 10^{-10} \exp(-150/T)$	[61]
R4	$N^+ + H_2O \rightarrow H_2O^+ + N$	2.8×10^{-9}	[62]
R5	$N^+ + NH_3 \rightarrow NH_3^+ + N$	2×10^{-9}	[62]
R6	$N^+ + NH_3 \rightarrow HN_2^+ + H_2$	2.2×10^{-10}	[62]
R7	$N^+ + NH_3 \rightarrow H_2N^+ + NH$	2.2×10^{-10}	[62]
R8	$O^+ + H \rightarrow H^+ + O$	6.8×10^{-10}	[1]
R9	$O^+ + H_2 \rightarrow OH^+ + H$	1.7×10^{-9}	[66]
R10	$O^+ + H_2O \rightarrow H_2O^+ + O$	3.2×10^{-9}	[66]
R11	$H_2^+ + H_2 \rightarrow H_3^+ + H$	2.1×10^{-9}	[63]
R12	$H_2^+ + N_2 \rightarrow HN_2^+ + H$	1.95×10^{-9}	[64]
R13	$N_2^+ + H_2 \rightarrow HN_2^+ + H$	1.7×10^{-9}	[65]
R14	$N_2^+ + H_2O \rightarrow H_2O^+ + N_2$	2.3×10^{-9}	[66]
R15	$N_2^+ + H_2O \rightarrow HN_2^+ + OH$	5×10^{-10}	[66]
R16	$N_2^+ + NH_3 \rightarrow NH_3^+ + N_2$	1.9×10^{-9}	[66]
R17	$O_2^+ + NH_3 \rightarrow NH_3^+ + O_2$	2×10^{-9}	[66]
R18	$NH^+ + H_2 \rightarrow NH_2^+ + H$	1.3×10^{-9}	[62]
R19	$NH^+ + H_2 \rightarrow H_3^+ + N$	2.3×10^{-10}	[62]
R20	$NH^+ + N_2 \rightarrow HN2^+ + N$	6.5×10^{-10}	[62]
R21	$NH^+ + O_2 \rightarrow O_2^+ + NH$	4.5×10^{-10}	[62]
R22	$NH^+ + O_2 \rightarrow NO^+ + OH$	2.1×10^{-10}	[62]
R23	$NH^+ + O_2 \rightarrow HO_2^+ + N$	1.6×10^{-10}	[62]
R24	$NH^+ + NO \rightarrow NO^+ + NH$	7.1×10^{-10}	[62]
R25	$NH^+ + NO \rightarrow HN_2^+ + O$	1.8×10^{-10}	[62]
R26	$NH^+ + H_2O \rightarrow H_2O^+ + NH$	1.1×10^{-9}	[62]
R27	$NH^+ + H_2O \rightarrow H_3O^+ + N$	1.1×10^{-9}	[62]
R28	$NH^+ + H_2O \rightarrow NH_2^+ + OH$	1.1×10^{-9}	[62]
R29	$NH^+ + H_2O \rightarrow HNO^+ + H_2$	3.5×10^{-10}	[62]
R30	$NH^+ + NH_3 \rightarrow NH_3^+ + NH$	1.8×10^{-9}	[62]
R31	$NH^+ + NH_3 \rightarrow NH_4^+ + N$	6×10^{-10}	[62]
R32	$H_3^+ + N_2 \rightarrow HN_2^+ + H_2$	1.8×10^{-9}	[67]
R33	$H_3^+ + NO \rightarrow NOH^+ + H_2$	1.4×10^{-9}	[1]
R34	$H_3^+ + NO_2 \rightarrow NO^+ + OH + H_2$	7×10^{-10}	[1]
R35	$H_3^+ + H_2O \rightarrow H_3O^+ + H_2$	3×10^{-9}	[1]
R36	$H_3^+ + NH_3 \rightarrow NH_4^+ + H_2$	5×10^{-9}	[68]

10.2 Ion–Molecule Reactions

Table 10.8. (Continued)

No.	Process	$k(\text{cm}^3/\text{s})$	Ref.
R37	$N_3^+ + H_2 \rightarrow HN_2^+ + NH$	2×10^{-13}	[66]
R38	$N_3^+ + H_2O \rightarrow H_2NO^+ + N_2$	3.3×10^{-10}	[66]
R39	$N_3^+ + NH_3 \rightarrow NH_3^+ + N_2 + N$	2.1×10^{-9}	[66]
R40	$HN_2^+ + N_2O \rightarrow N_2OH^+ + N_2$	7.9×10^{-10}	[1]
R41	$HN_2^+ + H_2O \rightarrow H_3O^+ + N_2$	5×10^{-10}	[1]
R42	$NH_2^+ + H_2 \rightarrow NH_3^+ + H$	1.6×10^{-9}	[62]
R43	$NH_2^+ + N_2 \rightarrow \text{products}$	$< 5 \times 10^{-13}$	[62]
R44	$NH_2^+ + O_2 \rightarrow H_2NO^+ + O$	1.2×10^{-10}	[62]
R45	$NH_2^+ + O_2 \rightarrow HNO^+ + OH$	2.1×10^{-11}	[62]
R46	$NH_2^+ + H_2O \rightarrow H_3O^+ + NH$	2.8×10^{-9}	[62]
R47	$NH_2^+ + H_2O \rightarrow NH_4^+ + O$	10^{-10}	[62]
R48	$NH_2^+ + NH_3 \rightarrow NH_3^+ + NH_2$	6.9×10^{-10}	[62]
R49	$NH_2^+ + NH_3 \rightarrow NH_4^+ + NH$	1.6×10^{-9}	[62]
R50	$H_2O^+ + O_2 \rightarrow O_2^+ + H_2O$	4.3×10^{-10}	[1]
R51	$H_2O^+ + H_2O \rightarrow H_3O^+ + OH$	1.7×10^{-9}	[1]
R52	$NH_3^+ + H_2 \rightarrow NH_4^+ + H$	$1.5 \times 10^{-9} \times \exp(-2570/T)$	[65]
R53	$NH_3^+ + N_2 \rightarrow \text{products}$	$< 5 \times 10^{-14}$	[62]
R54	$NH_3^+ + O_2 \rightarrow \text{products}$	$< 5 \times 10^{-13}$	[62]
R55	$NH_3^+ + H_2O \rightarrow NH_4^+ + OH$	$< 3 \times 10^{-11}$	[62]
R56	$NH_3^+ + NH_3 \rightarrow NH_4^+ + NH_2$	2.2×10^{-9}	[62]
R57	$N_4^+ + H_2 \rightarrow H_2^+ + 2N_2$	$3 \times 10^{-10} \exp(-1800/T)$	[66,69]
R58	$N_4^+ + H_2 \rightarrow HN_2^+ + H + N_2$	$6.7 \times k^{(R23)}$	[66,69]
R59	$N_4^+ + H_2O \rightarrow H_2O^+ + 2N_2$	3×10^{-9}	[66]
R60	$N_4^+ + NH_3 \rightarrow NH_3^+ + 2N_2$	1.8×10^{-9}	[66]
R61	$O_4^+ + H_2O \rightarrow O_3H_2^+ + O_2$	1.5×10^{-9}	[1]
R62	$O_2^+ N_2 + H_2O \rightarrow O_2^+ H_2O + N_2$	4×10^{-9}	[1]
R63	$O_2^+ N_2 + O_2 \rightarrow O_4^+ + N_2$	$> 5 \times 10^{-11}$	[1]
R64	$H_3O^+ OH + H_2O \rightarrow H_3O^+ H_2O + OH$	1.4×10^{-9}	[1]
R65	$O_2^+ H_2O + H_2O \rightarrow H_3O^+ + OH + O_2$	3×10^{-10}	[1]
R66	$NO^+(H_2O)_3 + H_2O \rightarrow H_3O^+(H_2O)_2 + HNO_2$	7×10^{-11}	[1]
R67	$NO^+ NO + H_2O \rightarrow NO^+ H_2O + NO$	1.4×10^{-9}	[1]
R68	$NO^+ NO + NH_3 \rightarrow NO^+ NH_3 + NO$	1.3×10^{-9}	[1]
R69	$NO^+ H_2O + NO \rightarrow NO^+ NO + H_2O$	9×10^{-14}	[1]
R70	$NO^+ H_2O + NH_3 \rightarrow NH_4^+ + HNO_2$	10^{-9}	[1]
R71	$NO_2^+ H_2O + NH_3 \rightarrow NH_4^+ + HNO_3$	1.1×10^{-9}	[1]
R72	$NO^+ NH_3 + NH_3 \rightarrow NH_4^+ + ON_2H_2$	9.1×10^{-10}	[1]
R73	$H_3O^+ + NH_3 \rightarrow NH_4^+ + H_2O$	2.1×10^{-9}	[1]
R74	$H_3O^+(H_2O) + NH_3 \rightarrow \text{products}$	2.6×10^{-9}	[1]
R75	$H_3O^+(H_2O)_2 + NH_3 \rightarrow \text{products}$	1.6×10^{-9}	[1]
R76	$H_3O^+(H_2O)_3 + NH_3 \rightarrow \text{products}$	2.1×10^{-9}	[1]
R77	$NH_4^+(H_2O) + NH_3 \rightarrow NH_4^+(NH_3) + H_2O$	1.2×10^{-9}	[1]
R78	$NH_4^+(H_2O)_2 + NH_3 \rightarrow NH_4^+(NH_3)(H_2O) + H_2O$	$> 9 \times 10^{-10}$	[1]
R79	$H_3O^+(H_2O)_3 + NH_3 \rightarrow \text{products}$	2.1×10^{-9}	[1]
R80	$NH_4^+(H_2O) + NH_3 \rightarrow NH_4^+(NH_3) + H_2O$	1.2×10^{-9}	[1]
R81	$NH_4^+(H_2O)_2 + NH_3 \rightarrow NH_4^+(NH_3)(H_2O) + H_2O$	$> 9 \times 10^{-10}$	[1]

No.	Three-body processes	$k(\text{cm}^6/\text{s})$	Ref.
R82	$H^+ + H_2 + M \rightarrow H_3^+ + M$	3.1×10^{-29}	[69]
R83	$O_2^+ + H_2 + He \rightarrow O_2^+ H_2O + He$	7.4×10^{-31} $(T = 80$ K$)$	[66]
R84	$O_2^+ + H_2O + He \rightarrow O_2^+ H_2O + He$	8.7×10^{-29}	[66]
R85	$NO^+ + H_2O + He \rightarrow NO^+ H_2O + He$	3.6×10^{-29}	[66]
R86	$NO^+ + NH_3 + N_2 \rightarrow NO^+ NH_3 + N_2$	3×10^{-28}	[66]
R87	$H_3^+ + H_2 + He \rightarrow H_5^+ + He$	2×10^{-29} $(T = 80$ K$)$	[66]
R88	$HN_2^+ + N_2 + N_2 \rightarrow HN_4^+ + N_2$	5.4×10^{-30}	[71]
R89	$HN_2^+ + N_2 + H_2 \rightarrow NH_4^+ + H_2$	$5.6 \times 10^{-30}(325/T)^{16}$	[72]
R90	$HN_2^+ + H_2 + H_2 \rightarrow H_3N_2^+ + H_2$	$3.3 \times 10^{-31}(300/T)^{4.5}$	[72]

Table 10.9. Reactions of negative oxygen and oxygen-nitrogen ions

No	Ion conversion	$k(\text{cm}^3/\text{s})$	Ref.
R1	$O^- + O_2(a) \to O_2^- + O$	10^{-10}	[47]
R2	$O^- + O_3 \to O_3^- + O$	8×10^{-10}	[73]
R3	$O^- + NO_2 \to NO_2^- + O$	1.2×10^{-9}	[74]
R4	$O^- + N_2O \to NO^- + NO$	2×10^{-10}	[53]
R5	$O^- + N_2O \to N_2O^- + O$	2×10^{-12}	[53]
R6	$O_2^- + O \to O^- + O_2$	3.3×10^{-10}	[47]
R7	$O_2^- + O_3 \to O_3^- + O_2$	3.5×10^{-10}	[75]
R8	$O_2^- + NO_2 \to NO_2^- + O_2$	7×10^{-10}	[21]
R9	$O_2^- + NO_3 \to NO_3^- + O_2$	5×10^{-10}	[47]
R10	$O_2^- + N_2O \to O_3^- + N_2$	$< 10^{-12}$	[1]
R11	$O_3^- + O \to O_2^- + O_2$	10^{-11}	[70]
R12	$O_3^- + N_2 \to NO_2^- + NO$	$< 5 \times 10^{-14}$	[53]
R13	$O_3^- + NO \to NO_3^- + O$	10^{-11}	[53,58]
R14	$O_3^- + NO \to NO_2^- + O_2$	2.6×10^{-12}	[47]
R15	$O_3^- + NO_2 \to NO_2^- + O_3$	7×10^{-11}	[58]
R16	$O_3^- + NO_2 \to NO_3^- + O_2$	2×10^{-11}	[58]
R17	$O_3^- + NO_3 \to NO_3^- + O_3$	5×10^{-10}	[58]
R18	$NO^- + O_2 \to O_2^- + NO$	5×10^{-10}	[1]
R19	$NO^- + NO_2 \to NO_2^- + NO$	7.4×10^{-10}	[1]
R20	$NO^- + N_2O \to NO_2^- + N_2$	2.8×10^{-14}	[1]
R21	$NO_2^- + H \to OH^- + NO$	3×10^{-10}	[77]
R22	$NO_2^- + O_3 \to NO_3^- + O_2$	1.8×10^{-11}	[75]
R23	$NO_2^- + NO_2 \to NO_3^- + NO$	4×10^{-12}	[1]
R24	$NO_2^- + NO_3 \to NO_3^- + NO_2$	5×10^{-10}	[58]
R25	$NO_2^- + N_2O_5 \to NO_3^- + 2NO_2$	7×10^{-10}	[47]
R26	$NO_3^- + NO \to NO_2^- + NO_2$	3×10^{-15}	[1]
R27	$O_4^- + M \to O_2^- + O_2 + M$	$10^{-10} \exp(-1044/T)(M=N_2, O_2)$	[60]
R28	$O_4^- + O \to O_3^- + O_2$	4×10^{-10}	[47]
R29	$O_4^- + O \to O^- + O_2 + O_2$	3×10^{-10}	[47]
R30	$O_4^- + O_2(a,b) \to O_2^- + O_2 + O_2$	10^{-10}	[47]
R31	$O_4^- + NO \to NO_3^- + O_2$	2.5×10^{-10}	[1]
	Electron detachment		
R32	$O^- + O \to O_2 + e$	1.4×10^{-10}	[78]
R33	$O^- + N \to NO + e$	2.6×10^{-10}	[53]
R34	$O^- + NO \to NO_2 + e$	2.6×10^{-10}	[53]
R35	$O^- + N_2 \to N_2O + e$	$10^{-13} - 10^{-12}$	[79]
R36	$O^- + O_2 \to O_3 + e$	5×10^{-15}	[80]
R37	$O^- + O_2(a) \to O_3 + e$	3×10^{-10}	[78]
R38	$O^- + O_2(b) \to O + O_2 + e$	6.9×10^{-10}	[81]
R39	$O^- + N_2(A) \to O + N_2 + e$	2.2×10^{-9}	[81]
R40	$O^- + N_2(B) \to O + N_2 + e$	1.9×10^{-9}	[81]
R41	$O^- + O_3 \to 2O_2 + e$	3×10^{-10}	[77]
R42	$O_2^- + O \to O_3 + e$	1.5×10^{-10}	[56]
R43	$O_2^- + N \to NO_2 + e$	5×10^{-10}	[56]
R44	$O_2^- + O_2 \to 2O_2 + e$	$2.7 \times 10^{-10}(T/300)^{0.5} \exp(-5590/T)$	[56]
R45	$O_2^- + O_2(a) \to 2O_2 + e$	2×10^{-10}	[56]
R46	$O_2^- + O_2(b) \to 2O_2 + e$	3.6×10^{-10}	[81]
R47	$O_2^- + N_2 \to O_2 + N_2 + e$	$1.9 \times 10^{-12}(T/300)^{0.5} \exp(-4990/T)$	[56]
R48	$O_2^- + N_2(A) \to O_2 + N_2 + e$	2.1×10^{-9}	[81]
R49	$O_2^- + N_2(B) \to O_2 + N_2 + e$	2.5×10^{-9}	[81]
R50	$O_3^- + O \to 2O_2 + e$	3×10^{-10}	[77]
R51	$O_3^- + N_2 \to N_2O + O_2 + e$	$< 10^{-15}$	[1]

Table 10.9. (Continued)

	Three-body collisions	$k(\text{cm}^6/\text{s})$	Ref.
R52	$O^- + O_2 + M \rightarrow O_3^- + M$	$1.1 \times 10^{-30}(300/T)$ (M=N_2,O_2)	[47]
R53	$O^- + NO + M \rightarrow NO_2^- + M$	10^{-29} (M=N_2,O_2)	[47]
R54	$O_2^- + O_2 + M \rightarrow O_4^- + M$	$3.5 \times 10^{-30}(300/T)$ (M=N_2,O_2)	[47]

Table 10.10. Reactions of negative oxygen–nitrogen–hydrogen ions

No	Ion conversion	$k(\text{cm}^3/\text{s})$	Ref.
R1	$H^- + H_2O \rightarrow OH^- + H_2$	3.8×10^{-9}	[1]
R2	$H^- + N_2O \rightarrow OH^- + N_2$	1.1×10^{-9}	[1]
R3	$H^- + NO_2 \rightarrow NO_2^- + H$	2.9×10^{-9}	[1]
R4	$H^- + NH_3 \rightarrow NH_2^- + H_2$	$1.6 \times 10^{-9} \exp(-2230/T)$	[82]
R5	$NH_2^- + H_2 \rightarrow H^- + NH_3$	$1.6 \times 10^{-9} \exp(-1260/T)$	[82]
R6	$O^- + H_2 \rightarrow OH^- + H$	3.3×10^{-11}	[1]
R7	$O^- + H_2O \rightarrow OH^- + OH$	1.4×10^{-9}	[1]
R8	$OH^- + NO_2 \rightarrow NO_2^- + OH$	1.9×10^{-9}	[1]
R9	$O_2^- + H \rightarrow$ products	1.5×10^{-9}	[1]
R10	$O_2^- + H_2 \rightarrow$ products	$< 10^{-12}$	[1]
R11	$NO_2^- + H \rightarrow OH^- + NO$	4×10^{-10}	[1]
R12	$NO_2^- + HNO_3 \rightarrow NO_3^- + HNO_2$	1.6×10^{-9}	[21]
R13	$O_3^- + H \rightarrow OH^- + O_2$	8.4×10^{-10}	[1]
R14	$O_4^- + H_2O \rightarrow O_2^- H_2O + O_2$	1.4×10^{-9}	[1]
R15	$O_2^- H_2O + O_2 \rightarrow O_4^- + H_2O$	2.5×10^{-15}	[1]
R16	$O_2^- H_2O + NO \rightarrow NO_3^- + H_2O$	3.1×10^{-10}	[1]
R17	$O_2^- H_2O + O_3 \rightarrow$ products	3×10^{-10}	[1]
R18	$O_2^- (H_2O)_2 + O_3 \rightarrow$ products	3.4×10^{-10}	[1]
	Electron detachment		
R19	$H^- + H \rightarrow H_2 + e$	1.8×10^{-9}	[1]
R20	$H^- + O_2 \rightarrow HO_2 + e$	1.2×10^{-9}	[1]
R21	$H^- + NO \rightarrow HNO + e$	4.6×10^{-10}	[1]
R22	$O^- + H_2 \rightarrow H_2O + e$	7×10^{-10}	[1]
R23	$OH^- + H \rightarrow H_2O + e$	1.8×10^{-9}	[1]
R24	$OH^- + O \rightarrow HO_2 + e$	2×10^{-10}	[1]
R25	$OH^- + N \rightarrow HNO + e$	$< 10^{-11}$	[1]
R26	$OH^- + H \rightarrow H_2O + e$	1.8×10^{-9}	[1]
R27	$OH^- + NO \rightarrow HNO_2 + e$	10^{-9}	[48]

The rate coefficients for ion–molecule reactions, as well as for reactions of neutral particles, may depend strongly on the degree of vibrational excitation of the reacting molecule. An example of this is the reaction

$$O^+ + N_2(v) \rightarrow NO^+ + N, \qquad (10.22)$$

which plays an important role in the ion composition of the Earth's ionosphere. The corresponding rate coefficient increases considerably with increasing N_2 vibrational excitation [83] and can be expressed as

$$k^{(22)} = C \times k^{(22)}(v=0) \equiv \sum_v k^{(22)}(v)f(v), \qquad (10.23)$$

where $k^{(22)}(v)$ is the rate coefficient for reaction from vibrational level v, $f(v)$ is the $N_2(v)$ vibrational distribution function, normalized to 1 ($\sum_v f(v) = 1$), C is a correcting factor that takes into account the contribution to the reaction of vibrationally excited N_2 molecules. The value of $k^{(22)}(v=0)$ is given in Table 10.7. The following approximate formula, based on measurements [83] and calculations [84,85] in the range T=300–2000 K, was proposed for $k^{(22)}(v)$ [86,87]:

$$k^{(22)}(v) = A_v T + B_v . \tag{10.24}$$

The parameters A_v and B_v for levels $v = 1$–8 are given in Table 10.11.

Table 10.11. Values of A_v and B_v in (10.24)

v	1	2	3	4	5	6	7	8
A_v; (10^{-14} cm^3/(s K))	0.339	2.33	3.02	-2.74	-0.384	1.6	-2.3	2.77
B_v; (10^{-10} cm^3/s)	3.72^{-3}	3.09^{-1}	1.92	2.9	0.585	1.59	1.19	1.36

Here 3.72^{-3} is 3.72×10^{-3}

Figure 10.3 shows calculations of the factor C as a function of T and T_V, using (10.23), (10.24) and assuming either Boltzmann (harmonic model) or Treanor (anharmonic model) vibrational distributions. It can be seen that vibrational excitation strongly influences the total rate coefficient $k^{(22)}$ for $T_V \geq 1000$ K. The results on this figure stress again the importance of state–selected rate coefficients and of the shape of the vibrational distribution in determining the total rates (see also [88]).

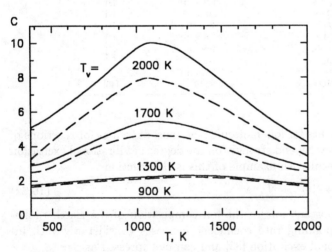

Fig. 10.3. Calculated values of the factor C in (10.23) as a function of gas temperature for different N_2 vibrational temperatures T_V. The dashed curves correspond to the harmonic model, the solid curves to the anharmonic model with a Treanor vibrational distribution [86,87]

Reactions involving positive and negative ions can also be of importance in plasmas. For example, ion-ion recombination reactions can sometimes control the concentration of charged particles. Ion–ion recombination processes and the corresponding rate coefficients are presented in Table 10.12 for nitrogen-oxygen plasmas. By comparing these data with those in Tables 10.7 and 10.8 it is seen that the rate coefficients for ion-ion recombination are larger by a factor of $\sim 10^2$, or even more, than the ion–molecule rate coefficients. This is explained by the strong Coulomb attraction between ions of opposite charges.

Table 10.12. Rate coefficients for reactions involving positive and negative ions

No	Process	Reactants, products	$k(\mathrm{cm}^3/s)$	Ref.
R1	$A^- + B^+ \to A+B$	$A^- = O^-, O_2^-, O_3^-,$ $NO^-, NO_2^-, NO_3^-, N_2O^-$ $B^+ = O^+, O_2^+, N^+,$ $N_2^+, NO^+, NO_2^+, NO_2^+$	2×10^{-7} $\times (300/T)^{0.5}$	[47]
R2	$A^- + (BC)^+ \to$ $A+B+C$	$A^- = O^-, O_2^-, O_3^-,$ $NO^-, NO_2^-, NO_3^-, N_2O^-$ $(BC)^+ = O_2^+, N_2^+, NO^+,$ $NO_2^+, N_2O^+, N_3^+, N_4^+, O_4^+$ $N_4^+, O_4^+, NO^+ N_2,$ $NO^+ O_2, NO^+ NO, O_2^+ N_2$	10^{-7}	[47]
R3	$O_4^- + C^+ \to 2O_2 + C$	$C^+ = O^+, O_2^+, N^+,$ $N_2^+, NO^+, NO_2^+, N_2O^+$	10^{-7}	[47]
R4	$O_4^- + (BC)^+ \to$ $2O_2 + B+C$	$B-C=N-N_2, N_2-N_2,$ $O_2-O_2, NO-N_2, NO-O_2,$ $NO-NO, O_2-N_2$	10^{-7}	[47]
R5	$A^- + B^+ + M \to$ $A+B+M$	$A^- = O^-, O_2^-;$ $B^+ = O^+, N^+,$ O_2^+, N_2^+, NO^+	2×10^{-25} $\times (300/T)^{2.5}$	[47]
R6	$O^- + B^+ + M \to$ $OB+M$	$B^+ = O^+, N^+, NO^+$ N_2, O_2	2×10^{-25} $(300/T)^{2.5}$ $M=N_2, O_2$	[47]
R7	$O_2^- + B^+ + M \to$ O_2B+M	$B^+ = O^+, N^+, NO$ O_2^+, N_2^+, NO^+	2×10^{-25} $(300/T)^{2.5}$ $M=N_2, O_2$	[47]

10.3 Vibrational Excitation of Reaction Products

Many of the reactions presented in Tables 10.1, 10.3, 10.4, 10.7–10.10, and 10.12 are exothermic. The energies released in these reactions may be converted not only into kinetic energy, but also into internal (electronic, vibrational and rotational) energy of the products. For example, associative reactions of neutral atoms lead to the production of electronically and vibrationally excited molecules. Such reactions are a source of electronically excited $N_2(A^3\Sigma_u^+, B^3\Pi_g)$, $O_2(a^1\Delta_g, b^1\Sigma_g^+, c^1\Sigma_u^-, A^3\Sigma_u^+)$ molecules. The corresponding rate coefficients were presented in Tables 9.3 and 9.4. Recombination between positive and negative ions can be also a source of electronically excited products. For example, according to calculations [89] from Landau-Zener theory, O atoms pairs in the states $2p^3P + 3p^5P$, $2p^3P + 3p^3P$, and $2p^1D + 3s^5S$ are created by the mutual neutralization of oxygen ions (see reaction (1) in Table 10.12)

$$O^- + O^+ \to O^* + O^{*'} . \tag{10.25}$$

The branching ratio for the formation of the above-mentioned atom pairs via this mutual neutralization is 0.42, 0.46 and 0.12 respectively.

Vibrational excitation of reaction molecular products is typical outcome of exothermic reactions. For example, the energy converted into vibrational excitation in the reactions

$$N + NO \to O + N_2^*(\bar{v} = 3.4) , \tag{10.26}$$

$$N + O_2 \to O + NO^*(\bar{v} = 2.73) , \tag{10.27}$$

$$N(^2D) + O_2 \to O + NO^*(\bar{v} = 4.75) , \tag{10.28}$$

$$N + NO_2 \to O + N_2O^*(\bar{v_3} = 2.47) \tag{10.29}$$

represents, respectively, 25, 39, 39 and 37% of the total energy released, according to measurements [90–93]. The mean number of excited vibrational quanta $\bar{v}$ produced in such reactions, according to the above authors, are given in brackets in (10.26)–(10.29). Processes (10.26)–(10.28) are important mechanisms for the excitation of N_2 and NO vibrations in laboratory discharges (mainly in the afterglow) and in the Earth's ionosphere. The corresponding total rate coefficients were given in Table 10.1

Measured relative rate coefficients [91,92], $\delta_v^{(27)}$ and $\delta_v^{(28)}$, for the production of $NO(v)$ via, respectively, reactions (10.27) and (10.28) are given in Table 10.13 as a function of v.

Table 10.13. Relative rate coefficients $\delta_v^{(27)}$ and $\delta_v^{(28)}$

v	0	1	2	3	4	5	6
$\delta_v^{(27)}$	0.21	0.104	0.218	0.116	0.163	0.085	0.074
$\delta_v^{(28)}$	0.08	0.1	0.1	0.1	0.11	0.12	0.1
v	7	8	9	10	11	12	
$\delta_v^{(27)}$	0.046	0	0	0	0	0	
$\delta_v^{(28)}$	0.09	0.08	0.05	0.03	0.02	0.02	

However, measurements of δ_v for many other reactions are lacking. In this case, to estimate δ_v, the fraction ξ of the reaction energy transferred into vibrations must be known. Using a statistical approach and information theory, one can obtain [94]:

$$\delta_v = \delta_0 \left(1 - \frac{E_v}{E_{\text{ch}}}\right)^{3/2} \exp\left\{-\frac{E_v}{kT_V}\right\}; \quad 1 \leq v \leq v_{\max}, \quad (10.30)$$

where E_v is the energy of vibrational level v, E_{ch} is the total energy released from the chemical reaction, T_V is a "temperature" characterizing the δ_v distribution which is determined by the equation

$$\frac{\sum_{v=0}^{v_{\max}} E_v \left(1 - \frac{E_v}{E_{\text{ch}}}\right)^{3/2} \exp\left\{-\frac{E_v}{kT_V}\right\}}{\sum_{v=0}^{v_{\max}} \left(1 - \frac{E_v}{E_{\text{ch}}}\right)^{3/2} \exp\left\{-\frac{E_v}{kT_V}\right\}} = \xi E_{\text{ch}}. \quad (10.31)$$

The limiting excitation level $v_{\max}$ for an anharmonic Morse-oscillator can be determined from

$$v_{\max} = \frac{E_{10}}{2\Delta E} + \frac{1}{2} - \sqrt{\left(\frac{E_{10}}{2\Delta E} + \frac{1}{2}\right)^2 - \frac{E_{\text{ch}}}{\Delta E}}, \quad (10.32)$$

while δ_0 can be obtained from the normalization condition

$$\sum_{v=0}^{v_{\max}} \delta_v = 1, \text{ that is } \delta_0 = \left[\sum_{v=0}^{v_{\max}} \left(1 - \frac{E_v}{E_{\text{ch}}}\right)^{3/2} \exp\left(-\frac{E_v}{kT_V}\right)\right]^{-1} \quad (10.33)$$

In conclusion, we wish to stress that the importance of vibration excitation in determining inelastic processes and chemical reactions is a topic of current interest as can be appreciated by the numerous publications on the subject. Recent calculations [95–101] using classical dynamics report state-to-state cross sections for the following processes:

$$N_2(v) + O \rightarrow NO + N; \quad (10.34)$$

$$N_2(v) + N \rightarrow 3N; \quad (10.35)$$

$$H_2(v) + H_2 \rightarrow 2H + H_2; \quad (10.36)$$

$$H_2(v) + H \rightarrow 3H; \quad (10.37)$$

$$O_2(v) + O \rightarrow O_2 + O; \quad (10.38)$$

$$O_2(v) + N \rightarrow NO + N; \quad (10.39)$$

References

1. Mc Ewan M. J. and Phillips L. F. (1975) *Chemistry of the Atmosphere*, Edward Arnold, New Zealand
2. Krivonosova O.E., Losev S.A., Nalivaiko V.P., Mukoseev Y.K. and Shatalov O.P. (1987) in *Plasma Chemistry*, ed. B.M. Smirnov, Energoatomizdat, Moscow, Vol. 14, p. 3 (*in Russian*)

3. AVOGADRO (1992-1995) *Data base on Rate Constants of Chemical and Plasmachemical Reactions, RRATE*, AVOGADRO Centre, Institute of Mechanics, Moscow State University, Moscow
4. Benson S.W., Golden D.M., Shaw R., Woolfolk R.W. and Lawrence R.W. (1975) Int. J. Chem. Kinet. **1**, 399
5. Baulch D.L., Drysdale D.D. and Horne D.C. (1973) *Evaluated Kinetic Data for High Temperature Reactions*, Vol. 2, Butterworths, London
6. Cline M.A. and Joko Ono (1982) Chem. Phys. **69**, 381
7. Boulch D.L., Cox R.A. and Hampson R.F. (1984) J. Phys. Chem. Ref. Data **13**, 1259
8. Stief L.J., Payne W.A., Lee J.H. and Michael J.V. (1979) J. Chem. Phys. **70**, 5241
9. Baulch D.L., Drysdale D.D., Duxbary J. and Grant S.L. (1976) *Evaluated Kinetic Data for High Temperature Reactions*, Vol. 3, Butterworths, London
10. Grahman R.A. and Johnston H.S. (1978) J. Phys. Chem. **82**, 254
11. Keiser E.W. and Japar S.W. (1978) Chem. Phys. Lett. **54**, 265
12. Baulch D.L., Cox R.A., Hampson R.F. et al. (1980) J. Phys. Chem. Ref Data **3**, 295
13. Mukkavilli S., Lee C.K., Varghese K. and Tavlarides L.L. (1988) IEEE Trans. Plasma Sci. **16**, 652
14. Dautov N.G. and Starik A.M. (1996) Kinetics and Catalysis **37**, 346 (*in Russian*)
15. Dautov N.G. and Starik A.M. (1997) Kinetics and Catalysis **38**, 207 (*in Russian*)
16. Pagsberg P.B., Eriksen J. and Christensen H.C. (1979) J. Phys. Chem. **83**, 582
17. Slovetsky D.I. (1980) *Mechanisms of Chemical Reactions in Non-equilibrium Plasmas*, Nauka, Moscow (*in Russian*)
18. Mackie J.C. and Dovlan K.R. (1984) Int. J. Chem. Kinet. **16**, 525
19. Dean A.M., Chou Mau-Song and Stern D. (1984) Int. J. Chem. Kinet. **16**, 633
20. Tsang W. and Hampson R.F. (1986) J. Phys. Chem. Ref. Data **15**, 1087
21. Person J.C. and Ham D.O. Radiat. Phys. Chem. **31**, 1
22. Davidson D.F., Kohse-Hoinghaus K., Chang A.Y. and Hanson R.K. (1990) Int. J. Chem. Kinet. **2**, 513
23. Maetzing H (1991) in *Chemical Kinetics of Flue Gas Cleaning by Irradiation with Electrons*, Adv. in Chem. Phys., ed. I.Prigogine and S.A.Rice, **LXXX**, John Wiley and Sons, New York
24. Baulch D.L. et al. (1992) J. Phys. Chem. Ref. Data **21**, 411
25. Zaslonko I.S., Tereza A.M., Kulish O.N. and Zheldakov D.Yu. (1992) Chemical Physics **11**, 1491 (*in Russian*)
26. Kuznetsov N.M. (1972) Rep. Acad. Sci. USSR **202**, 1367 (*in Russian*)
27. Losev S., Sergievskaya A., Starik A. and Titova N. (1997) *AIAA Paper* **97-2532**
28. Nikitin E.E. (1978) Rep. Acad. Sci. USSR **239**, 380 (*in Russian*)
29. Eremin A.V., Zibrov V.S. and Shumova V.V. (private communication)
30. Armenise I., Capitelli M., Garcia E., Gorse C., Lagana A. and Longo S. (1992) Chem. Phys. Lett. **200**, 597
31. Gordiets B.F. and Sergievskaya A.L. (1997) Chemical Physics **16**, 11 (*in Russian*)
32. Gershenzon U.M., Egorov V.I. and Rozenshein V.B (1977) Chem. High Energy **11**, 291 (*in Russian*)
33. Levitscky A.A. and Polak L.S. (1980) Chem. High Energy **14**, 3 (*in Russian*)

34. Levitscky A.A., Macheret S.O. and Fridman A.A. (1983) in *Chemical Reactions in Non-equilibrium Plasmas*, Nauka, Moscow, p. 2 (*in Russian*)
35. Le Roy R.L. (1969) J. Phys. Chem. **73**, 4338
36. Macheret S.O., Rusanov V.D. and Fridman A.A. (1984) Rep. Acad. Sci. USSR **276**, 1420 (*in Russian*)
37. Rusanov V.D. and Fridman A.A. (1984) *Physics of Chemically Active Plasmas*, Nauka, Moscow (*in Russian*)
38. Macheret S.O., Fridman A.A. and Elkin A.A. (1990) Sov. Chem. Phys. **9**, 174
39. Adamovich I.V., Macheret S.O., Rich J.W., Treanor C.E. and Fridman A.A. (1995) in *Molecular Physics and Hypersonic Flows*, ed. M.Capitelli, Kluwer, Dordrecht, NATO ASI Series, Vol. 482, p. 85
40. Rusanov V.D., Fridman A.A. and Sholin G.V. (1978) in *Plasma Chemistry*, ed. B.M. Smirnov, Atomizdat, Moscow, Vol. 5, p. 222 (*in Russian*)
41. Macheret S.O., Rusanov V.D. Fridman A.A. and Sholin G.V. (1980) J. Tech. Phys. **50**, 705 (*in Russian*)
42. Polak L.S., Çoldenberg M.Ya. and Levitsky A.A. (1984) in *Numerical Methods in Chemical Kinetics*, ed. L.S. Polak, Nauka, Moscow (*in Russian*)
43. Bucharin E.V. and Lobanov A.N. (1984) in *Proceed. VI Sympos. Plasmachemistry*, Dnepropetrovsk, Russia, p.52 (*in Russian*)
44. Dmitrieva I.K. and Zenevich V.A. (1984) Chemical Physics **3**, 1075 (*in Russian*); Dmitrieva I.K., Jdanok S.A. and Zenevich V.A. (1984) Preprint UTMO, Acad. Sci. Bielorussia **24** (*in Russian*)
45. Gordiets B., Ferreira C.M., Guerra V., Loureiro J., Nahorny J., Pagnon D., Touzeau M. and Vialle M. (1995) IEEE Trans. Plasma Sci. **23**, 750
46. Cacace F. and de Petris G. (2000) Inter. J. Mass Spectrometry **194**, 1; Cacace F., de Petris G., Pepi F., Rosi M., and Sgamellotti A. (1999) Angew. Chem. Int. Ed. **38**, 2408; Cacace F., de Petris G., Pepi F., and Troiani A. (2000) Angew. Chem. Int. Ed. **39**, 367; Cacace F., de Petris G. and Troiani A. (1999) Rapid Commun. Mass Spectrometry **13**, 1903
47. Kossiy I.A., Kostinsky Yu.A., Matveyev A.A. and Silakov V. (1992) Plasma Sources Sci. Technol. **1**, 207
48. Akishev Yu., Derugin A., Karalnik V., Kochetov I., Napartovich A. and Trushkin N. (1994) Plasma Physics **20**, 571 (*in Russian*)
49. O'Keefe A., Mauclaire G., Parent D. and Bowers M.T. (1986) J. Chem. Phys. **84**, 215
50. Zinn J., Sutherland C.D., Stone S.N. and Duncan L.M. (1982) J. Atm. Terr. Phys. **44**, 1143
51. Viehland H.A. and Mason E.A. (1971) J. Chem. Phys. **66**, 422
52. Maurice J.P. and Torr D.G. (1978) J. Geophys. Res. **83**, 969
53. Mc Daniel E.W., Cermak V., Dalgarno A., Ferguson E.G. and Friedman L. (1970) *Ion-Molecule Reactions*, John Wiley and Sons, New York
54. Dunckin D.V., McFarland M., Fehsenfeld F.C. and Ferguson E.E. (1971) J. Geophys. Res. **76**, 3820
55. Polak L.C. Sergeev P.A. and Slovetsky D.I. (1977) in *Plasma Chemical Reactions and Processes*, Nauka, Moscow, p. 81 (*in Russian*)
56. Kozlov S.I., Vlaskov V.A. and Smirnova N.V. (1988) Space Res. **26**, 738 (*in Russian*)
57. Fehsenfeld F.C. and Ferguson E.E. (1972) Radio Sci. **7**, 113
58. Niles F.E. (1970) J. Chem. Phys. **52**, 408
59. Guthrie J.A., Chaney R.C. and Cunningham A.J. (1991) J. Chem. Phys. **95**, 930
60. Smirnov B.M. (1983) *Complex Ions*, Nauka, Moscow (*in Russian*)
61. Adams N.G. and Smit D. (1985) Chem. Phys. Lett. **117**, 67

62. Adams N.G., Smit D. and Paulson J.F. (1980) J. Chem. Phys. **72**, 288
63. Theard L.P., and Huntress W.T. (1974) J. Chem. Phys. **60**, 2840
64. Ryan K.R. (1974) J. Chem. Phys. **61**, 1559
65. Kim J.K., Theard L.P. and Huntress W.T. (1975) J. Chem. Phys. **62**, 45
66. Smith D., Adams N.G. and Miller T.M. (1978) J. Chem. Phys. **69**, 308
67. Hemsworth R.S., Payzant J.D., Schiff H.I. and Bohme D.K. (1974) Chem. Phys. Lett. **26**. 417
68. Bei Hu Cheng, Bhowmik P.K. and Su T. (1989) J. Chem. Phys. **90**, 7046
69. Tichy M., Twiddy N.D., Wareing D.P., Admas N.G. and Smith D. (1987) Int. J. Mass. Spectrosc. **81**, 235
70. Graham E., James D.R., Keever W.C., Gatland I.R., Albritton D.L. and Mc Daniel E.W. (1973) J. Chem. Phys. **59**, 4648
71. Meot-Ner M. and Field F.H. (1974) J. Chem. Phys. **61**, 3742
72. Blades A.T. and Kebarle P. (1983) J. Chem. Phys. **78**, 783
73. Ichikawa Y. and Wu R.L.C. (1990) J. Appl. Phys. **67**, 108
74. Ferguson E.E. (1969) Can. J. Chem. **47**, 1815
75. Fehsenfeld F.G., Schmeltecopf A.L., Schiff H.I. and Ferguson E.E. (1967) Planet. Space Sci. **15**, 373
76. Fehsenfeld F.G. and Ferguson E.E. (1972) Planet. Space Sci. **20**, 295
77. Steinfeld J.J., Adler -Golden S.M. and Gallager J.M. (1987) J. Phys. Chem. Ref.Data **16**, 911
78. Eliasson B. (1983) Brown Boveri Rep. **KRL 83-40c.** Baden, Switzerland
79. Rayments S.W. and Moruzzi J.L. [1978] Int. Mass. Spectr. Ion. Phys. **26**, 321
80. Houg R., Bastien F. and Lecuiller M. (1975) J. Chimie Physique **72**, 105
81. Aleksandrov N.L. (1978) J. Tech. Phys. **48**, 1428 (*in Russian*)
82. Mackay G.I., Hemsworth S. and Bohme D.K. (1976) Can. J. Chem. **54**, 1624
83. Walker J.C.G., Stolarski R.S. and Nagy A.F. (1969) Ann. Geophys. **25**, 831
84. O'Malley T.E. (1970) J. Chem. Phys. **32**, 3269
85. Van Zandt T.E. and O'Malley T.E. (1973) J. Geophys. Res. **78**, 6818
86. Pavlov A.V. (1986) Geomagnetism and Aeronomy **26**, 152 (*in Russian*)
87. Pavlov A.V. and Namgaladze A.A. (1988) Geomagnetism and Aeronomy **28**, 705 (*in Russian*)
88. Capitelli M., Giordano D., Gorse C. and Longo S. (1996) Chem. Phys. Lett. **263**, 635
89. Olsen R.E., Peterson J.R. and Moseley J.T. (1971) J. Geophys. Res. **76**, 2516
90. Black G., Sharpless R. and Slanger T. (1973) J. Chem. Phys. **58**, 4792
91. Winkler I.C., Stachnik R.A., Steinfeld J.E. and Miller S.M. (1986) J. Chem. Phys. **85**, 890
92. Kennealy J.P., Del Greco F.P., Calldonia G.E. and Green B.D (1978) J. Chem. Phys. **69**, 1574
93. Clough P.N. and Thrush B.A. (1969) Proc. Roy. Soc. **A309**, 419
94. Ben-Shaul A., Levine R.D. and Bernstein R.B. (1972) J. Chem. Phys. **57**, 5427
95. Bose D. and Candler G. V. (1996) J. Chem. Phys. **104**, 2825
96. Esposito F. and Capitelli M. (1999) Chem. Phys. Lett. **302**, 49; Esposito F., Gorse C. and Capitelli M. (2000) Chem. Phys. (in press)
97. Caballos A., Garcia E., Rodriguez A. and Lagana A. (1999) Chem. Phys. Lett. **305**, 276
98. Esposito F., Gorse C. and Capitelli M. (1999) Chem. Phys. Lett. **303**, 636
99. Lagana A., Riganelli A., Ochoa de Aspuro G., Garcia E. and Martinetz T. (1998) Chem. Phys. Lett. **288**, 616

100. Lagana A., Garcia E. and Martinez T. (1999) *Quasiclassical and Quantum Rate Coefficients for $O+O_2$ Reaction*, ed. A. Lagana, University of Perugia, Perugia
101. Gilibert M., Gonzalez M., Sayos R., Aguilar A., Gimenez X., and Hijazo J. (1996) in *Molecular Physics and Hypersonic Flows*, ed. M.Capitelli, Kluwer, Dordrecht, NATO ASI Series, Vol. 482, p. 53

11. Interactions of Gas Phase Species with Surfaces

11.1 Balance Equations and Boundary Conditions at Walls

An important area of scientific research is the investigation of wall kinetic processes in low pressure gases and plasmas, since the interaction of different atomic and molecular species with walls can significantly influence the gas phase concentration of these species. For atmospheric N_2 and O_2 gases such processes are important, first of all, for long-lived species such as N and O atoms, vibrationally excited $N_2(X, v \geq 1)$ and $O_2(X, v \geq 1)$ molecules, and electronically excited metastables $N_2(A^3\Sigma_u^+, a'^1\Sigma_u^-, a''^1\Sigma_g^+)$, $O_2(a^1\Delta_g, b^1\Sigma_g^+)$, $N(^2D, ^2P)$, $O(^1D, ^1S)$. The interaction of these species with surfaces can lead to them being lost. On the other hand, for some atmospheric components (such as NO and NO_2), there can be significant surface production.

To investigate the influence of surfaces on the densities of gas phase species the transfer of these species to or from the surface must be studied. This transfer is usually determined by diffusion. Let us analyze, following [1], the balance equation for the concentration [A] of a given gas species A which diffuses to and interacts with a wall. This is the ordinary continuity equation for the species, that is

$$\frac{\partial [A]}{\partial t} = \nabla.(\mathcal{D}\nabla[A]) - \nabla.(\mathbf{v}[A]) + P - L . \tag{11.1}$$

The first and second terms on the right-hand side of (11.1) describe the change of [A] due to diffusion and convective flow respectively, $\mathcal{D}$ denoting the diffusion coefficient and v the flow velocity. The quantities P and L are the volume rates of creation and loss of A species. Later in this chapter we will be interested only in cylindrically symmetrical situations. We will also assume for simplicity that L is linear with respect to A, that is

$$L = \nu_{\text{vol}}[A] \tag{11.2}$$

and that the gas flow along the tube is laminar, with a flow velocity profile

$$v_z(r) = 2\overline{v_z}\left(1 - \frac{r^2}{R^2}\right) , \tag{11.3}$$

where $\overline{v_z}$ is the radially averaged flow velocity along the tube axis z; r is distance from the tube axis; R is the tube radius. Let us also assume $\mathcal{D}(z)$=constant and $\overline{v_z}$= constant. With the above assumptions and for cylindrical geometry, (11.1) takes the form

$$\frac{\partial [A]}{\partial t} = \mathcal{D}\left[\frac{1}{r}\frac{\partial}{\partial r}\left(r\frac{\partial [A]}{\partial r}\right) + \frac{\partial^2 [A]}{\partial z^2}\right] - 2\overline{v_z}\left(1 - \frac{r^2}{R^2}\right)\frac{\partial [A]}{\partial z} + P - \nu_{\mathrm{vol}}[A]. \tag{11.4}$$

Boundary conditions are required for (11.4). Below we will examine, for simplicity, only losses of A species. In this case, to obtain the condition on the wall (for $r = R$), consider a wall layer with thickness $\simeq l$, where l is the mean free path for the species. Since particle collisions do not occur in this layer, under stationary conditions ($\partial [A]/\partial t = 0$) one has

$$\Phi_- = (1 - \gamma)\Phi_+ , \tag{11.5}$$

where Φ_+ and Φ_- are, respectively, the particle flux towards the wall and the flux reflected from it, γ denoting the particle wall loss probability per collision with the surface. From elementary kinetic theory [2] one has

$$\Phi_- = \left.\frac{[A]\overline{c}}{4}\right|_{r=R} - \left.\frac{1}{2}\mathcal{D}\frac{\partial [A]}{\partial r}\right|_{r=R} ;$$

$$\Phi_+ = \left.\frac{[A]\overline{c}}{4}\right|_{r=R} + \left.\frac{1}{2}\mathcal{D}\frac{\partial [A]}{\partial r}\right|_{r=R} , \tag{11.6}$$

where $\overline{c}$ is the average speed of particles A. By inserting $\mathcal{D} = l\overline{c}/3$, one gets from (11.5),(11.6) the following boundary condition [3]:

$$\left.\mathcal{D}\frac{\partial [A]}{\partial r}\right|_{r=R} = -\frac{2\gamma}{2-\gamma}\left.\frac{[A]\overline{c}}{4}\right|_{r=R} \equiv \gamma'\left.\frac{[A]\overline{c}}{4}\right|_{r=R}. \tag{11.7}$$

Note that

$$\gamma' = \frac{2\gamma}{2-\gamma} \simeq \gamma \quad \text{for} \quad \gamma \ll 1. \tag{11.8}$$

Note that under conditions (11.8), the diffusion flux to the wall (the second term on the right-hand side of (11.6)) is not important.

Using (11.7), it is possible to solve (11.4) analytically in some simple cases. For instance, if $P = 0$ and for static conditions ($\overline{v_z} = 0$) the concentration after a sufficiently long time will be

$$[A](r, t) = [A](r, t = 0) \exp[-(\nu_\mathrm{w} + \nu_{\mathrm{vol}})t] J_0\left(r\sqrt{\frac{\nu_\mathrm{w}}{\mathcal{D}}}\right), \tag{11.9}$$

where ν_w is the smallest root of the equation

11.1 Balance Equations and Boundary Conditions at Walls

$$G \equiv \frac{\gamma' \bar{c} R}{4\mathcal{D}} = R\sqrt{\frac{\nu_w}{\mathcal{D}}} \frac{J_1\left(R\sqrt{\frac{\nu_w}{\mathcal{D}}}\right)}{J_0\left(R\sqrt{\frac{\nu_w}{\mathcal{D}}}\right)}, \qquad (11.10)$$

J_0 and J_1 denoting the well-known Bessel functions.

A simple one-dimensional approximation is frequently used in kinetic modelling to describe the densities along a tube. In this case, a simpler equation than (11.4) is usually used, considering the surface process as a volume source (negative or positive) of some species A. This equation, yielding the radially averaged density $\overline{[A]}$ in our example ($P = 0$; $\overline{v_z} = 0$), will be:

$$\frac{d\overline{[A]}}{dt} = -(\nu_w + \nu_{\text{vol}})\overline{[A]}. \qquad (11.11)$$

The root ν_w of (11.10), for the limiting cases indicated below, is

$$\nu_w \approx \begin{cases} \gamma \bar{c}/(2R) & \text{for} \quad \gamma \ll 1 \quad \text{when} \quad G \ll 1 \\ 5.78\mathcal{D}/(R^2) & \text{for} \quad \gamma \sim 1 \quad \text{when} \quad G \gg 1. \end{cases} \qquad \begin{matrix}(11.12a)\\(11.12b)\end{matrix}$$

In the general case, the following approximation can be used [4]:

$$\nu_w(s^{-1}) \approx \left(\frac{R^2}{5.78\mathcal{D}} + \frac{2R}{\gamma \bar{c}}\right)^{-1}. \qquad (11.13)$$

Note that (11.12),(11.13) can also be useful to analyze other regimes. For example, when $P \neq 0$, $\overline{v_z} \neq 0$, the equation for the average concentration $\overline{[A]}$ along the tube, as derived from (11.4) assuming an one-dimensional flow, is

$$\frac{\partial \overline{[A]}}{\partial t} = \mathcal{D}\frac{\partial^2 \overline{[A]}}{\partial z^2} - \overline{v_z}\frac{\partial \overline{[A]}}{\partial z} + P - (\nu_w + \nu_{\text{vol}})\overline{[A]}. \qquad (11.14)$$

This equation has the following simple analytical solution for quasistationary conditions ($\partial \overline{[A]}/\partial t = 0$), constant $\mathcal{D}(z)$, $\nu_w(z)$, and $\nu_{\text{vol}}(z)$, and $P = 0$:

$$\overline{[A]}(z) = \overline{[A]}(z = 0)\exp(-\beta \frac{z}{R}), \qquad (11.15)$$

where

$$\beta = \frac{(-\overline{v_z} + \sqrt{\overline{v_z}^2 + 4(\nu_w + \nu_{\text{vol}})\mathcal{D}})R}{2\mathcal{D}}. \qquad (11.16)$$

The solution (11.15),(11.16) describes the density distribution along the tube for post-discharge conditions. However, this solution must be used with care. since it applies only when $v_z(r) = $ const. across the tube. However, v_z usually has a parabolic profile (11.3), which can in some cases influence $[A](r)$ and the average $\overline{[A]}$. This influence is likely to be more important for post-discharge conditions, when $P = 0$. The case of a velocity profile like (11.3) was investigated by several authors [1,5–8]. It was found that the distribution $\overline{[A]}(z)$ has the exponential form (11.15) but with β different from (11.16). To illustrate this, the dimensionless parameter $\lambda^2 = \beta^2 \overline{v_z} R/\mathcal{D}$, for the case $\nu_{\text{vol}} = 0$, $\nu_w \neq 0$, is given in Fig. 11.1 as a function of $G \equiv \gamma'\bar{c}R/4\mathcal{D}$ (see (11.10)), assuming either a parabolic or a constant velocity profile. The

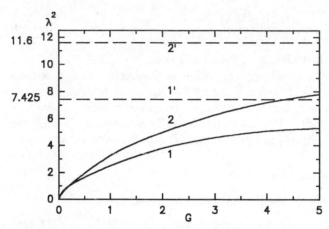

Fig. 11.1. Values of $\lambda^2 = \beta^2 \overline{v_z} R/\mathcal{D}$ as a function of the parameter $G \equiv \gamma' \bar{c} R / 4\mathcal{D}$. Curves 1, 1' – for a parabolic gas flow velocity profile (11.3); curves 2, 2' – for a uniform velocity profile $v_z = \overline{v_z}$. Curves 1', 2' are the limits of λ^2 as $G \to \infty$ [6]

results nearly coincide only for small γ', when $G \leq 0.3$. In other cases, the parabolic velocity profile reduces β, that is the surface loss rate decreases. In the limit of large G (that is for $\gamma' \sim 1$) $\nu_w \simeq 3.66 \mathcal{D}/R^2$. This value is ~ 1.6 times less than that for the one-dimensional case (11.12b). Such a difference can have considerable influence on the calculated average density $\overline{[A]}$ in the post-discharge due to the exponential dependence of $\overline{[A]}$ on β.

11.2 Wall Loss Probabilities γ

Interaction with surfaces can thus have a major influence on the volume concentration of some species, and information on wall loss probabilities γ is important for investigating and modelling kinetic processes in non-equilibrium gases and plasmas. Data in the literature on γ for some long-lived N_2, O_2, N, and O species are presented in Tables 11.1, 11.2 and in Figs. 11.2, 11.3.

The values of γ for deactivation of the metastables $N_2(A^3\Sigma_u^+)$, $N_2(a'^1\Sigma_u^-)$ and $O_2(b^1\Sigma_g^+)$ on glass are also worth mentioning: $\gamma_{a'} \leq 10^{-2}$ [22] and $\gamma_b \simeq (1\text{--}2) \times 10^{-2}$ [23,24] for deactivation of $N_2(a'^1\Sigma_u^-)$ and $O_2(b^1\Sigma_g^+)$ respectively; for $N_2(A^3\Sigma_u^+)$, $\gamma_A \sim 1$ and so diffusion to the wall usually controls the value of ν_w [25] (see (11.12b)). Fast wall deactivation is usually also assumed for the metastable atoms $N(^2D)$, $N(^2P)$, $O(^1D)$, and $O(^1S)$.

Many measurements of the probabilities γ_O and γ_N of wall losses of O and N atoms have been carried out. Results for silica-based surfaces are presented in Figs. 11.2 and 11.3. Values of γ_O and γ_N calculated from a surface kinetic model (see sect. 11.3) are also shown in these figures.

11.2 Wall Loss Probabilities γ 197

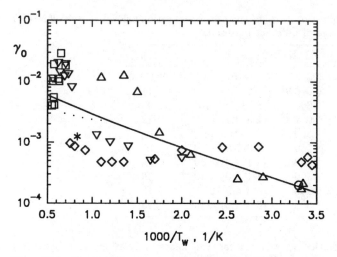

Fig. 11.2. Probability γ_O as a function of wall temperature. Points are measurements: $\triangle$–[26]; $\triangledown$–[27]; $\square$–[28]; $\circ$–[29]; $\Diamond$–[30]; $\star$–[31]; ······–[32]. The solid line is a calculation [51] from a model assuming an ordinary system of chemically active sites on the surface (post-discharge conditions)

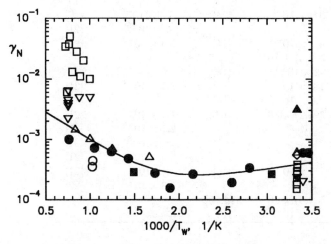

Fig. 11.3. Probability γ_N as a function of wall temperature. Points are measurements: $\square$–[33]; $\triangle$–[34]; $\triangledown$–[35]; $\circ$–[36]; $\blacksquare$–[37]; $\blacktriangle$–[38]; $\blacktriangledown$–[34]; $\bullet$–[30]. The solid line is a calculation [51] from a model assuming an ordinary system of chemically active sites on the surface (post-discharge conditions)

Table 11.1. Probability γ_v for deactivation of vibrational level $v = 1$ of $N_2(X)$ molecules on the surface of different materials

Surface	T_w, K	γ_v	Ref.
Pyrex	300–350	$(2-10) \times 10^{-4}$	[8,9,10]
molybdenum glass	300–600	$(1-3) \times 10^{-3}$	[11]
quartz	300	$(1.8-7) \times 10^{-4}$	[8]
quartz	300–700	$(2.3-3.1) \times 10^{-3}$	[12]
steel	300	$(1-3) \times 10^{-3}$	[8]
aluminium alloy	300	$(1.3-5.0) \times 10^{-3}$	[8]
Al_2O_3	300	$(1.1-1.4) \times 10^{-3}$	[8]
copper	300	$(1.1-4.0) \times 10^{-3}$	[8]
silver	295	1.4×10^{-2}	[13]
Teflon	300	$(0.4-2.0) \times 10^{-4}$	[13]

Table 11.2. Probabilities γ_O and γ_{a^1} for wall losses of O atoms and $O_2(a^1\Delta_g)$ metastables on the surface of different materials

Surface	γ_O	Ref.	γ_{a^1}	Ref.
glass	see Fig. 11.2		$(7.6-15) \times 10^{-6}$	[16–18]
			4×10^{-5}	[19]
steel	9.9×10^{-3}	[14]	$(7-13)10^{-6}$	[20]
			$4.4-10^{-3}$	[21]
aluminium	4.4×10^{-3}	[14]	5.9×10^{-5}	[21]
Al_2O_3	$(1.7-3.4) \times 10^{-3}$	[14,15]	-	
ceramic Al_2O_3	2×10^{-3}	[14]	-	
copper	1.7×10^{-1}	[15]	8.5×10^{-4}	[21]
	2.6×10^{-2}	[14]		
silver	$(3.2-5.2) \times 10^{-3}$	[14,15]	1.1×10^{-2}	[21]
nickel	-		2.7×10^{-3}	[21]
molybdenum	-		8×10^{-5}	[21]
titanium	-		6.5×10^{-5}	[21]
platinum	2.7×10^{-3}	[14]	4×10^{-4}	[21]
			1.3×10^{-3}	[22]
magnesium	$(2.3-2.6) \times 10^{-3}$	[14,15]	-	
Teflon	7.3×10^{-5}	[14]	-	

It is important to note that the experimental probabilities given in Figs. 11.2, 11.3 have, as a rule, been obtained from post-discharges in pure gases, that is, in N_2–N and O_2–O systems. Under such circumstances, these probabilities are dependent on surface temperature only. Most theories so far put forward to explain the experimental temperature dependence of γ_M (usually of a non-Arrhenius type over a wide temperature range) take into consideration only one kind of atom [29,39–45].

However, some experimental results have been published showing that the interaction with surfaces of different atoms and molecules in N_2–N–O_2–O gas mixtures can significantly influence γ_N and γ_O and lead to the production of heteronuclear gas phase molecules. For instance, optical and mass spectrometric measurements in space experiments [46,47] show that gas phase NO_2 molecules can be produced on the Space Shuttle surface in the ram direction in stationary orbit, where strong interaction with the surface takes place for ramming O atoms and N_2 molecules. Laboratory measurements [48,49] for low pressure ($\simeq 1\text{--}2$ Torr) O_2 or N_2 low temperature plasmas show a decrease in γ_O and γ_N with the addition of a small admixture ($\sim 1\%$) of the other gas (N_2 or O_2, respectively) in the discharge. A strong increase in γ_N, up to $\sim 10^{-2}$ as compared to values $\sim 10^{-4}$ for pure nitrogen, has been found [50] for discharges in $N_2\text{--}O_2$ mixtures with relative O_2 concentration $\geq 20\%$. Interpretation of the above experiments [46–50] requires the development of a kinetic theory for surface processes that takes into account different kinds of atoms and surface diffusion of physisorbed atoms. Such a theory has recently been developed [50,51] and is briefly presented in the next section.

11.3 Kinetic Model for Surface Processes

A kinetic model for surface processes was developed [50,51] in order to calculate the surface losses of gas phase N and O atoms and NO, NO_2 molecules, and the surface production of gas phase N_2, O_2, NO, NO_2 molecules, in discharges of binary molecular $N_2\text{--}O_2$ gas mixtures. This model takes into account:

(a) physical adsorption and desorption of N and O atoms;
(b) chemical adsorption and desorption of both types of atoms and NO, NO_2 molecules at vacant chemically active sites on the surface;
(c) surface diffusion of physisorbed N_f and O_f atoms;
(d) the reactions of chemisorbed N_S and O_S atoms both with gas phase N, O atoms and NO molecules (Eley–Rideal mechanism) and with physisorbed N_f and O_f atoms (Langmuir–Hinshelwood mechanism), leading to the formation of gas phase N_2, O_2 and chemisorbed $(NO)_S$, $(NO_2)_S$ molecules.

$(NO)_S$ molecules can be desorbed or can react with O, O_f and produce $(NO_2)_S$, which by desorption will produce gas phase NO_2 molecules.

The list of reactions taken into account in the model is presented in Table 11.3, where S_v and F_v denote vacant chemisorption (active) and physisorption sites, respectively.

This model for surface kinetics is simple and does not take into account a number of processes. The interaction between two physically adsorbed atoms or two chemically adsorbed particles, dissociative chemisorption of molecules, and processes involving physisorbed molecules have been ignored. Neglecting dissociative chemisorption and reactions between two chemisorbed particles

Table 11.3. List of surface processes

No.	Process	No.	Process
S1	$N+F_v \rightleftharpoons N_f$	S11	$N+O_S \rightarrow (NO)_S$
S2	$O+F_v \rightleftharpoons O_f$	S12	$N_f+O_S \rightarrow (NO)_S+F_v$
S3	$N+S_v \rightleftharpoons N_S$	S13	$O+N_S \rightarrow (NO)_S$
S4	$N_f+S_v \rightarrow N_S+F_v$	S14	$O_f+N_S \rightarrow (NO)_S+F_v$
S5	$N+N_S \rightarrow N_2+S_v$	S15	$NO+S_v \rightleftharpoons (NO)_S$
S6	$N_f+N_S \rightarrow N_2+S_v+F_v$	S16	$NO_2+S_v \rightleftharpoons (NO_2)_S$
S7	$O+S_v \rightleftharpoons O_S$	S17	$(NO)_S+O \rightarrow (NO_2)_S$
S8	$O_f+S_v \rightarrow O_S+F_v$	S18	$(NO)_S+O_f \rightarrow (NO_2)_S+F_v$
S9	$O+O_S \rightarrow O_2+F_v+S_v$	S19	$NO+O_S \rightarrow (NO_2)_S$
S10	$O_f+O_S \rightarrow O_2+S_v$		

is justified for a relatively low surface density of active sites (see below) and low wall temperatures ($T_w \leq 1000$ K), due to the high energy threshold of such reactions. The processes involving physisorbed molecules are also ignored here for simplification.

Note that a model accounting for surface diffusion of physisorbed atoms, ignoring the reactions of these atoms with vacant active sites and considering only atoms of a single kind, was investigated in [29]. Another model with two different atoms (N and O) has also been proposed [52,53], but this was only used for interpretation of high temperature experiments and did not include processes with NO_2 and physically adsorbed atoms.

The rate coefficients for processes involving gas phase particles (S1,S2, S3,S5,S7,S9,S11,S13,S15,S16,S17,S19 in Table 11.3) are in units of cm^{3}s^{-1}. The rate coefficients K_i for surface reactions (i=S4,S6,S8,S10,S12,S14,S18) are in units of cm^{2}s^{-1}. They are determined by surface diffusion coefficients D_N and D_O (in cm^{2}s^{-1}) for N_f and O_f physisorbed atoms:

$$K_i \left(\text{cm}^2\text{s}^{-1}\right) = \mathcal{D}_M K_i^0 \exp\left(-\frac{E_i}{T_w}\right), \qquad (11.17)$$

where K_i^0 are dimensionless steric factors (≤ 1) and E_i activation energies (in K) of the chemical processes i. The surface diffusion coefficients $\mathcal{D}_M$ can be expressed by the following equation [29]:

$$\mathcal{D}_M \simeq \frac{1}{4}a^2 \nu_{DM}^0 \exp\left(-\frac{E_{DM}}{T_w}\right) \simeq \frac{\nu_{DM}^0}{4[F]} \exp\left(-\frac{E_{DM}}{T_w}\right), \qquad (11.18)$$

where a is an elementary distance for the jump of a physisorbed atom from one physisorption site to another (that is, the distance between neighboring sites); ν_{DM}^0 is the frequency factor; and E_{DM} is the activation energy for surface diffusion of M atoms.

The desorption coefficients K_{-S1} and K_{-S2} of N_f and O_f physisorbed atoms read:

$$K_{-i}\left(\text{s}^{-1}\right) \simeq \nu_{dM}^0 \exp\left(-\frac{E_{dM}}{T_w}\right), \tag{11.19}$$

where ν_{dM}^0 is the frequency factor and E_{dM} is the activation energy for desorption of M atoms (M=N for i=1 and M=O for i=2). The coefficients of desorption for chemically adsorbed atoms and molecules (processes -S3,-S7,-S15,-S16) have the same dimensions and form as (11.19), but with the relevant frequency factors and activation energies for desorption.

From the master kinetic equations for the processes (S1)–(S19), with rate coefficients K_i, one readily obtains, under steady state conditions,

$$[N_f] = A_N[N]; \quad [O_f] = A_O[O], \tag{11.20}$$

where

$$A_N = K_{S1}\tau_N[F_v]; \quad A_O = K_{S2}\tau_O[F_v]; \tag{11.21}$$

$$[F_v] = \frac{[F]}{1 + K_{S1}\tau_N[N] + K_{S2}\tau_O[O]}; \tag{11.22}$$

$$[N_S] = \alpha_N[S_v]; \quad [O_S] = \alpha_O[S_v]; \tag{11.23}$$

$$[(NO)_S] = \alpha_{NO}[S_v]; \quad [(NO_2)_S] = \alpha_{NO_2}[S_v]; \tag{11.24}$$

$$[S_v] = \frac{[S]}{1 + \sum_M \alpha_M} \tag{11.25}$$

with M=N, O, NO, NO$_2$. The variables [X] in (11.20)–(11.25) are the volume or the surface densities of the X species; the values $\alpha_M/(1 + \sum_M \alpha_M)$ are the fraction of active sites occupied by chemisorbed particles M=O, N, NO, NO$_2$; τ_N and τ_O are the lifetimes of physically adsorbed N_f and O_f atoms, respectively:

$$\tau_N = (K_{-S1} + K_{S4}[S_v] + K_{S6}[N_S] + K_{S12}[O_S])^{-1}; \tag{11.26}$$

$$\tau_O = (K_{-S2} + K_{S8}[S_v] + K_{S10}[O_S] + K_{S14}[N_S] + K_{S18}[(NO)_S])^{-1}. \tag{11.27}$$

For a wide range of the concentrations [N] and [O] and for $T_w \geq 300$ K, we can further assume that

$$K_{S1}[N] \ll K_{-S4} \quad \text{and} \quad K_{S2}[O] \ll K_{-S2}, \tag{11.28}$$

so that $[F_v] \simeq [F]$. This condition corresponds to a small coverage of the surface by physisorbed atoms. Under this assumption, using dimensionless rate coefficients (see below) the α_M in (11.23)–(11.25) can be expressed as

$$\alpha_N = \frac{(K'_{S3} + P_{S4})\Phi_N}{K'_{-S3} + (K'_{S5} + P_{S6})\Phi_N + (K'_{S13} + P_{S14})\Phi_O}, \tag{11.29}$$

$$\alpha_O = \frac{(K'_{S7} + P_{S8})\Phi_O}{K'_{-S7} + (K'_{S9} + P_{S10})\Phi_O + (K'_{S11} + P_{S12})\Phi_N + K'_{S19}\Phi_{NO}},$$
(11.30)

$$\alpha_{NO} = \frac{(K'_{11} + P_{S12})\alpha_O\Phi_N + (K'_{S13} + P_{S14})\alpha_N\Phi_O + K'_{S15}\Phi_{NO}}{K'_{-S15} + (K'_{S17} + P_{S18})\Phi_O}, \quad (11.31)$$

$$\alpha_{NO_2} = \frac{(K'_{S17} + P_{S18})\alpha_{NO}\Phi_O + K'_{S19}\alpha_O\Phi_{NO} + K'_{S16}\Phi_{NO_2}}{K'_{-S16}}. \quad (11.32)$$

In (11.19)–(11.22), Φ_M are the fluxes of gas phase atoms or molecules M = N, O, NO, NO$_2$ to the surface:

$$\Phi_M = \frac{1}{4}\overline{c_M}[M], \quad (11.33)$$

where $\overline{c_M}$ denotes the mean velocity of the gas phase particle M near the surface. The rate coefficients K'_i and P_i are defined by the following expressions:

$$K'_i = \frac{4[S]}{\overline{c_M}}K_i; \quad K'_j = \frac{4[F]}{\overline{c_M}}K_j; \quad K'_n = [S]K_n \quad (11.34)$$

with $\overline{c_M} = \overline{c_N}$ for j=S1, i=S3,S5,S11; $\overline{c_M}=\overline{c_O}$ for j=S2, i=S7,S9,S13,S17; $\overline{c_M}=\overline{c_{NO}}$ for i=S15,S19; $\overline{v_M}=\overline{v_{NO_2}}$ for i=S16; n=-S3,-S7,-S15,-S16, and

$$P_i = K'_j K_i \tau_M [S] \quad (11.35)$$

with M=N for j=S1, i=S4,S6,S12; M=O for j=S2, i=S8,S10,S14,S18. All the coefficients K'_i and P_i, excepting K'_{-S3}, K'_{-S7}, K'_{-S15}, K'_{-S16} are dimensionless and, in fact, represent reaction probabilities. The units of K'_{-S3}, K'_{-S7}, K'_{-S15} and K'_{-S16} are cm^{-2}s^{-1}. They are proportional to the chemical desorption fluxes.

The dimensionless reaction probabilities (11.34) for processes involving gas phase atoms and molecules are conveniently expressed in the following form:

$$K'_i = \frac{[S]}{[F]}K^0_i \exp\left(-\frac{E_i}{T_{gw}}\right); \quad (11.36)$$

$$K'_j \equiv P_M = K^0_M \exp\left(-\frac{E^0_M}{T_{gw}}\right), \quad (11.37)$$

where i=S3,S5,S7,S9,S11,S13,S15,S16,S17,S19; M = N for j=S1, M=O for j=S2. The values K'_i are the probabilities of gas phase atoms or molecules being chemically adsorbed (processes S3,S7,S15,S16) or reacting with chemically adsorbed particles (processes S5,S9,S11,S13,S17,S19). The ratio $\frac{[S]}{[F]}$ represents the probability of direct impingement of gas phase particles on either vacant or occupied active sites; it is the fraction of the surface covered by such sites. The values $K'_j \equiv P_M$ are the probabilities of gas phase atoms M=N, O being physically adsorbed (processes S1, S2). The factors K^0_i, K^0_M

11.3 Kinetic Model for Surface Processes

in (11.36),(11.37) are the steric factors for surface processes involving gas phase particles. E_i is the activation energy of reaction i in absolute temperature units, E_M^0 is the activation energy for physical adsorption of M atoms, and T_{gw} is the gas temperature near the wall, which we will assume equal to the wall temperature T_w.

The values of P_i in (11.35) for surface processes involving physisorbed N_f and O_f atoms (i=S4,S6,S8,S10,S12,S14,S18) are determined by the following equations:

$$P_i = P_M P_i^0 \times [1 + \Psi_M]^{-1} ; \qquad (11.38)$$

$$P_i^0 = K_{Mi} \exp\left(\frac{E_{dM} - E_{DM} - E_i}{T_w}\right); \quad K_{Mi} = \frac{1}{4} K_i^0 \frac{[S]}{[F]} \frac{\nu_{DM}^0}{\nu_{dM}^0} ; \qquad (11.39)$$

$$\Psi_N = \delta \left(P_{S4}^0 + \alpha_N P_{S6}^0 + \alpha_O P_{S12}^0\right) ; \qquad (11.40)$$

$$\Psi_O = \delta \left(P_{S8}^0 + \alpha_O P_{S10}^0 + \alpha_N P_{S14}^0 + \alpha_{NO} P_{S18}^0\right) ; \qquad (11.41)$$

$$\delta = \frac{1}{(1 + \sum_M \alpha_M)} \qquad (11.42)$$

(with M= N, O, NO, NO$_2$). Equations (11.38)–(1.41) can be obtained from (11.17)–(11.19), (11.26),(11.27),(11.35). The factors Ψ_N and Ψ_O describe the influence on the lifetime of N_f and O_f physisorbed atoms of the surface chemical processes (in addition to desorption). However, for wall temperatures $T_w \geq 300$ K the lifetime of physisorbed atoms is controlled by desorption. In this case, $\Psi_M \ll 1$ and these factors have no influence on the probabilities P_i.

Let us now find the wall loss probabilities γ_M. Note that the diffusion rates of gas phase N and O atoms to the wall are much higher than the corresponding rates for wall destruction under typical low pressure discharge conditions ($p \leq 5$ Torr, $R \leq 2$ cm) (that is, the second term on the right-hand-side of (11.13) is the most important). This is known from experimental data. The same will also be assumed for NO and NO$_2$ molecules. Under such condition, the net rates for wall loss and production of gas phase N, O atoms and NO, NO$_2$ molecules in infinite cylindrical geometry are the following:

$$\left(\frac{d[N]}{dt}\right)_{wall} = -\frac{2}{R}\epsilon\{(K_{S1}[F_v] - K_{-S1}\frac{[N_f]}{[N]}$$
$$+ K_{S3}[S_v] + K_{S5}[N_S] + K_{S11}[O_S])[N] - K_{-S3}[N_S]\} , \qquad (11.43)$$

$$\left(\frac{d[O]}{dt}\right)_{wall} = -\frac{2}{R}\epsilon\{(K_{S2}[F_v] - K_{-S2}\frac{[O_f]}{[O]}$$
$$+ K_{S7}[S_v] + K_{S9}[O_S] + K_{S13}[N_S] + K_{S17}[(NO)_S])[O] - K_{-S7}[O_S]\} , \qquad (11.44)$$

$$\left(\frac{d[NO]}{d}\right)_{wall} = -\frac{2}{R}\epsilon\{(K_{S15}[S_v] + K_{S19}[O_S])[NO] - K_{-S15}[(NO)_S]\}, \tag{11.45}$$

$$\left(\frac{d[NO_2]}{dt}\right)_{wall} = -\frac{2}{R}\epsilon\{K_{S16}[S_v][NO_2] - K_{-S16}[(NO_2)_S]\}. \tag{11.46}$$

Here, $\epsilon \geq 1$ is the roughness factor, that is, the ratio of the real surface area to the geometric surface area, per unit length.

Using (11.20)–(11.27) and (11.29)–(11.32), the above equations can be rewritten in the form

$$\left(\frac{d[N]}{dt}\right)_{wall} = -\frac{\overline{v_N}}{2R}\gamma_N[N] + \frac{2}{R}K'_{-S3}\epsilon\delta\alpha_N; \tag{11.47}$$

$$\left(\frac{d[O]}{dt}\right)_{wall} = -\frac{\overline{v_O}}{2R}\gamma_O[O] + \frac{2}{R}K'_{-S7}\epsilon\delta\alpha_O; \tag{11.48}$$

$$\left(\frac{d[NO]}{dt}\right)_{wall} = -\frac{\overline{v_{NO}}}{2R}\gamma_{NO}[NO] + \frac{2}{R}K'_{-S15}\epsilon\delta\alpha_{NO}; \tag{11.49}$$

$$\left(\frac{d[NO_2]}{dt}\right)_{wall} = -\frac{\overline{v_{NO_2}}}{2R}\gamma_{NO_2}[NO_2] + \frac{2}{R}K'_{-S16}\epsilon\delta\alpha_{NO_2}, \tag{11.50}$$

where γ_M is the probability of wall losses given by the expression:

$$\gamma_N = \gamma_N^{(1)} + \gamma_N^{(2)}, \tag{11.51}$$

$$\gamma_N^{(1)} = \epsilon\delta(K'_{S3} + K'_{S5}\alpha_N + P_{S4} + P_{S6}\alpha_N), \tag{11.52}$$

$$\gamma_N^{(2)} = \epsilon\delta(K'_{S11} + P_{S12})\alpha_O, \tag{11.53}$$

$$\gamma_O = \gamma_O^{(1)} + \gamma_O^{(2)} + \gamma_O^{(3)}, \tag{11.54}$$

$$\gamma_O^{(1)} = \epsilon\delta(K'_{S7} + K'_{S9}\alpha_O + P_{S8} + P_{S10}\alpha_O), \tag{11.55}$$

$$\gamma_O^{(2)} = \epsilon\delta(K'_{S13} + P_{S14})\alpha_N, \tag{11.56}$$

$$\gamma_O^{(3)} = \epsilon\delta(K'_{S17} + P_{S18})\alpha_{NO}, \tag{11.57}$$

$$\gamma_{NO} = \epsilon\delta(K'_{S15} + K'_{S19}\alpha_O), \tag{11.58}$$

$$\gamma_{NO_2} = \epsilon\delta K'_{S16}. \tag{11.59}$$

Note that $\gamma_N^{(1)}$ and $\gamma_O^{(1)}$ are responsible for the production of, respectively, N_2 and O_2 gas phase molecules, and $\gamma_N^{(2)}$, $\gamma_O^{(2)}$ for the production of $(NO)_S$.

Equations (11.47)–(11.59) in principle enable the wall loss and production of gas phase atoms and molecules N, O, NO, NO_2 to be calculated. Of course, in practice the use of these equations presents difficulties due to the lack of information about rate coefficients and other parameters of the model. Some important parameters can be estimated, but it is clear that simplifications are necessary for practical purposes.

Such simplifications have been made in [36,37]. From comparisons with experimental data, the parameters and activation energies determining γ_N and γ_O in pure N_2-N, O_2-O post-discharge conditions have been estimated. The model presented above yields reasonable agreement with experimental data for silica-based surfaces using the following values: $E_{dO} - E_{DO} - E_O^0 \simeq 2000$ K; $K_O^0 K_{Oi} \simeq 1.5 \times 10^{-3}$ (for $i = S8, S10$), $E_{dN} - E_{DN} - E_N^0 \simeq 3100$ K; $K_N^0 K_{Ni} \simeq 1.5 \times 10^{-5}$ (for $i = S4, S6$), $E_{S4} \simeq 900$ K, $E_{S6} \simeq 2400$ K, $E_{S8} \simeq 0$ K, $E_{S10} \simeq 3100$ K, $[S]/[F] \simeq 3 \times 10^{-3}$, $\epsilon = 2.4$. A comparison of the model predictions with experimental data from different authors is shown in Figs. 11.2 and 11.3.

Note that the calculations of γ_O and γ_N in Figs. 11.2, 11.3 were carried out taking into account the reactions of chemisorbed atoms with both physisorbed atoms (Langmuir–Hinshelwood mechanism) and gas phase atoms (Eley–Rideal mechanism). For the latter (which becomes important for $T_w \geq 500$ K), it was assumed that all steric factors are equal to 1 and that the activation energies E_i are the same as for the reactions with physisorbed atoms. Taking into consideration both mechanisms allows one to explain the non-Arrhenius temperature dependence of γ_O and γ_N in a wide temperature range, from ~ 300 up to ~ 2000 K. This is particularly noticeable in Fig. 11.3 for γ_N.

It was mentioned above that γ_O and γ_N can change in N_2-N-O_2-O mixtures (as compared with pure N_2-N and O_2-O systems). They are moreover greater in discharge than in post-discharge [50–56]. The above model for surface processes has been used [50,51] to interpret these phenomena. An example of γ_O values, measured and calculated from the model, is presented in Fig. 11.4.

The physical reason for the decrease of γ_O with addition of nitrogen into a O_2 discharge is as follows. The N atoms (which are present in the discharge due to N_2 dissociation) occupy part of the chemically active sites and, as a result, the surface density of chemically adsorbed O_S atoms falls. This will reduce γ_O in cases where chemical reactions of N_S chemisorbed atoms with gas phase and physisorbed O and O_f atoms are not effective.

The reason for increased γ_O in the discharge as compared with post discharge, as found in [50,54–56], is so far unclear. It was quantitatively explained [50] by assuming the existence of an additional system of chemically active sites under discharge conditions. This could be connected with surface ions resulting from the electron and ion discharge fluxes to the surface. For example, the surface will likely be partly covered with negative atomic ions due to surface attachment of electrons to O_S atoms. Assuming that the lifetime of positive and negative surface ions is controlled by mutual ion-ion recombination, it is possible to estimate a value $\sim 10^{-3}$ for the relative surface density of positive and negative charges. This corresponds well to the

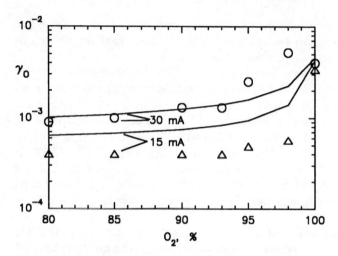

Fig. 11.4. Wall loss probabilities of O atoms in N_2-O_2 discharges versus O_2 percentage, for different currents. Pressure – 2 Torr; gas flow rate – 100 sccm; distance from gas entrance in discharge – 43 cm. Points – measurements [50], curves – model calculations [51]

values estimated to explain the high γ_O values observed under discharge conditions [50]. So far, this idea is only a working hypothesis which needs to be further developed in the future (see also Chap. 13).

11.4 Molecular Dynamics Approach

Recently, Billing [57] and Cacciatore and Billing [58] have developed a semi-classical approach to studying the reactive scattering of molecules and atoms impinging on a surface. Their model describes the motion of the impinging particles classically while an effective Hamiltonian is built to describe the coupling of internal molecular energy with the phononic and electronic structure of the surface. This model has been applied to the interaction of vibrationally excited $H_2(v)$ and atomic hydrogen on a copper surface [58]. More recently, it has been also applied to the recombination of atomic oxygen on silica [59]. These results were able to reproduce the experimental wall loss probabilities and provide the nascent vibrational distribution function of $O_2(v)$. This distribution was used to study hypersonic flows over catalytic walls [60].

Finally, Molinari and Tomellini [61] have discussed the influence of non-equilibrium vibrational kinetics on catalytic processes, with particular attention paid to heterogeneous isotope exchange reactions. This study together with the parallel development of gas–surface elementary processes from molecular dynamics can open interesting new perspectives in this field

References

1. Gershenzon Yu.M., Rosenshtein V.B. and Umansky S.Ya. (1977) in *Plasma Chemistry*, ed. B.M. Smirnov, Atomizdat, Moscow, **4**, 61 (*in Russian*)
2. Hirschfelder J.O., Curtiss C.F. and Bird R.B. (1964) *Molecular Theory of Gases and Liquids*, John Wiley and Sons, New York
3. Semenoff N.N. (1943) Acta Phys. USSR **18**, 93
4. Frank-Kamenetsky D.A. (1967) *Diffusion and Thermotransfer in Chemical Kinetics*, Nauka, Moscow (*in Russian*)
5. Gershenzon Yu.M., Rosenshtein V.B., Spassky A.I. and Kogan A.M. (1972) Rep. Acad. Sci. USSR **205**, 624 (*in Russian*)
6. Gershenzon Yu.M., Rosenshtein V.B., Spassky A.I. and Kogan A.M. (1972) Rep. Acad. Sci. USSR **205**, 871 (*in Russian*)
7. Gershenzon Yu.M. and Rosenshtein V.B. (1974) Theor. Exp. Chem. **10**, 769 (*in Russian*)
8. Black G., Wise H., Schechter S. and Sharpless R.L. (1974) J. Chem. Phys. **60**, 3526
9. Morgan J.E. and Schiff H.J. (1963) Can. J. Chem. **41**, 903
10. Abouaf R. and Legay F. (1966) J. Chimie Physique **63**, 1393 (*in French*)
11. Gershenzon Yu.M., Egorov V.I., Rosenshtein V.B. and Umansky S.Ya. (1973) Chem. Phys. Lett. **20**, 77
12. Gershenzon Yu.M. and Rosenshtein V.B. (1975) Chem. High Energy **9**, 413 (*in Russian*)
13. Gershenzon Yu.M., Kovalevsky S.A., Rosenshtein V.B. and Shub B.R. (1974) Rep. Acad. Sci. USSR **219**, 1400 (*in Russian*)
14. Wickramanayaka S., Meikle S., Kobayashi T., Hosokawa N. and Hatanaka Y. (1991) J. Vac. Sci. Technol. **A9**, 2999
15. Greaves J.C. and Linnett J.W. (1958) Trans. Faraday. Soc. **54**, 1323
16. Nugaev T.B.H., Popovich M.P., Filippov G.V. (1985) J. Phys. Chem. **59**, 1632 (*in Russian*)
17. Billington A.P. and Borrell P. (1986) J. Faraday Trans. **82**, 963
18. Giachrdi D.J., Harris G.W. and Wayne R.F. (1976) J. Faraday Trans. **72**, 619
19. Yaron M., von Engol A. and Vidaut P.H. (1976) Chem. Phys. Lett. **37**, 159
20. Becker K.H., Groth W. and Scurath U. (1971) Chem. Phys. Lett. **8**, 259
21. Riskin M.E. and Shub B.P. (1981) React. Kinet. Catal. Lett. **17**, 41
22. Gordiets B.F., Pinheiro M., Ferreira C.M. and Ricard A. (1998) Plasma Sources Sci.Technol. **7**, 363; 379
23. Fisk G.A. and Hays G.N. (1982) J. Chem. Phys. **77**, 4965
24. Hadwerk V. and Zellner R. (1986) Ber. Buns. Phys. Chem. **90**, 92
25. Cernogora G., (1981) *PhD Thesis*, Université de Paris-Sud, Orsay
26. Greaves J.C. and Linnett J.W. (1955) Trans. Faraday Soc. **55**, 1355
27. Steward D.A., Chen Y.K. and Henline W.D. (1991) AIAA Paper **91-1373**
28. Kolodziej P. and Steward D.A. (1987) AIAA Paper **87-1637**
29. Young C. Kim and Boudart M. (1991) Langmuir **7**, 2999
30. Marinelli W.J. and Campbell J.P. (1986) NASA Report **17565**
31. Zoby E.V., Gups R.N. and Simmonds A.L. (1984) AIAA Paper **84-0224**
32. Stewart D.A., Rakich J.V. and Lanfranco M.J. (1981) AIAA Paper **81-1143**
33. Breen J. et al. (1973) NASA Report **CR-134124**
34. Marshall T.C. et al. (1964) Report **RADC-TDR-63-275**, Griffis AFB, New York
35. Rosner D.E. and Cibrian R. (1974) AIAA Paper **74-755**
36. Back R.A. et al. (1959) Can. J. Chem. **37**, 2059
37. Kelly R. and Winkler C.A. (1959) Can. J. Chem. **37**, 62

38. Flower O. and Stewart D.A. (1985) NASA Report **TM-86770**
39. Wise H. and Wood B.J. (1967) Adv. Mol. At. Phys. **3**, 291
40. Scott C.D. (1980) AIAA Paper **80-1477**
41. Seward W.A. and Jumper E.J. (1991) Thermophysics and Heat Transfer **5**, 284
42. Jumper E.J., Newman M., Kitchen D.R. and Seward W.A. (1993) AIAA Paper **93-0477**
43. Barbato M. and Bruno C. (1995) in *Molecular Physics and Hypersonic Flows*, ed. M. Capitelli, Kluwer, Dordrecht, NATO ASI Series, Vol. 482, p. 139
44. Scott C.D. (1995) in *Molecular Physics and Hypersonic Flows*, ed. M. Capitelli, Kluwer, Dordrecht, NATO ASI Series, Vol. 482, p. 161
45. Jumper E.J. (1995) in *Molecular Physics and Hypersonic Flows*, ed. M. Capitelli, Kluwer, Dordrecht, NATO ASI Series, Vol. 482, p. 181
46. Swenson G.R., Mende S.B. and Clifton K.S. (1985) Geophys. Res. Lett. **12**, 97
47. Von Zanh U. and Murad E. (1986) Nature **321**, 147
48. De Souza A. R., Mahlmann C. M., Muzart J. H., and Speller C. V. (1993) J. Phys. D: Appl. Phys. **26**, 2164
49. Talsky A. and Zvonicek V. (1994) Proceed. 10th Symposium on Chemical Reactions in Low Temperature Plasmas, University of Bratislava, Slovakia
50. Gordiets B., Ferreira C.M., Nahorny J., Pagnon D., Touzeau M., and Vialle M. (1996) J. Phys. D: Appl. Phys. **29**, 1021
51. Gordiets B. and Ferreira C.M. (1997) AIAA Paper **97-2504**; (1998) AIAA Journal **36**, 1643
52. Nasuti F., Barbato M. and Bruno C. (1993) AIAA Paper **93-2840**
53. Kovalev V.L., Suslov O.N. and Tirskiy G.A. (1995) in *Molecular Physics and Hypersonic Flows*, ed. M. Capitelli, Kluwer, Dordrecht, NATO ASI Series, Vol. 482, p. 193
54. Gousset G., Panafieu P., Touzeau M. and Vialle M. (1987) Plasma Chem. Plasma Process. **7**, 409
55. Gousset G., Touzeau M., Vialle M., and Ferreira C. M., (1989) Plasma Chem. Plasma Process. **9**, 189
56. Magne L., Coitout H., Cernogora G. and Gousset G. (1993) J. Phys. III France **3**, 1871
57. Billing G. D. (1999) *Dynamics of Molecule Surface Interactions*, John Wiley and Sons, New York
58. Cacciatore M. and Billing G. D. (1996) Pure and Applied Chemistry **68**, 1075
59. Cacciatore M., Rutigliano and Billing G. D. (1999) J. Thermophys. and Heat Transfer **13**, 195
60. Armenise I., Cacciatore M., Capitelli M., Gorse C., and Rutigliano M. (1999) AIAA Paper **99-3631** (2000) J. Spacecraft and Rockets (in press)
61. Molinari E. and Tomellini C. (2000) Chem. Phys. **253**, 367

12. Discharges in Pure N_2 and O_2

Discharges and post-discharges in pure N_2 and O_2 or in their mixtures with inert gases are receiving increasing attention, due to their importance for understanding atmospheric and ionospheric physics and to their use as active laser media (ultraviolet N_2 lasers from pulsed discharges [1,2]) and as sources of active species (N and O atoms, metastable atoms and molecules, ions) for applications in plasma chemistry. Such applications include surface treatments, coating processes, metal nitriding [3,4], TiN deposition [5], oxidation and etching of polymers and semiconductors [6–8].

12.1 Discharge and Post-Discharge in N_2

Numerous aspects of nitrogen discharge physics have so far been investigated using both kinetic models and experiment. Calculations of electron energy distribution functions $f(\epsilon)$ (EEDF), transport parameters and other energy averaged properties (electron mobility μ_e and diffusion coefficients D_e, average $\bar{\epsilon}$ and characteristic ϵ_k energy, electron power balance), and of electron rate coefficients for excitation, ionization and dissociation of N_2 have been performed by many authors. From experimental work and theoretical modelling we are now able to understand the main mechanisms that control dissociation, ionization, gas heating, the populations of vibrational and electronic states, and the intensities of optical emissions in N_2 discharges and post-discharges, as well as the discharge instabilities and their volt-ampere characteristics.

The dissociation mechanisms of nitrogen molecules in a discharge were first investigated in [9–11]. In principle, two such mechanisms are possible. The first takes place through the excitation of unstable, dissociative electronic states by direct electron impact from the ground ($v = 0$) or excited ($v \geq 1$) vibrational levels of the ground electronic state. The second occurs through electron impact excitation of the lower vibrational levels of the ground electronic state and subsequent molecular transitions to the higher vibrational levels close to the dissociation limit from which dissociation occurs by V-V and V-T processes (this is the vibrational dissociation mechanism discussed in Chap. 3). The rates of N_2 dissociation by both mechanisms were calculated in [9–11], for typical discharge conditions ($N_e \sim 10^{12} cm^{-3}$, $kT_e \sim 2$ eV), using the anharmonic oscillator model with 45 vibrational levels. The master equations for the populations of all vibrational levels were solved assuming a Maxwellian EEDF and taking into account the vibrational-vibrational (V-V), vibrational-translational (V-T) and electron-vibrational (e-V) energy exchange processes. It was concluded that the vibrational dissociation mechanism is dominant under conditions of quasi-stationary vibrational level populations. However, the calculations of dissociation rates from the upper vibrational levels are rather inaccurate. This is mainly caused by inaccurate rate coefficients for V-V and V-T processes, especially for upper vibrational levels.

The actual form of the EEDF has a significant influence on the dissociation rate by direct electron impact. This was shown in later works [12–21], where the equations for the vibrational level populations were solved simultaneously with the Boltzmann equation. The time evolution of the vibrational distribution function and rate of dissociation was investigated in [12,13,15,21], while steady-state conditions were studied in [14,16–20]. In [21] direct electron impact dissociation from excited vibrational levels of the electronic ground state $N_2(X^1\Sigma_g^+)$ was included into the model. The effects of wall deactivation of N_2 vibrational levels on the vibrational distribution function and the rate of dissociation due to the vibrational mechanism were taken into account in [20].

The main difficulty in calculating accurately the vibrational dissociation rate is to predict correctly the populations of the upper $N_2(X)$ vibrational levels, since V-T relaxation by $N_2(v)-N$ collisions can strongly affect such populations. The influence of this process on the N_2 vibrational distribution function and the total dissociation rate has been investigated in [21]. It was found that $N_2(v)-N$ collisions can strongly affect the relative role of the various dissociation channels. Figure 12.1 illustrates the time evolution of the dissociation rates due to the different mechanisms after an external electric field is instantaneously applied. It can be seen that the dissociation rate by direct electron impact from levels $v > 0$ can considerably exceed those by electron impact from level $v = 0$ and through a pure vibrational mechanism. Electron impact dissociation from $N_2(X, v \geq 1)$ levels occurs through the excitation of different electronic excited states [13]. As an example, Fig. 12.2 reports the behavior of the dissociation cross–section through the following process

$$e + N_2(X, v \geq 1) \rightarrow e + N_2(B^3\Pi_g, v' \geq 12) \rightarrow e + N + N. \qquad (12.1)$$

This cross–section strongly increases with v, while the corresponding energy threshold decreases (see Fig. 12.2).

The calculated dissociation rate coefficient of N_2 molecules is also influenced by the vibrational energy spectrum and by the probabilities of V-V and V-T exchange (including the probability of two-quantum V-V exchanges) in $N_2(v)-N_2(v')$ collisions. This was studied in [18], where it was shown that

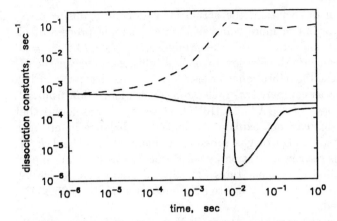

Fig. 12.1. Rate constants for different dissociation mechanisms of N_2 as a function of time. Dashed line - direct electron impact dissociation from all vibrational levels $N_2(X)$; full line – from level $v = 0$; dash-dotted line – through pure vibrational mechanism. These calculations are for $E/N = 3 \times 10^{-16}$ Volt cm^2, $N_e = 10^{11}$ cm^{-3}, $p = 5$ torr, $T = 500$ K [21]

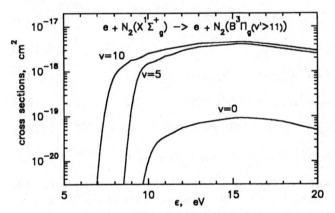

Fig. 12.2. Cross-sections for dissociation by processes (12.1) as a function of electron energy ϵ, for different vibrational levels of $N_2(X^1\Sigma_g^+)$ [13]

using the intramolecular potential (5.6) leads to an N_2 molecule model with 65 discrete vibrational levels. This can considerably decrease the dissociation rate coefficient as compared to the classical Morse-oscillator model with 45 levels. Inclusion of two-quantum V-V exchanges into the kinetic model leads to a sharp decrease in the calculated dissociation rate coefficient. Generally speaking, due to the lack of information on the vibrational energy spectrum and the probabilities of the different vibrational relaxation processes, only order-of-magnitude calculations of N_2 dissociation rates are possible at the present time.

As in the case of dissociation, the actual ionization mechanisms of N_2 in a discharge can be much more complex than the simple process of direct electron impact ionization from the ground state $N_2(X^1\Sigma_g^+, v = 0)$ [13,15,18,22–37]. associative ionization in binary collisions between the N_2 molecule in highly-excited vibrational levels, $v \geq 32$, was discussed in [22] and its importance was compared with electron impact ionization from different vibrational levels of $N_2(X^1\Sigma_g^+)$ [13,15]. However, it was found that the above mechanisms can not explain the ionization balance in nitrogen discharges over a wide range of gas pressure, reduced electric field E/N, and current. For this reason, ionization by such processes as electron impact ionization from excited electronic states [27–32] and associative ionization in collisions between electronically excited N_2 molecules [18,22,24,25,33–37] were also investigated.

According to present knowledge, the main N_2 ionization mechanisms in a discharge are the following:

$$e + N_2(X^1\Sigma_g^+, v \geq 0) \rightarrow 2e + N_2^+ ; \tag{12.2}$$

$$e + N_2(A^3\Sigma_u^+) \rightarrow 2e + N_2^+ ; \tag{12.3}$$

$$e + N_2(B^3\Pi_g) \rightarrow 2e + N_2^+ ; \tag{12.4}$$

$$e + N_2(a'^1\Sigma_u^+) \rightarrow 2e + N_2^+ ; \tag{12.5}$$

$$N_2(A^3\Sigma_u^+) + N_2(a'^1\Sigma_u^+) \rightarrow e + N_2 + N_2^+ \text{ or } e + N_4^+ ; \tag{12.6}$$

$$N_2(a'^1\Sigma_u^+) + N_2(a'^1\Sigma_u^+) \rightarrow e + N_2 + N_2^+ \text{ or } e + N_4^+ ; \tag{12.7}$$

$$N_2(A^3\Sigma_g^+) + N_2(X^1\Sigma_g^+, v \geq 30) \rightarrow e + N_2 + N_2^+ \text{ or } e + N_4^+ ; \tag{12.8}$$

$$N_2(a'''^1\Sigma_g^+) + N_2(X^1\Sigma_g^+, v \geq 13) \rightarrow e + N_2 + N_2^+ \text{ or } e + N_4^+ . \tag{12.9}$$

The rate coefficients for processes (12.6)–(12.9) were given in Table 9.3. For processes (12.2), these rates can be calculated from (8.16) (for $v = 0$), and by using (8.15) to estimate the contributions of levels $v \geq 1$.

As a matter of fact, not all processes (12.2)–(12.9) are equally important in N_2 discharges. To find the quantitative contributions of processes (12.2)–(12.9) to the ionization balance in a particular discharge, self-consistent calculations of the EEDF, vibrational distribution function and populations of the excited electronic states have to be carried out. Similar calculations are required for investigating optical emissions of N_2 in discharges and post-discharges. Such problems have been treated in [18,24,25,27–30,33,36–45], and the vibrational distributions of excited electronic states were calculated in [38–43,45,46]. Presently, the most important channels for N_2 ionization in steady-state discharges are considered to be (12.2),(12.6), and (12.7). The processes (12.8),(12.9) are less important, since the populations of $N_2(a'''^1\Sigma_g^+)$ and of the upper ($v \geq 30$) $N_2(X)$ vibrational levels are small (due to efficient deactivation by collisions $N_2(a'')+N_2$, $N_2(v)+N$).

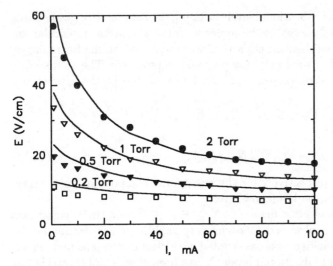

Fig. 12.3. Electric field E versus current I in a static N_2 DC discharge in a tube of 1 cm in radius, for different gas pressures. Points - measurements [48], curves – calculations [37]

Ionization by processes (12.6),(12.7) decreases the electric field E required to sustain a stationary discharge, the so-called maintaining field. Figure 12.3 shows the maintaining field E of a DC N_2 discharge as a function of gas pressure and discharge current. Good agreement between theory and experiment has been obtained by taking into account processes (12.6),(12.7) with the rate coefficients given in Table 9.3.

It is worth noting that the main excitation channel of the N_2 1st negative system, which is a strong emission corresponding to the transition $N_2^+(B) \to N_2^+(X)$, is the quasi-resonant V-E exchange

$$N_2(X, v \geq 12) + N_2^+(X) \to N_2(X, v - 12) + N_2^+(B) , \qquad (12.10)$$

as demonstrated in [36,37,42] for discharge conditions, and in [47] for post-discharge.

The kinetic modelling of N_2 discharges further enables the rate coefficients of various processes to be determined more accurately and the main gas heating mechanisms to be understood. To this end, self-consistent calculations of vibrational level populations and gas temperature have been carried out and compared with experiment [18,48-53]. For example, from the modelling of gas heating in a flowing nitrogen discharge, it has been found [49] that good agreement with experiment is obtained at $E/N \sim 10^{-16}$ volt cm^2 using the electron rotational excitation cross-section given in [54]. Comparison between the theoretical [18,50–52] and experimental [27,50] time dependence of N_2 vibrational level populations and gas temperature in a pulsed discharge leads to more accurate rate coefficients $Q_{n+1,n}^{k,k+1}$ of V-V processes. In fact, the V-V pumping of the upper vibrational levels is an important source of gas heating, due to the decreasing energy of the upper vibrational quanta (molecular anharmonicity). This mechanism is particularly important at the beginning of the pulse, when V-T processes as yet play no essential role. The volume rate of gas heating by V-V processes, P_{VV}, can be calculated by the expression

$$P_{VV} = \sum_{n=s}^{n_b} \sum_{m=1}^{s} (E_n - E_{n-1} + E_m + E_{m-1})$$

$$[Q_{n,n-1}^{m-1,m} N_n N_{m-1} - Q_{n-1,n}^{m,m-1} N_{n-1} N_m] \,, \tag{12.11}$$

where E_i, N_i stand respectively for the energy and the population of vibrational level "i"; n_b is the boundary level.

At high reduced electric fields ($E/N \geq 8 \times 10^{-16}$ volt cm^2), significant gas heating has been observed to occur very rapidly, which cannot be explained by electron energy losses associated with elastic collisions and excitation of rotational and vibrational levels. It has been shown [55] that this fast gas heating can be caused by electron impact excitation of electronic states (mainly, $N_2(A^3\Sigma_u^+)$ metastables) followed by fast deactivation of these states in collisions with heavy particles. According to [55], $\sim 30\%$ of the electronic state excitation energy is transferred into gas heating. Gas heating resulting from V-T processes with atomic nitrogen can now be investigated since a complete set of V-T rates for such processes is available.

Recently, a self–consistent model has been developed to describe the plasma properties of RF parallel plate discharges in N_2 using a PIC/MCC model (particle-in-cell with Monte Carlo collisions). This method provides the electron and the ion energy distribution functions as well as the self–consistent electric field from Poisson's equation [56,57]. For the first time, a complete vibrational and electronic kinetics has been coupled to the PIC/MCC model to show the mutual effects between free electron dynamics and heavy–particle kinetics. At the same time, experiments have been carried out [58,59] on stationary and transient RF parallel plate reactors using LIF and Langmuir probe techniques to monitor metastable concentrations and the electron energy distribution function in N_2 and N_2–O_2. Experimental EEDFs have also been obtained by deconvolution of spectroscopic quantities [60,61].

Spatio-temporal instabilities of gas discharges are processes whose investigation allows the most important energetic properties of discharges to be understood. The different types of instabilities in molecular gas discharges have been discussed in [62–67]. Instabilities in nitrogen plasmas have been investigated in [68–76].

A well-known instability mechanism is ionization overheating, due to the sharp (exponential) dependence of the ionization rate coefficient k_{ion}, and, consequently, of the electron concentration N_e, on the reduced electric field E/N. For example, when gas ionization by direct electron impact and electron-ion dissociative recombination (with a rate coefficient $k_{\rm d.r}$) are respectively the main creation and loss channels of charged particles, the electron density $N_{\rm e}$ in a steady-state N_2 discharge is determined by the expression

$$N_{\rm e} \simeq \frac{A}{k_{\rm d.r}}[N_2]\exp\left\{-\frac{B}{E/N} + \frac{C}{(E/N)^2}e^{-E_1/kT_V}\right\}, \qquad (12.12)$$

where N is the total gas density, T_V is the N_2 vibrational "temperature", and A, B, C are parameters. The exponential dependence (12.12) reflects the dependence (8.16) for the rate coefficient $k_{\rm ion}$. The fluctuating increase in gas temperature leads to a decrease of the gas density N at constant pressure. As seen from (12.12), this causes a strong increase of $N_{\rm e}$ (mainly due to the term $\frac{B}{E/N}$ in the exponent of (12.12)), which, in turn, increases gas heating and gas temperature. In this way, thermal perturbations can develop causing discharge instability and contraction. As seen from (12.12), instabilities caused by vibrational excitation can occur as well. In fact, an increase in $N_{\rm e}$ usually leads to a higher rate of vibrational excitation by electron impact, and thus to an increased vibrational "temperature" T_V. This results in additional growth of $N_{\rm e}$, due to the second term in the exponent of (12.12)), which constitutes an additional cause of instability [70]. Physically, the instability caused by vibrational excitation is related to the influence of vibrational excitation on the "tail" of the EEDF due to superelastic electron collisions with vibrationally excited molecules.

Other instability mechanisms associated with the ionization processes (12.3)–(12.9) involving vibrationally and electronically excited states can take place in N_2 discharges. To find out the actual role of each process in the development of instabilities, detailed self-consistent models accounting for the charge and vibrational kinetics and for the discharge thermal balance have to be worked out. Further, model predictions have to be compared with experimental data. Some aspects of these problems have been investigated in [71–77].

12.2 Discharge in O_2

Generally speaking, the same problems as for N_2 need to be solved for O_2 discharges, by combining theoretical and experimental investigations. However, the electronegativity of oxygen introduces quite different properties, since considerable concentrations of O^- and O_2^- ions, sometimes exceeding the concentration of plasma electrons, can exist due to the high rate of electron attachment to O atoms and O_2 molecules (see (8.27) and Table 8.12). The presence of such ions changes plasma properties such as conductivity, maintaining field, energy balance and spatial distributions of charged particles. Discharges in electronegative gases have long been the subject of theoretical investigation [78,79]. Detailed analyses of the oxygen positive column have

12.2.1 The Positive Column in Electronegative Gases

Let us consider an axially uniform positive column with cylindrical geometry and radius R. The DC electric field E is assumed uniform and directed along the axis. We will consider only three species of charged particles: electrons (mass m), positive ions of mass M_+, and negative ions of mass M_-. Electron impact ionization of neutral molecules creates free electrons and positive ions at a rate ν_{ion}, while negative ions are created at a rate ν_{att} by electron attachment to the gas molecules (e.g., two-body dissociative attachment) and lost at a rate ν_{det} by electron detachment on collisions with some molecular species, unspecified for the moment, uniformly distributed through the volume. The rates ν_{ion}, μ_{att} and ν_{det} are assumed spatially constant. The gas density, N, is assumed sufficiently high for the charged particle motion to be governed by their respective diffusion and mobility coefficients, D and μ, respectively, which are also assumed spatially constant. The charged particle concentrations are assumed to obey the quasi-neutrality condition $N_{\text{e}} + N_- = N_+$ and the total radial current density is assumed to vanish (ambipolar condition). Hereafter, subscripts e, + and − will be used to identify data related to electrons, and positive and negative ions, respectively.

The continuity equations for the three kinds of charged particles can be written

$$\nabla \Gamma_+ = \nu_{\text{ion}} N_{\text{e}} ; \qquad (12.13)$$

$$\nabla \Gamma_{\text{e}} = (\nu_{\text{ion}} - \nu_{\text{att}}) N_{\text{e}} + \nu_{\text{det}} N_- ; \qquad (12.14)$$

$$\nabla \Gamma_- = \nu_{\text{att}} N_{\text{e}} - \nu_{\text{det}} N_- , \qquad (12.15)$$

where the fluxes Γ are given by the expressions

$$\Gamma_j = -D_j \nabla N_j + z_j N_j \mu_j E_{\text{s}} \qquad (12.16)$$

with j holding for e, − or +, and z_j being the number of electron charges associated with each species; z_j is negative for negative charge carriers. For simplicity we shall consider singly charged ions only, so that $z_j = \pm 1$.

An expression for the space-charge field E_{s} can be easily derived from (12.16) and the ambipolar condition, i.e.,

$$\sum_j z_j \Gamma_j = 0 \qquad (12.17)$$

yielding

$$E_{\text{s}} = \sum_j z_j D_j \nabla N_j / \sum_j z_j^2 \mu_j N_j = 0 . \qquad (12.18)$$

Substituting (12.18) in (12.16) we obtain

$$\frac{\Omega}{\mu_e}\Gamma_e = -[\alpha\mu_- + (1+\alpha)\mu_+]\epsilon_k\nabla N_e - \mu_+T_+\nabla N_+ + \mu_-T_-\nabla N_-$$
$$-\mu_+T_+\nabla N_+ + \mu_-T_-\nabla N_- \; ; \tag{12.19}$$

$$\frac{\Omega}{\mu_e}\Gamma_- = \mu_-\epsilon_k\alpha\nabla N_e - \alpha\frac{\mu_-}{\mu_e}\mu_+T_+\nabla N_+$$
$$- \left[1 + (1+\alpha)\frac{\mu_+}{\mu_e}\right]\mu_-T_-\nabla N_- \; ; \tag{12.20}$$

$$\frac{\Omega}{\mu_e}\Gamma_+ = -\mu_+\epsilon_k(1+\alpha)\nabla N_e - \left(1 + \alpha\frac{\mu_-}{\mu_e}\right)\mu_+T_+\nabla N_+$$
$$-(1+\alpha)\frac{\mu_+}{\mu_e}\mu_-T_-\nabla N_- \; , \tag{12.21}$$

where $\epsilon_k = D_e/\mu_e$, $T_+ = D_+/\mu_+$ and $T_- = D_-/\mu_-$ are the electron characteristic energy, and the temperatures of positive and negative ions, respectively, expressed in eV. We have defined the quantity

$$\Omega = \mu_e + \alpha\mu_+(1+\alpha)\mu_+ \; . \tag{12.22}$$

Inspection of (12.19)–(12.21) reveals that in all these expressions the second and the third terms on the right-hand side are in the ratio T_+/ϵ_k and T_-/ϵ_k, respectively, to the first one. We can then write the expressions for the fluxes in the form $\Gamma_j = \Gamma_j^0 + \Gamma_j^1$, where Γ_j^0 and Γ_j^1 represent the zeroth and first-order terms with respect to the small parameters T_+/ϵ_k and T_-/ϵ_k, respectively. These parameters are indeed expected to be small under the conditions prevailing in the discharge positive column, and therefore several approximations can be made. The so-called cold ion approximation consists in neglecting Γ_j^1, and another useful one consists in neglecting $\alpha\mu_-$ and $(1+\alpha)\mu_+$ as compared to μ_e in (12.22), yielding $\Omega \simeq \mu_e$. This is valid provided that $\alpha \ll \mu_e/\mu_-, \mu_e/\mu_+$. Once the solution has been obtained these conditions can easily be checked *a posteriori*.

With the above simplifications the expressions for the fluxes Γ_j become

$$\Gamma_e \simeq \Gamma_e^0 = -\left[1 + \alpha\left(1 + \frac{\mu_-}{\mu_+}\right)\right]\mu_+\epsilon_k\nabla N_e \; ; \tag{12.23}$$

$$\Gamma_- \simeq \Gamma_-^0 = \alpha\mu_-\epsilon_k\nabla N_e \; ; \tag{12.24}$$

$$\Gamma_+ \simeq \Gamma_+^0 = -(1+\alpha)\mu_+\epsilon_k\nabla N_e \; . \tag{12.25}$$

Note that, in the absence of negative ions, i.e., when $\alpha = 0$, we simply recover the usual ambipolar flow expressions for the ordinary positive column, namely, $\Gamma_e = \Gamma_+ = -D_a\nabla N_e$, where $D_a \simeq \mu_+\epsilon_k$ is the approximate expression for the ambipolar diffusion coefficient if terms of order T_+/ϵ_k are neglected.

To get the final form of the continuity equations we must now insert (12.23)–(12.25) in (12.13)–(12.15). There exist only two independent equations since each of the three equations can be obtained from a linear combination of the other two. If one chooses (12.13) and (12.15) as the independent equations, the following system is obtained for cylindrical geometry

$$\frac{1}{r}\frac{d}{dr}\left(r(1+\alpha)\frac{dN_e}{dr}\right) + \frac{\nu_{ion}}{\mu_+ \epsilon_k} N_e = 0 ; \qquad (12.26)$$

$$\frac{1}{r}\frac{d}{dr}\left(r\alpha\frac{dN_e}{dr}\right) - \frac{\nu_{att} - \nu_{det}\alpha}{\mu_- \epsilon_k} N_e = 0 . \qquad (12.27)$$

Let us define the following dimensionless variables and parameters:

$$X = \frac{r}{R}; \quad g_e = \frac{N_e}{N_e^0}; \quad g_\alpha = \frac{\alpha}{\alpha_0} ; \qquad (12.28)$$

$$\lambda = \frac{\nu_{ion} R^2}{\mu_+ \epsilon_k}; \quad P = \frac{\mu_+}{\mu_-}\frac{\nu_{att}}{\nu_{ion}}; \quad Q = \frac{\mu_+}{\mu_-}\frac{\nu_{det}}{\nu_{ion}} , \qquad (12.29)$$

where the subscript zero denotes conditions at the axis.

In terms of these quantities, (12.26) and (12.27) can be written in dimensionless form as follows:

$$\frac{1}{X}\frac{d}{dX}\left(X(1+\alpha_0 g_\alpha)\frac{dg_e}{dX}\right) + \lambda g_e = 0 ; \qquad (12.30)$$

$$\frac{1}{X}\frac{d}{dX}\left(X\alpha_0 g_\alpha \frac{dg_e}{dX}\right) - \lambda(P - Q\alpha_0 g_\alpha)g_e = 0 . \qquad (12.31)$$

These equations involve four independent parameters, namely α_0, λ, P and Q, two of which are to be treated as eigenvalues of the boundary-value problem formulated below while the other two remain as independent parameters. Physically, the choice of P and Q as independent parameters seems obvious, since they mostly depend on atomic and collision data. Note that both P and Q depend on E/N through the electron rate coefficients $k_{ion} = \nu_{ion}/N$ and $k_{att} = \nu_{att}/N$. Besides, Q also depends on the relative concentration of the species producing electron detachment upon collisions with the negative ions. The parameters α_0 and λ are taken as eigenvalues and thus will be determined by numerical calculations as functions of P and Q. The normalized radial distributions $g_e(X)$ and $g_\alpha(X)$ are simply the eigenfunctions of this problem.

The set of boundary conditions can be chosen as

$$g_e(0) = g_\alpha(0) = 1 ; \qquad (12.32)$$

$$g_e'(0) = g_\alpha'(0) = 0 ; \qquad (12.33)$$

$$g_e(1) = 0 ; \qquad (12.34)$$

$$\Gamma_-(1) = 0, \tag{12.35}$$

where g' stands for dg/dX. The conditions (12.32) and (12.33) at the axis ($X = 0$) follow from the definition of g_e and g_α and from the problem symmetry. The negative ions are completely confined to the volume and virtually none will reach the wall. In this case, (12.35) (zero ion flux at the wall) is the physically correct condition. This condition together with (12.24) implies that $\alpha = 0$ at the wall. Therefore, $\Gamma_-(1) = 0$ and $\alpha(1) = 0$ are equivalent conditions.

The boundary condition (12.34), meaning that the electron density is assumed vanishingly small at the boundary, is the one usually employed in the classical ambipolar diffusion theory.

An analytical relationship between the parameters α_0, P and Q can also be derived from (12.30),(12.31) and the boundary conditions at the axis. To see this, we first subtract (12.31) from (12.30) yielding

$$\frac{1}{X}\frac{d}{dX}\left(X\frac{dg_e}{dX}\right) + \lambda(P - Q\alpha_0 g_\alpha)g_e = 0. \tag{12.36}$$

But, (12.31) can also be written as

$$\frac{g_\alpha}{X}\frac{d}{dX}\left(X\frac{dg_e}{dX}\right) + \frac{dg_e}{dX}\frac{dg_\alpha}{dX} - \lambda\left(\frac{P}{\alpha_0} - Qg_\alpha\right)g_e = 0. \tag{12.37}$$

At $X = 0$ we have $g_e(0) = g_\alpha(0) = 1$ and $g'_e(0) = g'_\alpha(0) = 0$. We can then conclude from (12.36) and (12.37) that

$$\nabla^2_X g_e|_{X=0} = -\lambda(1 + P - Q\alpha_0) = \lambda\left(\frac{P}{\alpha_0} - Q\right). \tag{12.38}$$

Hence,

$$P + (1 + P - Q)\alpha_0 - Q\alpha_0^2 = 0. \tag{12.39}$$

This is a quadratic equation for α_0 which for $P, Q > 0$ has one positive root

$$\alpha_0 = \frac{(1 + P - Q) + \sqrt{(1 + P - Q)^2 + 4PQ}}{2Q}. \tag{12.40}$$

Thus, we have determined the relative negative ion density at the axis of the discharge tube: $N_-^0/N_e^0 \equiv \alpha_0$. Figure 12.4 shows α_0 versus P for various values of Q. As is seen from this figure, the ratio of negative ion density to electron density at the axis can reach large values if the ratio P/Q is large.

The radial density distributions N_e and N_- must be obtained from numerical solutions of (12.36) and (12.37). When

$$Q(1 + \alpha_0)\alpha_0(1 - g_\alpha) \ll 1 \tag{12.41}$$

across a major portion of the column, an approximate analytical solution is

$$\frac{N_e}{N_e^0} \equiv g_e \simeq J_0(2.405X); \tag{12.42}$$

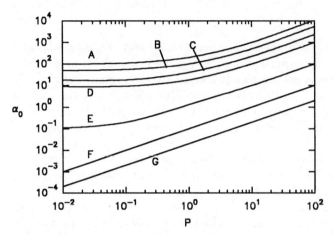

Fig. 12.4. Ratio $N_-^0/N_e^0 \equiv \alpha_0$ at the discharge tube axis as a function of P for the following values of Q: 0.01 (A), 0.02 (B), 0.05 (C); 0,1 (D); 1 (E); 10 (F) and 50 (G) [81]

$$\lambda \simeq (2.405)^2 (1 + \alpha_0) , \tag{12.43}$$

where J_0 is a Bessel function. In this case, the electron density profile in the low pressure cylindrical discharge in electronegative gases has the classical Bessel form.

The function $\lambda(P,Q)$, determined as a eigenvalue solution of (12.36), (12.37), must be equated to $\nu_{\text{ion}} R^2 / (\mu_+ \epsilon_k)$, according to the definition of λ (see (12.29)), i.e.,

$$\lambda(P,Q) = \frac{\nu_{\text{ion}} R^2}{\mu_+ \epsilon_k} . \tag{12.44}$$

When applied to a particular gas, and assuming that the relative concentration of detaching species can be independently known, (12.44) becomes a transcendental relation between E/N and NR, i.e., it gives the discharge characteristic E/N versus NR (to see this note that the right-hand side of (12.44) can be written in the form $k_{\text{ion}}(NR)^2 / (\mu_+^0 N_L \epsilon_k)$, where μ_+^0 is the reduced ion mobility and N_L is the Loschmidt number). Hence, from a physical point of view, (12.44) expresses the steady state discharge maintenance conditions. It constitutes a generalization of the well-known equation obtained in the case of an ordinary positive column, namely

$$\nu_{\text{ion}} = D_a / \Lambda^2 , \tag{12.45}$$

where $\Lambda = R/2.405$ is the characteristic diffusion length for cylindrical geometry. When the concentration of negative ions is vanishingly small ($\alpha_0 \ll 1$) $\lambda(P,Q) \simeq (2.405)^2 = 5.784$ (see (12.43)), and (12.44) becomes identical to (12.45).

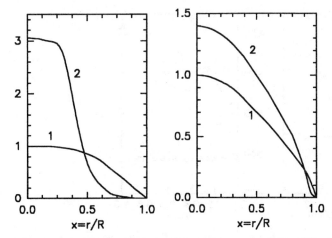

Fig. 12.5. Radial profiles N_e/N_e^0 (curves A) and N_-/N_e^0 (curves B) in an oxygen discharge ($I=50$ mA; $R=0.8$ cm). Full curves – pressure $p=0.22$ torr; dashed curves – $p=2.5$ torr [81]

The radial density distribution of negative ions depends on the range of P and Q values. For example, if $P \ll 1$ and $Q \leq 1$, $N_-(r)$ is a sharply peaked function at the axis. If $(P + Q) \gg 1$, the numerical solution of (12.30), (12.31) shows that $N_-/N_e \simeq N_-^0/N_e^0 \equiv \alpha_0$ over a major portion of the tube cross-section and decreases to zero only near the boundary.

Figure 12.5 illustrates an application of the above theory to the oxygen positive column. This figure shows the radial profiles N_e/N_e^0 and N_-/N_e^0 for the experimental conditions in [82].

From Fig. 12.5 one sees that, for a pressure of 0.22 torr, the radial profile of the negative ion density is far from a Bessel function. The axial concentration N_-^0 is about three times as large as the electron concentration.

12.2.2 Chemical and Charge Kinetics in O_2 Discharges

Oxygen dissociation in gas discharges by direct electron impact and pure vibrational mechanisms has been studied in [83,84]. However, as for N_2, the conclusions of this study are very sensitive to the choice of the kinetic model parameters, particularly the probabilities of V-V and V-T processes. In contrast with nitrogen discharges, large differences between vibrational and gas temperatures are difficult to achieve in oxygen discharges due to both high probabilities of V-T transitions in $O_2(v)-O$ collisions and small rate coefficients for electron excitation of O_2 vibrational levels. Experiment [85] confirms this fact (see Fig. 12.6). Therefore, it is clear that pure vibrational mechanisms are ineffective for O_2 dissociation in low-temperature plasma. Dissociation occurs mainly through electron impact excitation of $O_2(B^3\Sigma_u^-)$ and $O_2(c^1\Sigma_u^- + A'^3\Delta_u + A^3\Sigma_u^+)$ with formation of, respectively,

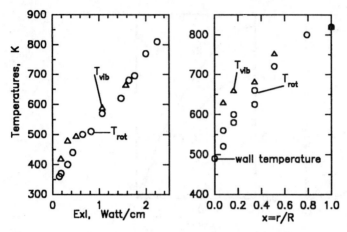

Fig. 12.6. Experimental rotational (T_{rot}) and vibrational (T_{vib}) temperatures as a function of discharge power $E \times I$ and radial profiles of these temperatures, for a DC oxygen discharge in a Pyrex tube with radius $R=0.8$ cm [85]

$O(^3P)+O(^1D)$ and $O(^3P)+O(^3P)$ atoms as dissociation products

The modelling of kinetic processes in low-temperature oxygen plasma was also carried out in [86–97]. Detailed data for rate coefficients of elementary processes in such plasmas have been given in [87,90,92,93]. A self-consistent kinetic model determining the EEDF and the ion and neutral compositions in a DC O_2 discharge was developed in [88,89], for the following discharge conditions: $NR = 10^{15}$–10^{18} cm^{-2}, $I/R = 0.1$–100 mA/cm. A more detailed kinetic model was developed in [87]. It includes 142 equations and describes the electron kinetics, power balance, and chemical and ionic composition of oxygen plasmas, for gas densities $N = (3\text{--}10) \times 10^{16}$ cm^{-3} and reduced electric fields $E/N = (4\text{--}10) \times 10^{-16}$ volt cm^2.

Kinetic models of DC O_2 discharges suitable to calculate the concentrations of $O_2(X^3\Sigma)$, $O_2(a^1\Delta_g)$, $O_2(b^1\Sigma_g^+)$, O, O$^-$, were developed in [91-93]. The model predictions were shown to be in good agreement with experiment. These results are presented in Fig. 12.7.

Modelling and experimental investigations of DC O_2 discharge properties were also carried out in [94,95]. In [95], a semi-empirical model describing electronic-vibrational relaxation of $O_2(b^1\Sigma_g^+, v = 1, 2)$ metastables in collisions with O, O_2, and O_3 was proposed.

Concerning microwave oxygen discharges, a simple model describing the power transfer from the microwave field to the electrons, and thence to neutral gas, was presented in [96]. Modelling of post-discharge conditions has also been carried out [97]. Finally, a detailed model for surface wave discharges in flowing oxygen has been developed in [98–101].

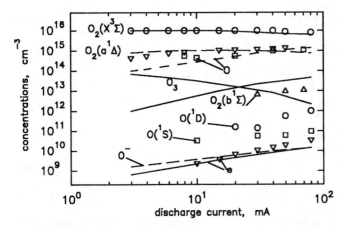

Fig. 12.7. Concentrations of active species in the positive column of a DC oxygen discharge versus discharge current, for $R = 0.8$ cm; $p = 0.38$ torr. Curves – calculations, points – experiment [93]

Following [92,93,95–101], the principal chemical and charge kinetic processes determining the concentrations of electrons and heavy particles $O_2(X^3\Sigma)$, $O_2(a^1\Delta_g)$, $O_2(b^1\Sigma)$, O_3, O, O^- in oxygen discharges, for pressures in the range 0.1–10 torr, are the following:

$$O_2(X^3\Sigma, v) + e \longleftrightarrow O_2(X^3\Sigma, v') + e; \tag{12.46}$$

$$O_2(X^3\Sigma) + e \longleftrightarrow O_2(a^1\Delta_g) + e; \tag{12.47}$$

$$O_2(X^3\Sigma) + e \longleftrightarrow O_2(b^1\Sigma) + e; \tag{12.48}$$

$$O_2(a^1\Delta_g) + e \longleftrightarrow O_2(b^1\Sigma) + e; \tag{12.49}$$

$$O_2(X^3\Sigma) + e \longrightarrow O_2^+ + 2e; \tag{12.50}$$

$$O_2^+ + e \longrightarrow 2O; \tag{12.51}$$

$$O_2(X^3\Sigma) + e \longleftrightarrow O^- + O; \tag{12.52}$$

$$O_2(a^1\Delta) + e \longrightarrow O^- + O; \tag{12.53}$$

$$O_2(a^1\Delta) + O^- \longrightarrow O_3 + e; \tag{12.54}$$

$$O_2(X^3\Sigma) + e \longrightarrow 2O + e; \tag{12.55}$$

$$O_2(a^1\Delta_g) + e \longrightarrow 2O + e; \tag{12.56}$$

$$O_3 + e \longrightarrow O + O_2 + e; \tag{12.57}$$

$$O_2 + O + M \longrightarrow O_3 + M; \quad M = O_2, \ O; \tag{12.58}$$

$$O_3 + O \longrightarrow 2O_2(X^3\Sigma); \tag{12.59}$$

$$O_3 + O \longrightarrow O_2(X^3\Sigma) + O_2(a^1\Delta_g) \;; \tag{12.60}$$

$$O_3 + O_2(a^1\Delta_g) \longrightarrow 2O_2(X^3\Sigma) + O \;; \tag{12.61}$$

$$O_3 + O_2(b^1\Sigma_g) \longrightarrow 2O_2(X^3\Sigma) + O \;; \tag{12.62}$$

$$2O_2(a^1\Delta_g) + O_2 \longrightarrow 2O_3 \;; \tag{12.63}$$

$$O + \text{wall} \longrightarrow \frac{1}{2}O_2(X^3\Sigma) \;; \tag{12.64}$$

$$O_2(a^1\Delta_g) + \text{wall} \longrightarrow O_2(X^3\Sigma) \;; \tag{12.65}$$

$$O_2(b^1\Sigma_g) + \text{wall} \longrightarrow O_2(X^3\Sigma) \;. \tag{12.66}$$

The rate coefficients for most of the processes (12.46)–(12.65) were given in Chaps. 8–11. Note that the heterogeneous processes (12.64)–(12.66) in discharge tubes with radius $R \sim 1$ cm, at pressures $p \leq 0.1$–10 torr, can play an important role in the kinetics of $O_2(a^1\Delta_g)$ and $O_2(b^1\Sigma_g)$ metastables and O atoms.

Collisions between electrons and O atoms can significantly affect the EEDF in oxygen discharges, since the degree of dissociation is usually high. For this reason, besides processes (12.46)–(12.66), self-consistent kinetic models should also include electron-atom collisions. Ionization of O atoms and excitation of the $O(^1D)$, $O(^1S)$, $O(^3S)$ states by electron impact were taken into account in [93], in addition to the processes (12.46)–(12.66).

Calculated and measured active species concentrations are shown in Fig. 12.7 for a typical low pressure oxygen plasma. It can be concluded that the theoretical modelling yields results that agree well with experiment.

Simplified kinetic schemes have been used to explain planar inductive oxygen discharges [102,103], while PIC/MCC has been used to describe capacitively coupled oxygen discharges [104].

References

1. Petrash G.G. (1971) Adv. Phys. Sci. **105**, 645 (*in Russian*)
2. Ischenko V.N., Lisitsin V.N., Razhev A.M. and Starinsky V.N. (1987) in *Gas Lasers*, eds. R.I. Solouhin and V.P. Chebotaev, Nauka, Novosibirsk, p. 224 (*in Russian*)
3. Lebrun J.P., Michel H. and Gantonis M. (1972) Mem. Sci. Rev. Metall. **69**, 727
4. Ricard A., Oseguera J., Michel H. and Gantois M. (1989) Plasma Surface Engineering **1**, 83
5. Jacquot P., Poirson J.M., Michel H., Gantonis M. and Pegre J.P. (1984) Traitement Thermique **184**, 25
6. Hudis M. (1979) *Plasma Treatment of Solid Materials in Techniques and Applications of Plasma Chemistry*, John Wiley and Sons, New York
7. Walkup R.E., Saenger K.L. and Selwyn G.S. (1986) J. Chem. Phys. **84**, 2668

8. Hopwood J., Dahimene M., Reingard D.K. and Asmussen J. (1988) J. Vac. Sci. Technol. **86**, 268
9. Capitelli M. and Dilonardo M. (1977) Chem. Phys. **24**, 417
10. Capitelli M. and Dilonardo M. (1978) Rev. Phys. Appl. **13**, 115
11. Capitelli M. and Molinari E. (1980) in *Topics in Current Chemistry*, ed. F.L. Boschke, Springer, Berlin Heidelberg, Vol. 90, p. 59
12. Capitelli M., Dilonardo M. and Gorse C. (1981) Chem. Phys. **56**, 29
13. Cacciatore M., Capitelli M. and Gorse C. (1982) Chem. Phys. **60**, 141
14. Rohlena K. and Masek K. (1985) Acta Phys. Slovaka **35**, 141
15. Cacciatore M., Capitelli M., De Benedictis S., Dilonardo M. and Gorse C. (1986) in *Non-equilibrium Vibrational Kinetics*, ed. M. Capitelli, Springer, Berlin Heidelberg, p. 5
16. Loureiro J. and Ferreira C.M. (1986) J. Phys. D: Appl. Phys. **19**, 17
17. Loureiro J. and Ferreira C.M. (1989) J. Phys. D: Appl. Phys. **22**, 67
18. Porshev P.I. (1989) *PhD Thesis*, Inst. TMO, Minsk (*in Russian*)
19. Loureiro J., Ferreira C.M., Capitelli M., Gorse C. and Cacciatore M. (1990) J. Phys. D: Appl. Phys. **23**, 1371
20. Loureiro J. (1991) Chem. Phys. **157**, 157
21. Armenise I., Capitelli M., Garcia E., Gorse C., Lagana A. and Longo S. (1992) Chem. Phys. Lett. **200**, 597
22. Polak L.S., Sergeyev P.A. and Slovetsky D.I. (1977) Thermophys. High Temp. **15**, 15 (*in Russian*)
23. Slovetsky D.I. (1980) *Mechanisms of Chemical Reactions in Non-equilibrium Plasma*, Nauka, Moscow (*in Russian*)
24. Brunet H., Vincent P. and Rocca-Serra J. (1983) J. Appl. Phys. **54**, 4951
25. Brunet H. and Rocca-Serra J. (1985) J. Appl. Phys. **57**, 1574
26. Golubovsky Yu.B. and Telezhko B.M. (1984) Thermophys. High Temp. **22**, 428 (*in Russian*)
27. Baidze K.B., Vetsko V.M., Lopantseva J.B. et al., (1985) Plasma Phys. **11**, 352 (*in Russian*)
28. Akishev Yu.S., Baidze K.V., Vetsko V.M. et al., (1985) Plasma Phys. **11**, 999 (*in Russian*)
29. Pivovar V.A. and Sidorova T.D. (1985) J. Techn. Phys. **55**, 519 (*in Russian*)
30. Paniccia F., Gorse C., Cacciatore M. and Capitelli M. (1987) J. Appl. Phys. **61**, 3123
31. Gorse C. and Capitelli M. (1987) J. Appl. Phys. **62**, 4072
32. Mnatsakanyan A.H. and Naidis G.V. (1987) in *Plasma Chemistry*, ed. B.M. Smirnov, Energoatomizdat, Moscow Vol. 14, p. 227 (*in Russian*)
33. Berdyshev A.V., Kochetov I.V. and Napartovich A.P. (1988) Plasma Phys. **14**, 741 (*in Russian*)
34. Bol'shakova L.G., Golubovsky Yu.B., Telezhko V.M. and Stoyanov D.G. (1990) Sov. Phys. Tech. Phys. **35**, 665
35. Nahorny J., Pagnon D., Touzeau M., Vialle M., Gordiets B. and Ferreira C.M. (1995) J. Phys. D: Appl. Phys. **28**, 738
36. Gordiets B., Ferreira C.M., Guerra V., Loureiro J., Nahorny J., Pagnon D., Touzeau M. and Vialle M. (1995) IEEE Trans. Plasma Sci. **23**, 750
37. Gordiets B., Ferreira C.M., Nahorny J., Pagnon D., Touzeau M. and Vialle M. (1996) J. Phys. D: Appl. Phys. **29**, 1021
38. Cacciatore M., Capitelli M., Gorse C., Massabiaeux B. and Ricard A. (1982) Lettere Nuovo Cimento **34**, 417
39. Massabieaux B., Plain A., Ricard A., Capitelli M. and Gorse C. (1983) J. Phys. B: At. Mol. Phys. **16**, 1863

40. Cernogora G., Ferreira C.M., Hochard L., Touzeau M. and Loureiro J. (1984) J. Phys. B: At. Mol. Phys. **17**, 4429
41. Ferreira C.M., Touzeau M., Hochard L. and Cernogora G. (1984) J. Phys. B: At. Mol. Phys. **17**, 4439
42. Plain A., Gorse C., Cacciatore M., Capitelli M., Massabieaux B. and Ricard A. (1985) J. Phys. B: At. Mol. Phys. **18**, 843
43. Capitelli M., Gorse C. and Ricard A. (1986) in *Non-equilibrium Vibrational Kinetics*, ed. M. Capitelli, Springer, Berlin Heidelberg, p. 315
44. Matveev A.A. and Silakov V.P. (1999) Plasma Sources Sci. Technology **8**, 162
45. Gorse C. and Capitelli M. (1996) in *Molecular Physics and Hypersonic Flows*, ed. M. Capitelli, Kluwer, Dordrecht, NATO ASI Series, Vol. 482, p. 437
46. Magne L., Cernogora G., Loureiro J. and Ferreira C.M. (1991) J. Phys. D: Appl. Phys. **24**, 1758
47. Golubovsky Yu.B. and Telezhko V.M. (1993) J. Appl. Spectroscopy **39**, 1429 (*in Russian*)
48. Blois D., Supiot P., Baj M., Chaput A., Foissac C., Dessaux O., and Goudman P. (1998) J. Phys. D: Appl. Phys. **31**, 2521
49. Napartovich A.N., Naumov B.G. and Shashkov V.M. (1977) Rep. Acad. Sci. USSR **232**, 570 (*in Russian*)
50. Akishev Yu.S., Demianov A.B., Kochetov I.B. at al. (1982) Thermophys. High Temp. **20**, 818 (*in Russian*)
51. Valyansky S.I., Vereschagin K.A., Vernke B. et al. (1984) Quant. Electron. **11**, 1833 (*in Russian*)
52. Deviatov A.A., Dolenko S.A., Rahimov A.T. at al. (1986) J. Exp. Theor. Phys. **90**, 429 (*in Russian*)
53. Baeva M., Luo X., Pfelzer B., and Uhlembusch J. (1999) Plasma Sources Sci. Technol. **8**, 404; Baeva M., Luo X., Pfelzer B., Schafer J. H. Uhlembusch J., and Zhang Z. (1999) Plasma Sources Sci. Technol. **8**, 142
54. Morrison M.A. and Lane N.F. (1977) Phys. Rev. A **16**, 975
55. Berdyshev A.V., Viharev A.O., Gitlin M.S. et al. (1988) Thermophys. High Temp. **26**, 661 (*in Russian*)
56. Longo S. and Capitelli M. (1994) Phys. Rev. E **49**, 2302
57. Longo S., Hassouni K., Iasillo D., and Capitelli M. (1997) J. Physique III, **7**, 707; Longo S., Hassouni K. and Capitelli M. (1997) J. Physique IV **7**, C4–271; Longo S., Capitelli M. and Hassouni K. (1998) J. Thermophys. Heat Transfer **12**, 473
58. De Benedictis S. and Dilecce G. (1997) J. Chem. Phys. **107**, 6129
59. De Benedictis S., Dilecce G. and Simek M. (1998) J. Phys. D: Appl. Phys. **31**, 1197; De Benedictis S., Dilecce G., Simek M., and Vigliotti M. (1998) Plasma Sources Sci. Technol. **7**, 557; Dilecce G. and De Benedictis S. (1999) Plasma Sources Sci. Technol. **8**, 266; Dilecce G., Capitelli M. and De Benedictis S. (1991) J. Appl. Phys. **69**, 121
60. Bibinov N. K., Kokh D. B., Kolokolov N. B., Kostenko V. A., Meyer D., Vinogradov I. P., and Wiesemann K. (1998) Plasma Sources Sci. Technol. **7**, 298; Vinogradov I. P. (1999) Plasma Sources Sci. Technol. **8**, 295
61. Fischer R. and Dose V. (1999) Vinogradov I. P., and Wiesemann K. (1998) Plasma Phys. Control. Fusion **41**, 1109
62. Nighan W.H. and Wiegand J.W. (1974) Phys. Rev. A **A10**, 922; Nighan W.H. (1977) Phys. Rev. A **A15**, 1701
63. Velihov E.P., Pismenniy V.D. and Rahimov A.T. (1977) Adv. Phys. Sci. **122**, 419 (*in Russian*)
64. Napartovich A.P. and Starostin A.N. (1979) in *Plasma Chemistry*, ed. B.M. Smirnov, Atomizdat, Moscow, Vol. 6, p. 153 (*in Russian*)

65. Raizer Yu.P. (1991) *Physics of Gas Discharges*, Springer, Berlin Heidelberg
66. Eletsky A.V. and Rahimov A.T. (1977) in *Plasma Chemistry* ed. B.M. Smirnov, Atomizdat, Moscow, Vol. 4, p. 123 (*in Russian*)
67. Velihov E.P., Kovalev A.S. and Rahimov A.T. (1987) *Physical Phenomena in Gas Discharge Plasma*, Nauka, Moscow (*in Russian*)
68. Menahin L.P., Eroschenko E.K. and Ulianov K.N. (1975) J. Tech. Phys. **45**, 1346 (*in Russian*)
69. Menahin L.P., Eroschenko E.K., Sibiryak I.O. and Ulianov K.N. (1976) J. Tech. Phys. **46**, 2429 (*in Russian*)
70. Osipov A.P. and Rahimov A.T. (1977) Plasma Phys. **3**, 644 (*in Russian*)
71. Vysikailo F.I., Napartovich A.P. and Son E.E. (1978) Plasma Phys. **4**, 1383 (*in Russian*)
72. Baidze K.B., Vetsko V.M., Zhdanok S.A., Napartovich A.P. and Starostin A.N. (1979) Rep. Acad.Sci. USSR **249**, 832 (*in Russian*)
73. Zhdanok S.I., Napartovich A.P. and Starostin A.N. (1979) Lett. J. Tech. Phys. **5**, 155 (*in Russian*)
74. Koroleva I.L., Napartovich A.P. and Starostin A.N. (1982) Plasma Phys. **8**, 561 (*in Russian*)
75. Koroleva I.L., Napartovich A.P. and Starostin A.N. (1983) Thermophys. High Temp. **21**, 852 (*in Russian*)
76. Akishev J.S., Koroleva I.L., Napartovich A.P., Ponomarenko V.V. and Starostin A.N. (1986) Thermophys. High Temp. **24**, 26 (*in Russian*)
77. Berdyshev A.B., Golovin A.C., Gurashvili A.V. et al. (1989) Plasma Phys. **15**, 335 (*in Russian*)
78. Holm R. (1932) Z. Phys. **75**, 171
79. Seeliger R. (1949) Ann. Phys. Lpz. **6**, 93.
80. Edgley P.D. and von Engel A. (1980) Proc. R. Soc. **A370**, 375.
81. Ferreira C.M., Gousset G. and Touzeau M. (1988) J. Phys. D: Appl. Phys. **21**, 1403
82. Gousset G., Panafieu P., Touzeau M. and Vialle M. (1987) Plasma Chem. Plasma Process. **6**, 409
83. Capitelli M. and Dilonardo M. (1978) Chem. Phys. **30**, 95
84. Capitelli M., Dilonardo M. and Gorse C. (1980) Beitr. Plasmaphys. **20**, 83
85. Lefebre M., Pealat M., Gousset G., Touzeau M. and Vialle M. (1990) Proceed. ESCAMPIG 90, Orléans, France, **14E**, 246
86. Masek K., Laska L. and Ruzicka T. (1978) Czech. J. Phys. **B28**, 1321
87. Dettmer J.W. (1978) *PhD Thesis*, Air Force Institute of Technology, Wright Patterson Air Force Base, Dayton, Ohio
88. Laska L., Masek K. and Ruzicka T. (1979) Czech. J. Phys. **B29**, 498
89. Masek K. and Laska L. (1980) Czech. J. Phys. **B30**, 805
90. Eliasson B. (1983) *Electrical Discharge in Oxygen*, Brown Boveri Report **KRL83-40C, CH-5405**, Baden, Switzerland
91. Gousset G., Touzeau M., Vialle M. and Ferreira C.M. (1989) Plasma Chem. Plasma Process. **9**, 189
92. Gousset G., Ferreira C.M., Pinheiro M., Sa P.A., Touzeau M., Vialle M. and Lourciro J. (1001) J. Phys. D: Appl. Phys **24**, 290
93. Touzeau M., Gousset G., Jolly J., Pagnon D., Vialle M., Ferreira C.M., Loureiro J., Pinheiro M. and Sa P.A. (1991) in *Non-equilibrium Processes in Partially Ionized Gases*, eds. M. Capitelli and I. N. Bardsley, Plenum Press, New York
94. Yukimi Ichikawa, Richard L.C. Wu and Teruo Kaneda (1990) J. Appl. Phys. **67**, 108
95. Yankovsky V.A. (1991) Chemical Physics **10**, 291 (*in Russian*)

96. Granier A., Pasquiers S., Boisse-Laporte C., Darchicourt B., Leprince P. and Marec J. (1989) J. Phys. D: Appl. Phys. **22**, 1487
97. Ta-Chin Wei and Phillips J. (1993) J. Appl. Phys. **74**, 825
98. Ferreira C.M., Alves L.L., Pinheiro M. and Sa A.B. (1991) IEEE Trans. Plasma Sci. **19**, 229
99. Pinheiro M.J., Ferreira C.M. and Gousset G. (1995) in *Molecular Physics and Hypersonic Flows*, ed. M.Capitelli, Kluwer, Dordrecht, NATO ASI Series, Vol. 482, p. 485
100. Pinheiro M. J., Gousset G. , Granier A. and Ferreira C.M. (1998) Plasma Sources Sci. Technol. **7**, 524
101. Pinheiro M., Gordiets B.F. and Ferreira C.M. (1999) Plasma Sources Sci. Technol. **8**, 31
102. Gudmundsson J.T. and Lieberman M.A. (1998 Plasma Sources Sci. Technol. **7**, 1
103. Gudmundsson J. T., Kimura T. and Lieberman M.A. (1999) Plasma Sources Sci. Technol. **8**, 22
104. Wang Z., Lichtenberg A.J. and Cohen R.H. (1999) Plasma Sources Sci. Technol. **8**, 151

13. Discharges in N_2-O_2 Mixtures

Interest in discharges in N_2-O_2 mixtures is furthered by research on air breakdown, optical characteristics of the atmosphere under perturbed non-equilibrium conditions, chemical processes for surface treatments, synthesis of nitrogen oxides, nitrogen isotope separation, cleaning of polluted atmospheric gases, and so on.

Discharges in N_2-O_2 mixtures can differ considerably from discharges in pure N_2 or O_2 in a number of characteristics (principally, in chemical and ionic composition). Theoretical modelling of kinetic processes in N_2-O_2 discharges has been developed in several works. For example, the EEDF, the rate coefficients for excitation, dissociation and ionization, and the electron energy balance in DC and high frequency electric fields were calculated in [1–5]. In [5], useful analytical approximations have been obtained for different rate coefficients as well as for plasma energy and transport characteristics, as a function of the reduced electric field E/N, N_2 vibrational temperature and the O_2 percentage in N_2-O_2 mixtures (see also Chaps. 4 and 8). The EEDF, energy balance and relative electron energy loss rates for different processes of interaction of secondary electrons with the main gas species (elastic collisions, excitation of rotational, vibrational and electronic states, dissociation, ionization, dissociative ionization, electron attachment) have been calculated in [6–12] for the interaction of high energy ($\epsilon \geq 1$ keV) electron beams with atmospheric gases. For this case, analytical approximations for the energy loss rates and energetic costs of various processes were obtained in [10–12] (see also Chap. 8).

13.1 Modelling of Low Pressure N_2-O_2 Discharges

The non-equilibrium kinetics of low pressure plasmas in N_2-O_2 mixtures is an important field of research, in the search for a better understanding of the processes occurring in the atmosphere and ionosphere and in a variety of modern plasma technologies. In particular, knowledge of the volume and the surface kinetics of active species such as N and O atoms, $N(^2D)$, $O_2(a^1\Delta)$, and $N_2(A^3\Sigma)$ metastables and NO molecules is important for understanding the workings of plasma reactors used for chemical synthesis and surface treatments of various materials. The reentry heating of the Space Shuttle (surface recombination of O and N atoms) and emissions near the surface of space vehicles in stationary orbits (surface production of gas phase NO and NO_2 excited molecules) are other examples of situations where further attention is needed.

Modelling of kinetic processes in low pressure N_2-O_2 discharges has been carried out in [13–29]. Special attention was given to the study of reaction

$$N_2(v \geq 12) + O \rightarrow NO + N \tag{13.1}$$

in low-temperature plasmas with high non-equilibrium N_2 vibrational energy as a promising method for plasmachemical synthesis of nitrogen oxides. In [13,15,16,21,22] this modelling was carried out for the specific conditions of the Earth's ionosphere (altitudes 90–200 km). Processes in laboratory plasmas were analyzed in [14,17–20,23–30] with the purpose of interpreting experiments, investigating the synthesis of nitrogen oxides, the production of N and O atoms, and the interaction of these atoms with surfaces.

Post-discharges in N_2-O_2 mixtures were investigated in [23,31]. For example, in experiment [31], discharge optical emissions and concentrations of O and N atoms and NO molecules were measured in the flowing afterglow of a microwave N_2-O_2 discharge with a total power of 120 Watt, at 3 Torr pressure, for different O_2 percentages. It was found that, with the addition of O_2 in the discharge, the smooth decrease in N density in the post-discharge turned into a sharp decrease for some critical O_2 percentage (8–10%). The variation in NO density is quite the opposite. The kinetics of such post-discharges has been developed in [23], and used for interpretation of this experiment. As $[O_2]$ grows, the decrease in N atom density and the sharp increase in NO density beyond the critical point are due to the decrease in the production rate of N atoms by electron collisions, and to the increased role of the reactions

$$N + O_2 \rightarrow NO + O \tag{13.2}$$

in the discharge and

$$N + NO \rightarrow N_2(\bar{v} = 3.4) + O \tag{13.3}$$

in the post-discharge. The situation is very similar to that in the well-known titration method for the measurement of absolute N atom densities. It was also found that the direct reaction (13.1) and the reverse one (13.3)) constitute an effective V-T multi-quantum process for N_2 vibrational relaxation.

The most detailed self-consistent model has been developed in [25,27,29,30]. This model has been successfully used for the interpretation of a detailed experimental study [24,25,27] on DC discharges in N_2-O_2 mixtures with different O_2 (0–100%) percentages, for currents (15–80 mA), in a tube of 0.8 cm inner radius, at 2 Torr pressure. In the next subsection this model will be described briefly.

13.1 Modelling of Low Pressure N_2-O_2 Discharges

13.1.1 General Description of the Model

This model is one-dimensional and self-consistent, and applies to a DC flowing glow discharge and post-discharge. Its input parameters are the following: pressure p (Torr), radius R (cm) and length L (cm) of the discharge tube, electric current I (mA), gas flow rate Q (sccm) and initial gas temperature and composition (*i.e.*, the relative oxygen concentration $X(\%)$ in the binary mixture $N_2 + X(\%)O_2$ at the gas inlet into the discharge zone). The model predicts the following properties of the bulk plasma, as a function of the axial coordinate z: concentrations of ground state N_2, O_2, NO, N_2O, NO_2, NO_3, N_2O_5, O_3 molecules and N, O atoms, populations of the electronically excited states $N_2(A^3\Sigma_u^+, B^3\Pi_g, a'^1\Sigma_u^-, a^1\Pi_g, C^3\Pi_u, a''^1\Sigma_g^+)$, $O_2(a^1\Delta_g, b^1\Sigma_g^+)$, $N(^2D, ^2P)$, $O(^1D, ^1S)$, concentrations of N_2^+, $N_2^+(B)$, N_4^+, O^+, O_2^+, NO^+, O^- ions, populations of vibrational levels of N_2 molecules in the ground electronic state $X^1\Sigma_g^+$, density N_e of plasma electrons, their average kinetic $\frac{3}{2}kT_e$ and characteristic energy ϵ_k (that is, the ratio of the free diffusion coefficient to mobility), electron drift velocity v_d, discharge maintaining field E, radially averaged gas temperature T, and wall temperature T_w.

To determine the above plasma properties, a solution was found of a coupled system of equations describing the kinetics of free electrons, the vibrational kinetics of N_2 molecules, the kinetics of electronic states of molecules and atoms, the chemical kinetics of heavy neutrals and charged particles, the energy balance of the gas, and the charged particle balance determining the electric field needed to maintain the discharge. Empirical formulas have been derived and used for the calculation of the wall temperature T_w. Further, a set of equations for the surface kinetics of N and O atoms has been coupled to the gas phase chemical kinetic equations, which renders this model fully self-consistent.

13.1.2 Kinetics of Free Electrons

The electron kinetics has been investigated using the quasi-stationary electron Boltzmann equation. The degree of ionization was assumed small, and electron–electron collisions were therefore neglected. The collision integral, describing the collisions of electrons with heavy particles in the Boltzmann equation, has also been simplified. Due to the relatively small concentrations of the species involved, one can neglect electron collisions with excited electronic states of N_2, O_2, N, O, vibrationally excited $O_2(v)$ and other neutral and ionic species. The only exception concerns collisions with $N_2(X, v > 0)$ molecules, whose relative concentration can be large enough to affect the EEDF. The EEDF is a solution of the Boltzmann equation, and the electron transport parameters and rate coefficients, as calculated from the obtained EEDF, are functions of the relative composition of the main gas components N_2 and O_2, reduced electric field E/N (N is the total density of N_2 and O_2 molecules) and relative population of N_2 vibrational levels. The electron cross-sections used in the collision terms of the Boltzmann equation are the same as those used in [32] for N_2 and in [33] for O_2.

13.1.3 Vibrational Kinetics

The populations of $N_2(X,v)$ vibrational levels have been determined from a coupled system of master equations which take into account: a) vibrational excitation and de-excitation by electron collisions (e-V processes); b) vibration-vibration (V-V) energy exchanges; c) vibrational exchanges between N_2 and O_2, NO molecules (V-V' processes); d) vibration-translation (V-T) energy exchanges in collisions $N_2(v)$ with N_2, O_2, NO, N, O; e) chemical reactions involving vibrationally excited N_2 molecules; f) single-quantum energy exchanges in collisions of $N_2(v)$ with the wall. A list of these processes and the corresponding rate coefficients are given in [25]. Multi-quantum V-T processes in N_2–N collisions play an important role in the populations of the upper vibrational levels. The simple analytical approximation (7.11), (7.12) has been used to describe the total rate coefficients k_v for transitions from a vibrational level v to all lower levels $v - \Delta v$, $1 \leq \Delta v \leq v$ and the rate coefficients $k_{v,v-\Delta v}$ for V-T transitions $v \to v - \Delta v$ due to $N_2(v)$+N collisions. Note that N_2–N collisions are important only for the N_2 vibrational kinetics in pure nitrogen. In N_2–O_2 mixtures, where O atoms are present in large amounts due to O_2 dissociation, the main reaction affecting the N_2 upper vibrational level populations is (13.1). Reaction (13.1) is also very important for the production of NO. An expression similar to (10.17) has been used for the rate coefficient of process (13.1). Note that, due to process (13.1), the N_2 vibrational distribution function is strongly depleted for $v \geq 12$ levels in N_2–O_2 discharges with $\geq 1\%$ of $[O_2]$ (see Fig. 13.3).

13.1.4 Kinetics of Electronic States and Chemical Kinetics

A large number of chemical reactions involving atoms, molecules and ions, and processes for excitation and deactivation of N_2 and O_2 electronic states are used in the model. A list of these processes with the corresponding rate coefficients is given in [25,29,30]. Note that excitation of the electronic states $N_2(A, B, a'')$ upon collisions between two vibrationally excited $N_2(X)$ molecules has been ignored in [29,30]. This differs from [25] but such processes are not very important under discharge conditions. It was found that neglecting such processes yields more accurate calculated populations of these states in the post-discharge.

13.1.5 Interaction with the Wall

The interaction with the tube wall has a significant effect for some species in low pressure discharges. This interaction is approximately described using (11.13) for the radially averaged rate coefficients for losses on the wall,

ν_w. According to experimental evidence (see also Chap. 11), the wall loss probabilities for the species

$$N_2(A); \quad N(^2D); \quad N(^2P); \quad O(^1D); \quad O(^1S) \qquad (13.4)$$

are quite high (large γ values). For pressures ~ 1 Torr and tube radius $R \sim 1$ cm, the wall loss rates of the species $N_2(A)$, $N(^2D)$, $N(^2P)$, $O(^1D)$ and $O(^1S)$ are therefore determined by the corresponding rates of diffusion to the wall (first term on the right-hand side of (11.13)). The diffusion coefficients have been taken from the literature. For electronically excited particles, they have been assumed the same as for the corresponding ground state species. In the N_2–O_2 mixture, the diffusion coefficients were calculated using the well-known Blanc's law. In order to describe transport at low pressures more accurately, when the ion mean free path is not small compared to the tube radius, an effective ambipolar diffusion coefficient has been used for charged particles by the same procedure as in [34,35].

For the probabilities γ of single-quantum deactivation of $N_2(v)$ and deactivation of $N_2(a')$ metastables at the wall, the values 4.5×10^{-4} [36] and 10^{-3} [37], respectively, have been used.

Wall recombination of N and O atoms is the main loss channel of these species in low pressure discharges. The probabilities γ_N and γ_O are small, this being the main reason why the corresponding loss rates are controlled by these probabilities rather than by the diffusion time. The values γ_N and γ_O are modified in gas mixtures. For this reason, a kinetic model of surface processes involving N and O atoms has been developed in [27,29,30] to calculate γ_N, γ_O, γ_{NO} and γ_{NO_2} in N_2–N–O_2–O mixtures. This model was presented in Chap. 11.

13.1.6 Gas and Wall Temperatures

The radially averaged gas temperature has been calculated from the gas thermal balance equation, self-consistently with the vibrational and chemical kinetic equations. The most important sources of gas heating are V-T relaxation of N_2 and O_2 molecules, dissociation of N_2 and O_2 with production of "hot" N and O atoms, and deactivation of N_2 electronic states. Heating due to elastic electron collisions with heavy particles, excitation of N_2 and O_2 rotational levels (assuming fast rotational-translational relaxation) and exothermic chemical reactions have also been taken into account. Thermal conduction to the tube wall is the main gas cooling mechanism. In a one-dimensional model, the power density lost by this process can be taken into account in the thermal balance equation by a term of the form $8\lambda(T-T_w)/R^2$, where T is the radially averaged gas temperature, T_w is the wall temperature and λ is the thermal conductivity of the gas. This type of law results from assuming a parabolic gas temperature profile across the tube, the temperature on the axis being given by $T_a = 2T - T_w$. The wall temperature T_w is an important parameter as it influences the gas temperature and the rate coefficients of surface processes. A fitting empirical formula for T_w was obtained in [37] for a stationary N_2 DC discharge in a Pyrex tube with radius $R \sim 1$ cm, for different pressures and electric currents, in the absence of forced external cooling of the tube walls. This formula reads:

$$T_w(K) = T_0 + C\left(\frac{EI}{R}\right)^\beta, \qquad (13.5)$$

where T_0 is the ambient temperature, EI (watt/cm) is the discharge power per unit length (E (volt/cm) is the electric field and I (A) is the discharge electric current), R is the tube radius in cm, and C and β are fitting parameters. The values $T_0 = 296$ K, $C = 82$, $\beta = 0.87$ (in the corresponding appropriate units, taking into account the other units specified above) have been obtained by fitting (13.5) to experiment. These values have been used in the model for calculations of T_w. Note that (13.5) and the above fitting parameters can apply to a variety of gases and mixtures, including N_2–O_2 mixtures. In fact, from energy conservation, it is clear that the wall temperature must be nearly the same for identical discharge tubes and deposited powers per unit discharge tube length $(EI)/(2\pi R)$, irrespective of the nature of the gases, provided that a similar fraction (for example, a negligible one) of the total discharge power is dissipated via radiation that is not absorbed by the tube walls. This is the case with nitrogen, hydrogen, oxygen, and their mixtures, for reduced electric fields typically $\leq 2 \times 10^{-15}$ Volt. cm^2. As confirmation, using (13.5) to fit the experimental values of T_w in pure oxygen DC discharges in a Pyrex tube with $R = 0.8$ cm [38] yields the fitting values $T_0 = 296$ K, $C = 84$, $\beta = 0.75$.

The gas dynamic treatment used in the one-dimensional flowing discharge model was very simple. The pressure was assumed constant along the tube axis z. This is a good approximation under the present conditions of relatively small gas velocity, since viscosity is therefore small and the relaxation times for the gas temperature and the species concentrations are large as compared to the characteristic time for establishing the pressure along the tube, $\tau_p \simeq L/c_s$ (L denoting the length of the discharge tube and c_s the speed of sound). The average gas velocity $v(z)$ can easily be determined from conservation of the mass flow rate along the tube.

To close the model, the equation relating the current density to the electric field E must also be taken into consideration. This equation reads

$$J = ev_d(E)N_e(E), \qquad (13.6)$$

where J and v_d denote the current density and the electron drift velocity. The latter was calculated from the Boltzmann equation. The electron density N_e was determined as the difference between the densities of positive and negative ions (that is, from the condition of plasma quasi-neutrality). Note that (13.6) is strongly nonlinear in E, due to the strong nonlinear dependence on E of the ionization rate coefficients determining the charged particle densities.

It is interesting to compare model calculations and measurements of different plasma properties for N_2-O_2 mixtures. Detailed measurements have been carried out [24–26] for a low pressure glow N_2-O_2 discharge, with O_2 percentages $X_{O_2} = 0-100\%$, of gas temperature T, N_2 vibrational temperature, electron density N_e, electric field E, relative populations of $N_2(A)$, $N_2(C)$, $N_2^+(B)$, concentration of N atoms (for $X_{O_2} \leq 10\%$) and concentrations of O and NO. Examples of such comparisons are given in Figs. 13.1–13.4. It can be seen that the measurements and the calculations [25,27,29,30] are in reasonable agreement.

Two different excitation processes of the $NO(A)$ state have been investigated:
(i) electron excitation of ground state NO molecules

$$e + NO(X) \to NO(A) + e \tag{13.7}$$

(ii) E-E energy exchange due to collisions with $N_2(A)$ metastables

$$N_2(A) + NO(X) \to N_2(X) + NO(A) . \tag{13.8}$$

From comparison of the $NO(A)$ production rates by the above processes with the intensity of the $NO(\gamma)$ band versus O_2 percentages (see Fig. 13.1), it can

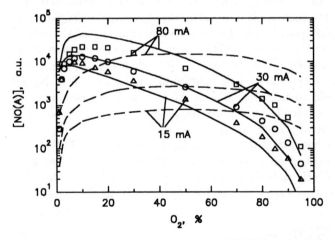

Fig. 13.1. Variation of the population of $NO(A)$ excited state (in arbitrary units) determined from the emission of the $NO(\gamma)$ at 237 nm versus the O_2 percentage for $I = 15, 30$ and 80 mA. The full lines represent the calculation of the excitation rate of $NO(A)$ by the reaction $N_2(A)+NO(X) \to NO(A)+N_2$ and the dashed lines by direct electronic excitation: $e+NO \to NO(A) + e$. The calculated and experimental values are adjusted for an O_2 percentage of 50% and a discharge current $I = 30$ mA. The measurements are represented by symbols; triangles: 15 mA; circles: 30 mA and diamonds: 80 mA. The calculated values are represented by solid lines. A: 15 mA; B: 30 mA; C: 80 mA [25]

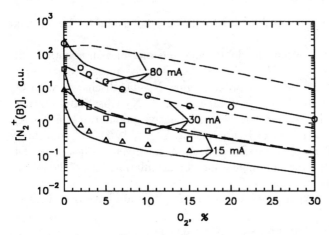

Fig. 13.2. Variation of the population of the $N_2^+(B)$ state (in arbitrary units) determined from the emission of the 1th negative system at 391 nm versus the O_2 percentage for $I = 15, 30$ and 80 mA. The experimental and theoretical values are adjusted for pure N_2 and $I = 30$ mA. The solid lines represent values calculated with large rate coefficients and dashed lines with rate coefficients smaller by two orders of magnitude. The symbols are as in the caption to Fig. 13.1 [25]

be concluded that E-E energy exchange is the main process for excitation of NO(A). The increase in the NO(γ) band intensity with the addition of 0–10% of oxygen is due to the increase in concentration of NO molecules (see Fig. 13.4) and in the $N_2(A)$ population. The decrease in intensity of the NO(γ) band for oxygen percentages above 10% is mainly due to the decrease in the $N_2(A)$ population, in spite of the growth of [NO] when the %O_2 is varied up to 50%. This decrease in $N_2(A)$ population is due to the fall in $N_2(X)$ concentration and reduced electric field E/N, and to the increased quenching by O atoms and NO molecules (which are the main destruction processes of this state in this range of O_2 percentages).

To explain the behaviour of the relative intensities of the $N_2^+(B \to X)$ 391.4 nm band versus O_2 percentage (see Fig. 13.2), three different processes of excitation of the $N_2^+(B)$ state have been investigated. The first is ionization of $N_2(X)$ molecules by electron impact leading directly to the creation of $N_2^+(B)$:

$$e + N_2(X) \to N_2^+(B) + 2e \,. \tag{13.9}$$

The second is excitation of $N_2^+(X)$ ions by electron impact

$$e + N_2^+(X) \to N_2^+(B) + e \,. \tag{13.10}$$

However, taking into account the smooth change in electron density and the increase in the electron rate coefficients for excitation of $N_2^+(X)$ as E/N increases, the calculated rates of production of $N_2^+(B)$ by the above processes

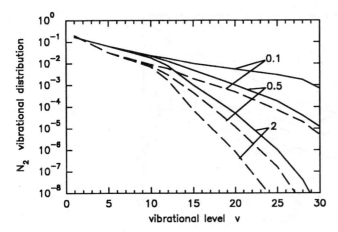

Fig. 13.3. Vibrational distribution of $N_2(X,v)$ molecules calculated for different O_2 percentages 0.1, 0.5 and 2%. Solid lines correspond to a discharge current $I = 80$ mA and dashed lines to $I = 30$ mA [25]

cannot explain the rapid decrease in intensity of the $N_2^+(B)$ 391.4 nm band with increasing %O_2.

The third excitation process investigated is the exothermic quasi-resonant V-E energy exchange in collisions of $N_2^+(X)$ ions with vibrationally excited N_2 molecules:

$$N_2(v > 12) + N_2^+(X) \to N_2(v-12) + N_2^+(B) \ . \tag{13.11}$$

This reaction is an important process for excitation of $N_2^+(B)$ in pure nitrogen post-discharges [39] and it also seems to be the main process for production of $N_2^+(B)$ in discharges.

The flux $[N_2^+(X)]\Sigma_{v>12}[N_2(v)]$, which is proportional to the rate of this reaction, has been calculated as a function of %O_2 for two values of the rate coefficient for (13.1): one, using (10.17), which gives large values ($\geq 10^{-11}$ cm^3/s for $v \geq 12$); and a two-order of magnitude smaller value. The results of these calculations, reported in Fig. 13.2, show that using the large value obtained from (10.17) leads to better agreement with the observed variation of the $N_2^+(B)$ 391.4 nm band.

Further evidence for the above conclusions is provided in Fig. 13.3 where the calculated vibrational distribution function of $N_2(X)$ is given for discharge currents of 30 and 80 mA and various percentages of O_2. As the latter increases, reaction (13.1) causes a steep decrease in the populations of the levels $v > 12$ which can thus explain the rapid decay observed in the 1$^-$ system emission.

Moreover, for reaction (13.11) to be much faster than reaction (13.10), the rate coefficient of the former has to be much larger than 10^{-12} cm^{3}s^{-1}.

For [O] and [NO], agreement between calculations and measurements could be achieved by using values of γ_O, γ_N considerably different from those

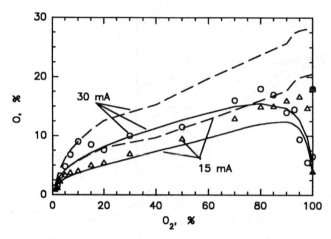

Fig. 13.4. Relative concentration of O atoms vs. O_2 percentage in an N_2-O_2 DC discharge in a tube 0.8 cm in radius, for a pressure 2 Torr, gas flow rate 100 sccm and at a distance of 43 cm from gas entrance into discharge. Data points are measurements [27]. Solid lines are calculations. Dotted lines – calculations using only the "ordinary" system of surface chemically active sites to determine γ_O [29,30]

obtained from afterglow experiments [40–43]. The values of γ_O and γ_N under discharge conditions can be obtained from a comparison between theoretical and experimental O and NO concentrations under different discharge conditions (note that [NO] is controlled by the fast reaction (13.3), and therefore depends on [N] and γ_N). Such comparisons were made in [27,29,30,44–46] for pure O_2 discharges and in [27,29,30] for N_2-O_2 mixtures. The results obtained for γ_O [29,30] have been presented in Fig. 11.4 where it can be seen that γ_O is relatively large and decreases with a small amount of N_2 admixed into an O_2 discharge. Such high γ_O values are required to interpret experimental O atom densities in discharges. Similarly, high γ_N values in N_2-O_2 discharges, with $X_{O_2} \geq 20\%$, are necessary to explain the observed NO densities.

Note, however, that much smaller γ_O and γ_N values have been derived from experiments in pure gases and afterglows [40–43]. To solve this apparent paradox and to explain the experiments [25,27] it is reasonable to assume that, under discharge conditions, there exist additional chemically active sites on the wall. Such an assumption was made in [27,29,30] and the parameters determining γ_O and γ_N in expressions (11.51)–(11.57) of Chap. 11 have been estimated, for relatively low wall temperatures $T_w \leq 400$ K, when surface diffusion and reactions of physisorbed N_f and O_f atoms control the surface losses of gas phase N and O atoms.

Calculations of O and NO concentrations, taking into consideration either two systems of chemically active sites or a single ordinary system (second system according to the designation adopted in [27]), are shown in Figs. 13.4

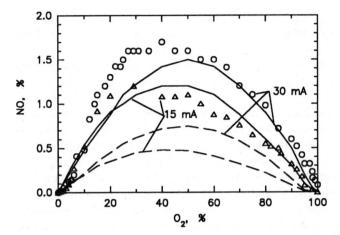

Fig. 13.5. Relative NO density vs. O$_2$ percentage. Data points are measurements [27]. Solid lines are calculations. Dotted lines – calculations with small $\gamma_N \leq 10^{-4}$ (using only the "ordinary" system of surface chemically active sites). Discharge parameters are the same as in Fig. 13.4 [29,30]

and 13.5. It can be seen that the calculations with only one system do not agree with experiment.

The nature of the additional system of chemically active sites under discharge conditions is unclear so far. It could be connected with surface ions resulting from discharge electron and ion fluxes to the surface. For example, the surface will likely be partly covered with negative atomic ions due to surface attachment of electrons to O$_S$ atoms. If one assumes that the lifetime τ_{ion} of surface ions is controlled by mutual ion-ion recombination, from the balance equation for the concentration [S^{ion}] of surface ions

$$\Phi^{\text{ion}} = \frac{1}{\tau_{\text{ion}}}[S^{\text{ion}}] \equiv K^{\text{ion}}_{\text{rec}}[S^{\text{ion}}]^2 \equiv \frac{\nu_{\text{ion}}}{4[F]} \exp\left(-\frac{E_{\text{ion}}}{T_{\text{w}}}\right) [S^{\text{ion}}]^2 \qquad (13.12)$$

it is easy to obtain the following expression for the relative surface density of positive and negative charges:

$$\frac{[S^{\text{ion}}]}{[F]} \simeq 2\sqrt{\frac{\Phi^{\text{ion}}}{[F]\nu_{\text{ion}}}} \exp\left(\frac{E_{\text{ion}}}{2T_{\text{w}}}\right). \qquad (13.13)$$

In (13.12) and (13.13), $[F] \sim 1/a^2$ is the surface density of all sites which can physisorb the gas phase species; $K^{\text{ion}}_{\text{rec}}$ is the rate coefficient (in cm^2/s) of surface ion-ion recombination; ν_{ion} and E_{ion} are, respectively, the frequency factor (in s^{-1}) and the activation energy (in K) for surface diffusion of charged particles; a is the elementary distance (in cm) for a jump of physisorbed ions; Φ_{ion} is the ambipolar diffusion flux (in cm^{-2}s^{-1}) of plasma electrons and ions to the surface, which is the source of surface ions. For the typical values $T_{\text{w}} \sim 350$ K, $\Phi^{\text{ion}} \sim 10^{14}$ cm^{-2}s^{-1} and physically acceptable values $a \sim 10^{-8}$

cm; $\nu_{\text{ion}} \sim 10^{13}$ s^{-1}; $E_{\text{ion}} \sim 7500$ K, one can estimate from (13.13) that $[S^{\text{ion}}]/[F] \sim 10^{-3}$. This value corresponds to the order of magnitude of the relative density of chemically active sites $[S]/[F]$ estimated in [27] to explain the large γ_O and γ_N values observed under discharge conditions. So far, the above explanation is just a working hypothesis which needs to be further developed.

The production mechanisms of NO_2 in low pressure plasmas have been investigated in [29,30]. An understanding of such a mechanism is especially important for interpreting the NO_2 visible emission generation on the Space Shuttle surface in the ram direction in stationary orbits, where strong interaction with the surface takes place for ramming O atoms and N_2 molecules [47,48] (see also Chap. 15). Calculations of NO_2 concentrations in low pressure laboratory plasmas using the self-consistent model [29,30] for volume and surface processes in N_2-O_2 discharges show that the surface production of NO_2 can be a very important mechanism for pressures $p \leq 1$ Torr.

It is worth mentioning that a similar self-consistent kinetic model to that for N_2-O_2 has also been developed for low pressure N_2-H_2 discharges and post-discharges [37]. It was found that, for example, the wall loss probabilities γ_N and γ_H of N and H atoms on a Pyrex wall were reduced by a factor of about 2, as a few % of H_2 or N_2 were introduced into N_2 or H_2, respectively. The calculations were compared with data from experiments (measurements of relative changes in the electric field and $N_2(C)$, $N_2^+(B)$ concentrations as a function of the H_2 percentage). From this comparison, rate coefficients for associative ionization upon collisions between two excited N_2 molecules and deactivation of $N_2(a')$ and $N_2(X,v)$ by H atoms were estimated from the model. It was also shown that surface processes are important for the production of gas phase NH_3 molecules.

In conclusion, self-consistent kinetic models combining volume and surface processes are an important improvement in the numerical modelling of non-equilibrium plasmas in low pressure molecular gas mixtures. The surface kinetic models developed in [25,27,29,30,37] enable us to obtain the dependence of the probabilities γ_M on the relative concentration of gas phase atoms and to calculate the rates for surface production of gas phase NO, NO_2 and NH_3 molecules. An important peculiarity of these models is the small number of required input parameters – only those externally controlled in experiments. This self-consistent approach provides quantitative interpretation of different experiments and an understanding of the role of different processes. It also enables us to estimate or define more accurately some important rate coefficients and parameters. This was achieved in [25,27,29,30,37] for the rate coefficients and parameters determining: the surface kinetics of O and N atoms; the surface kinetics of H and N atoms; associative ionization of N_2 molecules from $N_2(A,a')$ metastables; excitation of $N_2(a')$ from $2N_2(v \geq 16)$; deactivation of $N_2(a')$ on the walls; deactivation of $N_2(a')$ by H atoms; V-T relaxation of $N_2(X,v)$ molecules by $N_2(X,v)-H$ collisions;

chemical reaction $N_2(X, v \geq 12)+O \rightarrow NO+N$; V-E resonant exchanges via $N_2(X, v \geq 12)+N_2^+(X) \rightarrow N_2(X, v-12)+N_2^+(B)$; V-E resonant exchanges via $N_2(X, v \geq 14)+NH(X) \rightarrow N_2(X, v-14)+NH(A)$.

13.2 Modelling of High Pressure N_2-O_2 Discharges

The main interest in the investigation of high-pressure (atmospheric or nearly atmospheric pressures), low temperature non-equilibrium plasmas in N_2-O_2 mixtures, with small admixtures of other gases, is connected with ecological problems. Examples of such problems are the influence of artificial microwave discharges in the stratosphere on the ozone layer, the cleaning from the atmosphere of chlorofluorocarbons through artificial microwave discharges, and the removal of NO, NO_2, CO, SO_2 and other polluting admixtures by using corona, surface, DC glow, or dielectric barrier discharges and discharges generated by a high energy electron beam.

The charge and the chemical kinetics in air plasmas generated at atmospheric pressure by an electron beam were theoretically studied in [49]. Self-consistent calculations have been carried out of the concentrations of electrons, N_2^+, N^+, N_3^+, N_4^+, O^+, O_2^+, $O_2^+N_2$, NO^+, O^-, O_2^-, O_3^-, O_4^-, NO_2^-, NO_3^- ions, N_2, O_2, N, O, NO, O_3 neutral particles, several electronically excited states of ions and neutrals (such as $N_2(A)$, $N_2(B)$, $N_2(C)$, $O_2(a\Delta)$, $N(^2D)$, $N_2^+(A)$, $N_2^+(B)$, $O_2^+(X)$, $O_2^+(a^4\pi)$, $O_2^+(b^4\pi)$, $O_2^+(A^2\pi)$), vibrational excitation of N_2, and electron and gas temperatures. Figure 13.6 illustrates schematically the various processes determining the kinetics of positive and negative ions in the model [49].

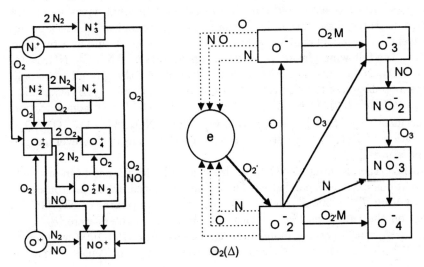

Fig. 13.6. Positive and negative charge flow chart [49]

13.2.1 Cleaning of Polluted Atmospheric Gases

The kinetic processes involved in the cleaning of polluted atmospheric gases by irradiation with electron beams have been modelled in [50–52]. A theoretical investigation of the efficiency of dry air cleaning from NO, NO_2 polluting admixtures has been carried out in [52]. The developed kinetic model determines the rate of electron energy losses along the beam assuming an energy distribution function (EDF) of the primary beam electrons of the form $f(\epsilon) \sim \epsilon^\alpha \exp\left\{-(\epsilon/\epsilon_0)^\beta\right\}$, where α, β, ϵ_0 are working parameters. The other input parameters of the model are the following: total energy flux F (W/cm^2) and duration of the beam τ(s) (or gas residence time in the irradiation zone); gas pressure p(Torr); initial gas temperature T_g^0 and composition N_2–O_2 with small impurities; columnar gas mass $\overline{X}$ (gram/cm^2) transmitted by the beam from the boundary (where $\overline{X}=0$) up to the point investigated. The value $\overline{X}$ is determined by the total gas density [N_{tot}] (cm^{-3}), the average molecular mass μ (gram), and the distance X (cm) from the boundary: $\overline{X}=\mu[N_{\text{tot}}]X$. The value $X = R$ has been chosen, where R is the radius of the discharge tube (which is irradiated perpendicularly to its surface). The distribution of the beam energy losses as a function of $\overline{X}$ (or of X, Z) is determined by a universal dimensionless energy dissipation function $\lambda(\overline{X},\epsilon)$. An analytical expression for $\lambda(\overline{X},\epsilon)$ is given, for example, in [10,53] (see also formula (8.45) in Chap. 8). The primary electrons ionize the main air components, and therefore the electron energy distribution function - degradation spectrum - consists of secondary electrons. This spectrum was not calculated in [52] and analytical approximations [11] (see also Chap. 8) have been used to determine the relative electron energy loss rates and the "energetic costs" associated with different interaction processes of secondary electrons with the main gas species (elastic collisions, excitation of rotational, vibrational and electronic states, dissociation, ionization, dissociative ionization, electron attachment). The model enables one to calculate, at point $\overline{X}$ and as a function of time, the following plasma properties: concentration of N_2, O_2, NO, N_2O, NO_2, NO_3, N_2O_5, O_3 molecules and N, O atoms in the ground electronic states; population of the electronically excited states $N_2(A^3\Sigma_u^+, B^3\Pi_g, a'^1\Sigma_u^-, a^1\Pi_g, C^3\Pi_u, a''^1\Sigma_g^+)$, $O_2(a^1\Delta_g, b^1\Sigma_g^+)$, $N(^2D,^2P)$, $O(^1D,^1S)$; concentration of the ions N^+, N_2^+, $N_2^+(B)$, N_4^+, O^+, O_2^+, NO^+, NO_2^+, O^-, O_2^-, O_3^-, O_4^-; population of $N_2(X,v)$ vibrational levels; electron density N_e; average gas temperature T_g.

In [52], the cleaning of NO from dry atmospheric gas has been studied. Calculations have been carried out for dry air irradiated by an electron beam ($p = 1$ atm; $T_g^0 = 300$ K), with different initial admixtures of NO (10^{-2}–10 %) on the tube axis ($R = 1$ cm). The beam parameters were: $\epsilon_0 = 30$ keV, $\alpha=25$, $\beta=25$ (nearly monoenergetic beam); $F = 100$–1000 watt/cm^2; $\tau = 10^{-4}$–10^{-2} s. The post-discharge time was 1 s. NO molecules are partly removed during irradiation. The main "useful" channels for NO removal are the reactions (13.3) and

$$NO + O_2^+ \rightarrow NO^+ + O_2 \tag{13.14}$$

involving respectively N atoms and O_2^+ ions created by electron impact N_2 dissociation and O_2 ionization. The reactions

$$NO + O + M \rightarrow NO_2 + M, \tag{13.15}$$

$$NO + O_3 \rightarrow NO_2 + O_2 \tag{13.16}$$

are, by contrast, "bad" NO loss channels, because they produce the bad admixture NO_2.

Examples of calculated NO and NO+NO_2 removal efficiencies, $\eta(NO) = ([NO]_0 - [NO]_{\text{fin}})/[NO]_0$, $\eta(NO_x) = ([NO]_0 - [NO]_{\text{fin}} + [NO_2]_{\text{fin}})/[NO]_0$, are given in Fig. 13.7.

It can be seen that full cleaning can be achieved, if $E_b/[NO]_0 \geq 80$ eV/molecule and the total absorbed energy is $E_b \leq 0.1$ J/cm^3. The analysis shows that, for $E_b \geq 0.4$ J/cm^3, full cleaning cannot be attained and that the removal efficiencies $\eta(NO)$ and $\eta(NO_x)$ start to fall as $E_b/[NO]$ rises. This is due to the increase in gas temperature, during the electron pulse, up to values ≥ 750 K. Under such gas heating conditions, the "bad" reaction (13.2) starts to be important.

Experimentally, the NO and NO_2 removal efficiencies in electron beam generated air plasmas have been measured in [54]. Note that, in real situations, polluted atmospheric gases are frequently not dry and contain an H_2O

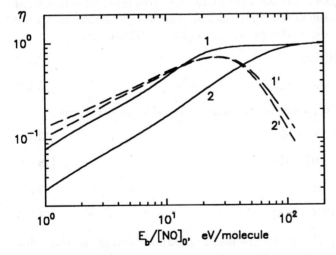

Fig. 13.7. Calculated NO and NO+NO_2 removal efficiencies $\eta(NO)$ (1, 1') and $\eta(NO_x)$ (2, 2') as a function of $E_b/[NO]_0$, the ratio of the absorbed beam energy E_b on the tube axis to the initial $[NO]_0$ concentration. Full lines – for $E_b \leq 0.1$ J/cm^3; broken lines – for $E_b = 0.45$ J/cm^3; $\tau = 10^{-3}$ s [52]

vapour admixture. In addition, other bad admixtures such as CO or SO_2 may also be present. It is clear that kinetic models describing air cleaning by electron beam irradiation under such conditions must be extended to include reactions with species containing H, C, and S atoms (see, for example, Tables 10.4, 10.8, 10.10 for hydrogen-oxygen-nitrogen reactions with neutral and ion species). Such extended kinetic models have been developed in [50,51]. For example, the model in [50] includes 748 gas phase reactions involving species with O, N, H, C and S atoms and ions. Moreover, the production of aerosol particles and the heterogeneous chemistry on their surface have been included in this model.

The use of corona and dielectric barrier discharges is highly promising for air cleaning. A dielectric barrier discharge is created between electrodes separated by a small (~ 1 mm) gap and covered with a dielectric. A large number of microdischarges, statistically spread in space and time, are created in the gas, when an alternating (from $\sim$ 50 Hz up to several kHz) high voltage is applied between the electrodes. Every microdischarge (discharge filament) develops in space and time like streamers in ordinary gas discharge breakdown. A model of individual discharge filaments was used in [55] to describe the dielectric barrier discharge. This model includes the kinetics of free electrons during the microdischarge phase and heavy particle charge and chemical kinetics during the subsequent reaction phase. The physical-chemical kinetics, reactions and processes considered are the same as in [56]. They include the chemistry of the hydrogen-oxygen-nitrogen components, to account for the presence of water vapours in the air.

The evolution of NO_x bad admixtures during the filament pulse and the interpulse reactive phase has been calculated and compared with measured values. The analysis shows that, in humid air, the NO molecules are mainly converted into NO_2 and HNO_2 by reactions (13.15), (13.16) and

$$NO + OH \rightarrow HNO_2 ; \tag{13.17}$$

$$NO + HO_2 \rightarrow NO_2 + OH . \tag{13.18}$$

The O_3, OH and HO_2 molecules are mainly created by the processes

$$O_2 + O + M \rightarrow O_3 + M ; \tag{13.19}$$

$$e + H_2O \rightarrow e + OH + H ; \tag{13.20}$$

$$O + H_2O \rightarrow 2 \, OH ; \tag{13.21}$$

$$O_2 + H + M \rightarrow HO_2 + M. \tag{13.22}$$

Moreover, the O, H, OH species are mainly produced during the short discharge phase, while for O_3 and HO_2 – this happens after the pulse.

An example of calculated and measured NO and NO_2 densities in a dielectric barrier discharge is given in Fig. 13.8. for an initial mixture 79%N_2 – 1%O_2 – 20%H_2O–500 ppm NO. It can be seen that a NO+NO_2 removal

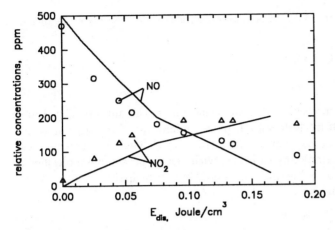

Fig. 13.8. Calculated and measured NO and NO_2 concentrations as a function of E_{dis}, the discharge energy per unit volume [55]

efficiency $\eta \simeq 40\%$ can be achieved for a discharge specific energy $\simeq 0.1$ J/cm^3.

The NO+NO_2 removal efficiency η can be improved and reach $\sim 100\%$ by adding ammonia (NH_3) to the initial gas mixture. To obtain $\eta(NO_x) \simeq 100\%$, an input energy of discharge $E_{dis}/[NO]_0 \simeq 70$ eV/(NO molecule) is needed. The most important processes involving NH_3 are

$$NO_2 + NH_3 \rightarrow HNO_2 + NH_2 ; \tag{13.23}$$

$$NO + NH_2 \rightarrow N_2 + H_2O . \tag{13.24}$$

Reaction (13.24) is particularly interesting because it yields ecologically pure products.

Kinetic models have also been developed for NO reduction from flue atmospheric gases excited either by a pulsed corona discharge [57] or a DC glow discharge [58,59]. The model [57] involves 39 species undergoing 287 gas phase reactions. The gas composition prior to the discharge includes N_2, O_2, CO_2, H_2O as the main species and a small NO admixture (of a few ppm). The kinetic model [58,59] is similar to the model [60], aimed at explaining the glow phase of spark breakdown in air.

An interesting method for cleaning chlorofluorocarbon from air admixtures in the troposphere has been suggested in [61]. There is great concern today about preserving the Earth's ozone layer, which is Earth's life natural screen from ultraviolet solar radiation. The chlorofluorocarbons (CFC: molecules – $CFCl_3$, CF_2Cl_2, $C_2F_3Cl_3$ and so on) gas admixtures are important anthropogenic products that destroy the ozone layer. These chemically stable species move through the troposphere (altitudes 0–10 km) and reach the stratosphere, where they can be dissociated by ultraviolet solar radiation, with production of free Cl atoms. These atoms can destroy O_3 molecules by reactions such as

$$Cl + O_3 \rightarrow ClO + O_2 ; \tag{13.25}$$

$$ClO + O \rightarrow Cl + O_2 . \tag{13.26}$$

For removing CFC molecules from the atmosphere, the authors of [61] have suggested the use of freely localized, pulsed microwave discharges at tropospheric altitudes, with electron density $\sim 10^{12}$ cm^{-3}. The CFC molecules can be destroyed, for example, by collisions with $O_2(a)$ metastable molecules or $O(^1S)$ atoms formed by electron collisions, but, principally, by dissociative attachment reactions, such as

$$CFCl_3 + e \rightarrow Cl^- + CFCl_2 ; \tag{13.27}$$

$$CF_2Cl_2 + e \rightarrow Cl^- + CF_2Cl . \tag{13.28}$$

These processes have very large cross-sections [62] for low energy electrons. For a Maxwellian EEDF with electron temperature $T_e \simeq 0.25$ eV, the rate coefficients of (13.27),(13.28) are 2.9×10^{-8} and 1.9×10^{-9} cm^3/s, respectively. Estimations [61] show that the removal of CFCs by (13.27) and (13.28) can be quite effective, if an electron density $\sim 10^{12}$ cm^{-3} can be achieved by a discharge pulse with energy $\sim 2 \times 10^{-5}$ J/cm^3, and if the decay of electron density in the post-discharge is controlled by recombination and is relatively slow (dissociative recombination, with a rate coefficient $k_{rec} \simeq 10^{-7}$ cm^3/s). Laboratory experiments [63] have shown such a type of slow air plasma decay. Its behaviour is associated with O atoms, which control electron detachment from negative ions, and thus compensate for the electron losses due to fast attachment. However, under optimal conditions for the removal of CFC molecules, the actual air plasma decay mechanisms are not yet clear. Further research, including detailed modelling, is necessary.

13.2.2 N^{14} and N^{15} Isotope Separation

Discharges in N_2–O_2 can also be used for separating the isotopes N^{14} and N^{15}. This separation is based on the difference in the rates of the chemical reactions (13.1) of the molecules $N^{14}N^{14}$ and $N^{14}N^{15}$ (excited in vibrational levels ($v \geq 12$), with O atoms [64–66]. As a result of collisions and intermolecular energy exchanges in the vibrational modes of isotopic molecules, quasi-stationary vibrational distribution functions are formed with different vibrational "temperatures". The isotopic separation effect can be illustrated by considering as example a binary mixture of isotopic molecules A and B, modelled by harmonic oscillators with vibrational quantum energies E_{10}^A and E_{10}^B, respectively. Let N_A, N_B and N_A', N_B' be the concentrations of A and B before and after the chemical reaction. The enrichment coefficient γ of isotope A in the reaction products is determined by the relation

$$\gamma = \frac{N'_A/N'_B}{N_A/N_B} - 1. \tag{13.29}$$

Since at the initial stage of the reaction $N'_i \sim k_i N_i$ (i =A,B; k_i is the reaction rate coefficient), one has

$$\gamma = k_A/k_B - 1. \tag{13.30}$$

If only molecules in vibrational levels, say r_A and r_B, with energy equal to or greater than the activation energy, E_{act}^A and E_{act}^B, can react, then γ depends on the ratio of the populations of these levels. Since the quantum energies E_{10}^A and E_{10}^B differ only slightly, and $E_{act}^A \simeq E_{act}^B$, in practice $r_A \simeq r_B \equiv r$ (except in the case of hydrogen isotopes). During the fast V-V' exchange, quasi-equilibrium is established with vibrational temperatures T_V^A and T_V^B related by (3.58). There will be enrichment of the heavier isotopic molecules, which have a lower vibrational quantum energy, and consequently a higher vibrational temperature. Considering (3.57), (3.58) and the fact that $k_i = k_i^0 \exp(-rE_{10}^i/kT_V^i)$ (where i =A,B), we get from (13.30)

$$\gamma = \frac{k_A^0}{k_B^0} \exp\left\{\frac{r(E_{10}^B - E_{10}^A)}{kT}\right\} - 1 \simeq \frac{k_A^0}{k_B^0} \exp\left\{\frac{E_{act}\Delta E_{AB}}{kTE_{10}^B}\right\} - 1. \tag{13.31}$$

It can be seen from (13.31) that a high enrichment coefficient can be obtained when the activation energy and isotopic frequency shift are such that $r(E_{10}^B - E_{10}^A) \equiv r\Delta E_{AB} > kT$. For example, for $T = 300$ K, $E_{act} = 2.5$ eV, $k_A^0 \simeq k_B^0$, $\Delta E_{AB}/E_{10}^B = 1/40$ (isotopes with mass number of the order of 20), one has $\gamma \simeq 10$. This value far exceeds the enrichment coefficient that can be obtained by traditional chemical methods under thermal equilibrium conditions, where usually $\gamma \leq 0.1$. Of course, to obtain a more rigorous value of γ and to assess the effectiveness of this method, a more detailed investigation must be carried out. A more accurate anharmonic model for the vibrations of chemically active isotopes has been used in [65] to estimate γ. Kinetic models for discharge isotope separation of N^{14}, N^{15} were developed in [67–71].

The most detailed of these models has been developed in [71] for N_2-O_2 discharges. It computes the populations of the electronically excited states $N_2(A^3\Sigma_g^+, B^3\Pi_g^+, a'^1\Sigma_u^-, C^3\Pi_u, O_2 a^1\Delta_g, b^1\Sigma_g^+, A^3\Sigma_u^+, N(^2D), N(^2P), O(^1D), O(^1S)$, of vibrationally excited $N_2(X)$, $O_2(X)$ molecules and of ground state of N, O, NO species. Of course, all species containing nitrogen with different N^{14} and N^{15} isotopes have been included in the model. Nonselective (for the isotopes) production of $N(^2D)$ metastables by electron impact dissociation of N_2 followed by the nonselective reaction

$$N(^2D) + O_2 \rightarrow NO + O \tag{13.32}$$

has been taken into account in calculating the isotopic enrichment coefficient γ concerning the products $N^{14}O$, $N^{15}O$. The input parameters of the model are the gas temperature $T = 200$–500 K and pressure $p = 10^{-3}$–1 atm, the ionization degree $y_e = 10^{-10}$–10^{-6}, the reduced electric field $E/N = (1\text{--}10)^{-16}$ Volt cm^2 and the gas composition (O$_2$ percentage in the N$_2$–O$_2$ mixture). It was found that, at $T \sim 300$ K, optimal conditions occur for $p \sim 1$ atm, $y_e \sim 10^{-9}$, $E/N \sim (1\text{--}5)^{-16}$ V cm^2, yielding an enrichment coefficient $\beta \sim 5$, a NO yield $\sim 0.1\%$, and an energy cost per N^{15}O molecule ~ 6 keV/mol.

To conclude, the examples discussed above as well as other papers continuously appearing on the subject [72–79] show that high pressure discharges in N$_2$–O$_2$ mixtures have several interesting and useful applications, and that the modelling of kinetic and plasma chemical processes in such discharges is quite useful for understanding the different phenomena and for choosing optimal conditions for specific applications

References

1. Aleksandrov N.L., Konchakov A.M. and Son E.E. (1979) High Temp. **17**, 179
2. Aleksandrov N.L, Vysikailo F.I., Islamov R.Sh., Kochetov I.V., Napartovich A.P. and Pevgov V.G. (1981) Thermophys. High Temp. **19**, 22 (*in Russian*)
3. Aleksandrov N.L, Vysikailo F.I., Islamov R.Sh., Kochetov I.V., Napartovich A.P. and Pevgov V.G. (1981) Thermophys. High Temp. **19**, 485 (*in Russian*)
4. Aleksandrov N.L., Bazyelan E.M., Kochetov I.V., and Dyatko N.A. (1997) J. Phys. D: Appl. Phys. **30**, 1686
5. Guerra V., Pinheiro M.J., Gordiets B.F., Loureiro J. and Ferreira C.M. (1997) Plasma Sources Sci. Technol. **6**, 220
6. Medvedev Yu.A. and Hohlov V.D. (1979) J. Techn. Phys. **49**, 309 (*in Russian*)
7. Lappo G.B., Prudnikov M.M. and Checherin V.G. (1980) Thermophys. High Temp. **18**, 677 (*in Russian*)
8. Konovalov V.P. and Son E.E. (1980) J. Techn. Phys. **50**, 300 (*in Russian*)
9. Konovalov V.P. and Son E.E. (1987) in *Plasma Chemistry*, ed. B.M. Smirnov, Energoizdat, Moscow, Vol. 17, p.194 (*in Russian*)
10. Ivanov V.E. and Sergienko T.I. (1992) *Interaction of Aurora Electrons with Atmospheric Gases*, Nauka, St. Petersburg (*in Russian*)
11. Gordiets B.F. and Konovalov V.P. (1990) in *Plasma Jets in the Development of New Materials Technology*, eds. O.P. Solonenko and A.I. Fedorehenko, VSP, Utrecht Tokyo, p.617
12. Gordiets B.F. and Konovalov V.P. (1991) *Geomagnetism and Aeronomy* **31**, 649 (*in Russian*)
13. Gordiets B.F., Markov M.N. and Shelepin L.A. (1976) *Preprint Lebedev Phys. Inst.* **N84, N85**, Moscow (*in Russian*)
14. Rusanov V.D. and Fridman A.A. (1976) Rep. Acad. Sci. USSR **231**, 1109 (*in Russian*)
15. Gordiets B.F. (1977) Geomagnetism and Aeronomy **17**, 871 (*in Russian*)
16. Gordiets B.F., Markov M.N. and Shelepin L.A. (1978) Proceed. Lebedev Phys. Inst. Acad. Sci. USSR (FIAN) **105**, 7 (*in Russian*)
17. Cartry G., Magne L. and Cernogora G. (1999) J. Phys. D: Appl. Phys. **32**, 1894
18. Rusanov V.D. and Fridman A.A. (1984) *Physics of Chemically Active Plasmas*, Nauka, Moscow (*in Russian*)

19. Krasheninnikov S.I. and Nikiforov V.A. (1982) in *Plasma Chemistry*, ed. B.M. Smirnov, Energoizdat, Moscow, Vol. 9, p. 179 (*in Russian*)
20. Macheret S.O., Rusanov V.D., Fridman A.A. and Sholin G.V. (1980) J. Tech. Phys. **50**, 705 (*in Russian*)
21. Vlasov M.N., Mishin E.V. and Telegin V.A. (1980) *Geomagnetism and Aeronomy* **20**, 57 (*in Russian*)
22. Mishin E.V., Ruzhin Yu.Y. and Telegin V.A. (1989) *Interaction of Electron Fluxes with Ionospheric Plasma*, Gidrometeoizdat, Leningrad (*in Russian*)
23. Gordiets B. and Ricard A. (1993) Plasma Sources Sci. Technol. **2**, 158
24. Nahorny J., Pagnon D., Touzeau M., Vialle M., Gordiets B. and Ferreira C.M. (1995) J. Phys. D: Appl. Phys. **28**, 738
25. Gordiets B., Ferreira C.M., Guerra V., Loureiro J., Nahorny J., Pagnon D., Touzeau M., and Vialle M. (1995) IEEE Trans. on Plasma Sci. **23**, 750
26. Guerra V. and Loureiro J. (1995) J. Phys. D: Appl. Phys. **28**, 1903
27. Gordiets B., Ferreira C.M., Nahorny J., Pagnon D., Touzeau M., and Vialle M. (1966) J. Phys. D: Appl. Phys. **29**, 1021
28. Guerra V. and Loureiro J. (1997) Plasma Sources Sci. Technol. **6**, 373
29. Gordiets B.F. and Ferreira C.M. (1997) AIAA Paper **97-2504**
30. Gordiets B.F. and Ferreira C.M. (1998) AIAA Journal **36**, 1643
31. Ricard A., Berton B., Bonnet B. and Bacos M.P. (1990) Proceedings ESCAMPIG XX, Orléans, France, p. 195
32. Loureiro J. and Ferreira C.M. (1986) J. Phys. D: Appl. Phys. **19**, 17
33. Gousset G., Ferreira C.M., Pinheiro M., Sa P.A., Touzeau M., Vialle M. and Loureiro J. (1991) J. Phys. D: Appl. Phys. **24**, 290
34. Self S.A. and Ewald H.N. (1966) Phys. Fluids **9**, 2486
35. Ferreira C.M. and Loureiro J. (1984) J. Phys. D: Appl. Phys. **17**, 1175
36. Black G., Wise H., Schechter S. and Sharpless R.L., (1974) J. Chem. Phys. **60**, 3526
37. Gordiets B.F., Pinheiro M., Ferreira C.M. and Ricard A. (1998) Plasma Sources Sci. Technol. **7**, 363; 379; Guerra V. and Loureiro J. (1999) Plasma Sources Sci. Technol. **8**, 111
38. Touzeau M., Vialle M., Zellagui A., Gousset G., Lefebvre M. and Pealat M. (1991) J. Phys. D: Appl. Phys. **24**, 1
39. Golubovsky Yu. B. and Telezhko V.M. (1983) J. Appl. Spectroscopy **39**, 1429 (*in Russian*)
40. Scott C.D. (1980) AIAA Paper **80-1477**
41. Young C. Kim and Boudart M. (1991) Langmuir **7**, 2999
42. Seward W.A. and Jumper E.J. (1991) Thermophysics and Heat Transfer **5**, 284
43. Jumper E.J., Newman M., Kitchen D.R. and Seward W.A. (1993) AIAA Paper **93-0477**
44. Gousset G., Panafieu P., Touzeau M. and Vialle M. (1987) Plasma Chem. Plasma Process. **6**, 409
45. Gousset G., Touzeau M., Vialle M. and Ferreira C.M. (1989) Plasma Chem. Plasma Process. **9**, 189
46. Magne L. Coitout H., Cernogora G. and Gousset G. (1993) J. Phys. III **3**, 1871
47. Swenson G.R., Mende S.B. and Clifton K.S. (1985) Geophys. Res. Lett. **12**, 97
48. Von Zanh U. and Murad E. (1986) Nature **321**, 147
49. Ali A.W. (1990) in *Non-equilibrium Processes in Partially Ionized Gases*, eds. M. Capitelli and J.N. Bardsley, Plenum Press, New York, p.651
50. Matzing H. (1991) Adv. Chem. Phys. **LXXX**, 315
51. Petrusev A.S., Potapkin B.V., Shulakova E.V. and Fridman A.A. (1995) Proceed. 12th ISPC, Minnesota, USA, Vol. 2, p. 1009

52. Gordiets B.F., Stepanovich A.N. and Ferreira C.M. (1998) Proceed. 14th ESCAMPIG-98, Malahide, Ireland, p. 444
53. Lazarev V.I. (1967) Geomagnetism and Aeronomy **7**, 278 (*in Russian*)
54. Willibald U., Platzer K.-H. and Witting S. (1990) Radiat. Phys. Chem. **35**, 422
55. Klein M., Lins G., Rombheld M. and Seebock R.J. (1995) Proceed. 12th ISPC, Minnesota, USA, Vol. 2, p. 671
56. Yamamoto T., Lawless P.A., Owen M.K., Ensor D.S. and Boss C. (1993) in *Non-Thermal Plasma Techniques for Pollution Control*, eds. B. M. Penetrante and S. E. Schultheis, Springer, Berlin Heidelberg, NATO ASI Series, Vol. 34B, p. 223
57. Yousfi M., Hamani A., Hennad A., Benabdessadok M. and Eichwald O. (1995) Proceed. 12th ISPC, Minnesota, USA, Vol. 2, 1107
58. Akishev Yu.S., Deryugin A.A., Kochetov I.V., Napartovich A.P. and Trushkin N.I. (1993) J. Phys. D: Appl. Phys. **26**, 1630
59. Deryugin A, Napartovich A., Gorse C., Paniccia F. and Capitelli M. (1997) Plasma Chem. Plasma Process. **17**, 79
60. Rodriguez A.E., Morgan W.L., Touryan K.J. and Moeny W.M. (1991) J. Appl. Phys. **70**, 2015
61. Askaryan G.A., Batanov G.M., Barkhudarov A.E., Gritsinin S.I., Korchagina E.G., Kossyi I.A., Silakov V.P. and Tarasova N.M. (1994) J. Phys. D: Appl. Phys. **27**, 1311
62. McCorkle D.L., Christodoulides A.A., Christophorou L.G. and Szamrej I. (1980) J. Chem. Phys. **72**, 4049
63. Vikharev A.L., Ivanov O.A. and Stepanov A.N. (1984) Sov. J. Tech. Phys. **54**, 1617
64. Belenov E.M., Markin E.P., Oraevsky A.N. and Romanenko V.I. (1973) Letters J. Exper. Theor. Phys. **18**, 116 (*in Russian*)
65. Gordiets B.F. and Mamedov Sh.S. (1975) J. Quantum Electron. **5**, 1082 (*in Russian*)
66. Gordiets B.F., Osipov A.I. and Shelepin L.A. (1988) *Kinetic Processes in Gases and Molecular Lasers*, Gordon and Breach, New York
67. Dolinina V.I., Oraevsky A.N., Suchkov A.F., Urin V.M. and Shebeko Yu.N. (1978) J. Techn. Phys. **48**, 983 (*in Russian*)
68. Akulintsev V.M., Gorshunov N.M. and Neshchimenko Yu.P. (1979) High Energy Chemistry **13**, 165; 441 (*in Russian*)
69. Baiadze K.V. and Vetsko V.M. (1983) Chemical Physics **2**, 1185 (*in Russian*)
70. Baiadze K.V. and Vetsko V.M. (1987) Chemical Physics **6**, 1093 (*in Russian*)
71. Adamovich I.V., Borodin V.I., Chernukho A.P., Rich J.W. and Zhdanok S.A. (1995) AIAA Paper **95-1988**
72. Penetrante B.M. and Schulteiss S.E., eds. (1993) *Non-Thermal Plasma Techniques for Pollution Control*, Springer, Berlin Heidelberg, NATO ASI Series G: Ecological Sciences, Vol. 34
73. Penetrante B.M., Hsiao M.C., Bardsley J.N., Merritt B.T., Vogtlin G.E., Kuthi A., Burkhart C.P., and Bayless J.R. (1997) Plasma Sources Sci. Technol. **6**, 251
74. Vinogradov I.P. and Wiesemann K. (1997) Plasma Sources Sci. Technol. **6**, 307
75. Van Veldhuizen E.M., Zhou L.M. and Rutgers W.R. (1998) Plasma Chem. Plasma Process. **18**, 91
76. Zheleznyak M.B. and Filimonova E.A.(1998) High Temperature **4**, 533
77. Yan K., Kanazawa S., Okhubo T. and Nomoto Y. (1999) Plasma Chem. Plasma Process. **19**, 421
78. Stratton B.C., Knight R., Mikkelsen D.R., Blutke A., and Vavruska J. (1999) Plasma Chem. Plasma Process. **19**, 191
79. Aleksandrov N.L. and Bazelyan E.M. (1999) Plasma Sources Sci. Technol. **8**, 285

14. Kinetic Processes in the Ionosphere

The Earth's ionosphere (at altitudes higher than ~ 60 km) is a natural low-temperature, non-equilibrium plasma. The investigation of its structure and physical properties is one of the main purposes of geophysics and atmospheric physics. The analysis of physical-chemical processes in the Earth's ionosphere is also fundamental to understanding processes occurring in low-temperature laboratory plasmas and in technological devices.

Ultraviolet solar radiation and fluxes of high-energy charged particles (usually electrons, with energies ≥ 1 keV) "precipitating" from the magnetosphere are the main ionization sources in the upper Earth's atmosphere. These fluxes reach a maximum in the polar regions during perturbations, in auroras. The low-temperature plasma (ionosphere) formed by these ionization sources can reach an absolute concentration of charged particles of 10^6 cm^{-3}, and a degree of ionization from 10^{-11}–10^{-10} (at altitudes $Z \sim 90$ km) up to $\sim 10^{-3}$ (at altitudes ~ 400 km). The ionosphere has a rather complicated chemical composition. This includes atomic and molecular ions such as O^+, N^+, O_2^+, N_2^+, NO^+ and neutral components such as O, N, O_2, N_2, CO_2, NO, O_3, and OH. After ionization of N_2, O_2 and O, and dissociation of O_2 by primary electrons, the formation of the ionosphere's chemical composition results from the cooling processes of secondary electrons to temperatures 200–2000 K and different physical–chemical processes. Through such processes, the primary energy absorbed from ultraviolet radiation and "precipitating" fast particles is transformsed into other types of energy, namely: internal energy of atoms, molecules and ions; chemical and thermal energy; and energy radiated from the ionosphere.

The investigation and theoretical modelling of processes involving neutral and charged particles and electronically excited species in the ionosphere has a long history (see, for example, [1–8]). Such an interest is first of all due to the strong influence of charged species on the propagation of radio waves in the ionosphere, and also to the fact that optical measurements of atmospheric emissions, determined by the populations of atomic and molecular electronic states, are easily accessible using ground–based devices, especially in the visible spectrum.

14. Kinetic Processes in the Ionosphere

In this chapter, the main results obtained from the modelling of chemical, charge, and vibrational kinetic processes at altitudes $z \simeq 90$ km will be presented and discussed.

14.1 Probabilities and Rates of Dissociation and Ionization

Absorption of energy in the ionosphere is mainly associated with the dissociation of oxygen molecules and ionization of the basic neutral components N_2, O_2 and O. The volume rates q_s for dissociation or ionization of component "s" by ultraviolet solar radiation at altitude z can be calculated by the formula:

$$q_s(z) = N_s(z)$$
$$\times \int_0^\infty \sigma_s(\lambda)\Phi_\infty(\lambda) \exp\left\{-Ch(Z^\odot)\sum_M \sigma_M(\lambda)\int_z^\infty N_M(z')dz'\right\} d\lambda, \tag{14.1}$$

where $N_s(z)$, N_M are, respectively, the concentrations of the component "s" and of each component M absorbing the radiation, at altitude z, σ_s is the cross-section of the process (dissociation or ionization) in the wavelength range λ, $\lambda + d\lambda$; $\sigma_M(\lambda)$ is the total cross-section for absorption of radiation by component M in this range; $\Phi_\infty(\lambda)$ is the flux of ultraviolet solar radiation at the upper atmosphere boundary, in the range λ, $\lambda + d\lambda$. The exponential term in (14.1) describes the decrease in the rates $q_s(z)$ due to absorption of solar radiation by the atmospheric components M at altitudes $z' > z$. The magnitude of this absorption is determined by the Chapman function $Ch(Z^\odot)$, which depends on the Sun's zenith angle $Z^\odot$:

$$Ch(Z^\odot) \simeq \begin{cases} 1/\cos(Z^\odot) & \text{for } Z^\odot \leq 80° \\ \sqrt{\frac{\pi R_E}{2H}}e^{\tau_0}[1 - \text{erf}\sqrt{\tau_0}] & \text{for } 80° < Z^\odot \leq 90° \\ \sqrt{\frac{\pi R_E}{2H}}e^{\tau_0}[1 + \text{erf}\sqrt{\tau_0}] & \text{for } Z^\odot > 90° \end{cases} \tag{14.2}$$

Here, R_E is the Earth's radius, $\tau_0 = (R_E+z)/[(1-\sin Z^\odot)H]$; $H = k_B T/\mu g$ is a height scale for a gas with molecular weight μ; T is the gas temperature; g is the acceleration of gravity. Sometimes, instead of the complicated expression (14.1), it is possible to use for ionization rates the following approximated expressions [9], convenient for altitudes $z > 90$ km:

$$q_{N_2}^{\text{Ph.ion}}(z)\left(\frac{1}{\text{cm}^3\text{s}}\right) \simeq W_{N_2}^{\text{Ph.ion}}(\infty)[N_2]\left\{1 + X_1 + 2.91 X_1^2\right\}^{-1} ;$$

14.1 Probabilities and Rates of Dissociation and Ionization

$$q_{O_2}^{Ph.ion}(z)\left(\frac{1}{cm^3 s}\right)$$
$$\simeq W_{O_2}^{Ph.ion}(\infty)[O_2]\left\{1 + 0.91 X_1 + 1.182 X_1^{3/2}\right\}^{-1} ; \quad (14.3)$$

$$q_{O}^{Ph.ion}(z)\left(\frac{1}{cm^3 s}\right) \simeq W_{O}^{Ph.ion}(\infty)[O]\left\{1 + 1.39 X_1 + 0.435 X_1^{2.4}\right\}^{-1} ;$$

$$X_1 = 10^{-12}\left\{H_{N_2}[N_2] + H_{O_2}[O_2] + H_O[O]\right\} Ch(Z^\odot) .$$

Here, the concentrations [S] and the height scales H_s are related to the altitude z and are expressed in cm^{-3} and km, respectively; $W_S^{Ph.ion}(\infty)$ is the probability of photoionization of component "s" outside the atmosphere. These probabilities depend on solar activity and are approximately proportional to an index F characterizing this activity, which determines the intensity of solar radiation at the wavelength of 10.7 cm (in 10^{-22} W/(m^{2}Hz)):

$$W_{N_2}^{Ph.ion}(\infty) \simeq 3 \times 10^{-9} F ;$$

$$W_{O_2}^{Ph.ion}(\infty) \simeq 6.1 \times 10^{-9} F; \quad W_O^{Ph.ion}(\infty) \simeq 2.9 \times 10^{-9} F . \quad (14.4)$$

The values $W_S^{Ph.ion}(\infty)$ are given in s^{-1}. The magnitude of F varies from $\simeq 70$ (low solar activity) to $\simeq 270$ (high solar activity).

The following approximation can also be used to describe the O$_2$ dissociation rate by ultraviolet solar radiation [10], which constitutes an important ionospheric process:

$$q_{O_2}^{Ph.dis}(z)\left(\frac{1}{cm^3 s}\right) \simeq [O_2] A \exp(-X_2) + \frac{[O_2]}{X_2}$$
$$\times \left\{ B\left[\exp(-aX_2) - \exp(-bX_2)\right] + C\left[\exp(-cX_2) - \exp(-dX_2)\right]\right\} ;$$
$$(14.5)$$

$$X_2 = 10^{-12} H_{O_2}[O_2] Ch(Z^\odot) . \quad (14.6)$$

The coefficients A, B, C, a, b, c, d are given in Table (14.1).

Table 14.1. Values of the parameters in (14.5)

A	B	C	a	b	c	d
8×10^{-7}	1.07×10^{-6}	2.08×10^{-6}	0.151	1.1	0.029	0.154

To obtain the rates of volume ionization, dissociation and excitation of different states of the main components N$_2$, O$_2$, and O by fast electrons precipitating from the magnetosphere, the analytical approximations (8.42)–(8.45) can be used. However, it should be kept in mind that (8.44) gives the energy loss rate of a monoenergetic electron beam with electron energy ϵ_b. If the energy spectrum of the beam has the form

$$n(\epsilon) \sim \epsilon^\alpha \exp\left\{-\left(\frac{\epsilon}{\epsilon_0}\right)^\beta\right\} \tag{14.7}$$

and the total vertical energy flux of the beam on the upper atmospheric boundary is Φ_∞^e, (8.44) must be integrated over this spectrum to obtain the total rate of volume energy losses Q. As a result, instead of (8.44) one gets for Q:

$$Q\left(\frac{\text{ev}}{\text{cm}^3\text{s}}\right) \simeq 2.27 \times 10^{-7} \frac{\Phi_\infty^e \mu \beta [N_{\text{tot}}]}{\Gamma((\alpha+2.5)/\beta)(\epsilon_0)^{\alpha+2.5}}$$

$$\times \int_0^\infty \frac{\lambda(\overline{Z}/R)}{R} \epsilon^{\alpha+3/2} \exp\left\{-\left(\frac{\epsilon}{\epsilon_0}\right)^\beta\right\} d\epsilon . \tag{14.8}$$

Here, Γ is the gamma function, ϵ_0 is the characteristic energy of the energy spectrum (14.7), expressed in keV, α and β are the parameters determining this spectrum; $[N_{\text{tot}}]$ is the total gas density: $\lambda(\overline{Z}/R)$, $\overline{Z}$ and R are the same as in (8.45). The energy flux Φ_∞^e is expressed in erg cm^{-2}s^{-1}. The values of Φ_∞^e depend on the latitude, local time, and degree of geomagnetic perturbation. This flux is always small (≤ 0.1 erg cm^{-2}s^{-1}) for middle and low latitudes. However, it can sharply increase in polar areas (particularly in the aurora oval). For example, class I, II, III, IV auroras correspond to energy fluxes of 1, 10, 10^2, and $\geq 10^3$ erg cm^{-2}s^{-1}, respectively.

14.2 Chemical and Charge Kinetics

The atoms and ions O, O$^+$, N, N$^+$, O$_2^+$, N$_2^+$ (in ground and electronically excited states) resulting from the primary interactions of N$_2$, O$_2$, N, O with ultraviolet solar radiation and fast particles, are precursors of different ion-molecule reactions.

Modelling of chemical and charge kinetics in the ionospheric gas at altitudes $z > 90$ km has been carried out in detail in [11–22]. The following reactions are the main ion-molecule reactions at altitudes $z = 90$–500 km that determine the charge kinetics of nitrogen-oxygen components:

$$O^+ + N_2^* \to NO^+ + N ; \tag{14.9}$$

$$O^+ + O_2 \to O_2^+ + O ; \tag{14.10}$$

$$O_2^+ + N \to NO^+ + O ; \tag{14.11}$$

$$O_2^+ + NO \to NO^+ + O_2 ; \tag{14.12}$$

$$N_2^+ + O \to NO^+ + N(^4S, ^2D) ; \tag{14.13}$$

14.2 Chemical and Charge Kinetics

$$N_2^+ + O_2 \to O_2^+ + N_2 ; \tag{14.14}$$

$$N^+ + O_2 \to O_2^+ + N ; \tag{14.15}$$

$$N^+ + O_2 \to NO^+ + O ; \tag{14.16}$$

$$O_2^+ + e \to 2O(^3P, ^1D, ^1S) ; \tag{14.17}$$

$$N_2^+ + e \to 2N(^4S, ^2D) ; \tag{14.18}$$

$$NO^+ + e \to O + N(^4S, ^2D) . \tag{14.19}$$

The rate coefficients of reactions (14.9)–(14.16) were presented in Table (10.7), and those of processes (14.17)–(14.19) in Table (8.11). Note that the rate coefficient of reaction (14.9) depends on N_2 vibrational excitation. All the processes (14.9)–(14.19) are exothermic, and thus electronically excited O, N atoms and vibrationally excited molecular ions can be formed by these processes.

The reactions

$$N + O_2 \to NO + O , \tag{14.20}$$

$$N(^2D) + O_2 \to NO + O , \tag{14.21}$$

$$N + NO \to N_2 + O , \tag{14.22}$$

$$N(^2D) + O \to N + O \tag{14.23}$$

are the main chemical processes that control the concentration of the atomic nitrogen states $N(^4S)$, $N(^2D)$ and NO molecules. The rate coefficients of these reactions were given in Table (10.1). The processes (14.9), (14.13), (14.18) and (14.19), and also dissociation and dissociative ionization of N_2 by fast electrons (see Table (8.15)) play an important role in the production of N and $N(^2D)$. According to [23], the relative rates of production of N, $N(^2D)$, $N(^2P)$ and N^+ by dissociation and dissociative ionization of N_2 by electrons with an energy of 100 eV are ~ 0.4, ~ 0.3, ~ 0.16 and ~ 0.14, respectively.

Note that, due to low pressure, the characteristic times for chemical processes in the upper atmosphere are rather large and can exceed the characteristic particle diffusion times. Such a situation occurs for practically all neutral and charged particles in the ionospheric plasma at altitudes $z \geq 250$–300 km, and for NO molecules at $z \leq 100$ km. For this reason, the continuity equations must be used to calculate the concentrations of the above-mentioned species at those altitudes, in order to take into account, besides volume production and loss of species, transport processes due to molecular or turbulent diffusion. However, at altitudes $z \simeq 100$–250 km, the chemical lifetimes for oxygen-nitrogen neutral species (besides the main N_2, O_2 and O components), and all charged particles, are short when compared with diffusion times, and therefore they control the densities of these species. This considerably simplifies calculation of the concentrations of minor ionospheric constituents.

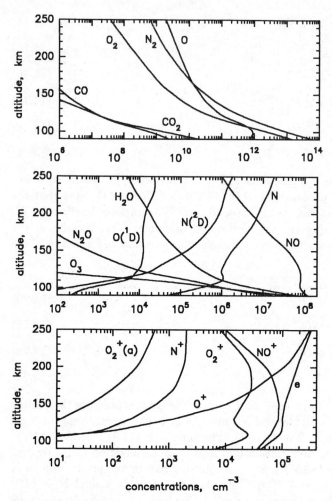

Fig. 14.1. Typical altitude profiles of the concentrations of ions, electrons and neutral species in daytime, middle latitude ionosphere, at middle solar activity [17]

Figure 14.1 illustrates typical altitude profiles of the concentrations of ions, electrons and neutral species.

It can be seen that the density of atomic oxygen equals the concentration of molecular oxygen at an altitude ~ 120 km and becomes dominant at altitudes $z \geq 200$ km. The main ions at altitudes $z \leq 180$ km are NO^+ and O_2^+, and O^+ at altitudes $z \geq 180$ km. The concentration of $N(^2D)$ metastables at all altitudes is 10^2–10^3 times smaller than the concentration of unexcited $N(^4S)$ atoms. However, the reaction (14.21) is the main source of NO in the region $z \leq 140$ km (due to its large rate coefficient; see Table 9.5), and process (14.20) plays a major role only at higher altitudes. This process, together with the main reaction for NO losses (14.22), yields the quasi-equilibrium concentration

$$[NO] \simeq \frac{k_{20}}{k_{22}}[O_2] \qquad (14.24)$$

NO molecules, which are a small impurity in the lower atmosphere, do however play an important role in the formation of the gas temperature altitude profile (and, consequently, in the gas density profile), due to their infrared radiation in the 5.3 μm vibrlational–rotational band.

14.3 Vibrational Kinetics

Molecular vibrational kinetics of both major and minor components plays an important role in the transformation of the energy entering the ionosphere.

Vibrationally excited molecules can influence the chemical and ionic composition, and also gas thermal conditions. This influence is due to the efficiency of certain reactions involving vibrationally excited levels, to energy exchanges between $N_2(v)$ molecules and electrons ($e-V$ processes), and to the infrared radiation of vibrlational–rotational bands.

Historically, vibrational kinetics of OH radicals was the first to attract attention owing to the discovery and identification [24–26] by ground–based devices of overtone vibrlational–rotational OH emission bands (in the visible spectrum) at altitudes 80–95 km. Since then, special attention has been given to experimental and theoretical research on atmospheric hydroxil. The results of those investigations were presented in [1,2,27]. Presently, the monitoring of OH atmospheric emissions is providing important information about dynamic processes at altitudes 80–100 km.

Investigation of vibrational kinetics and infrared emissions of other atmospheric molecules, under the non-equilibrium conditions of the upper atmosphere, only could begin with development of rocket and space technology, and of high-altitude balloon flights. This was so because of the impossibility of ground–based observations, since vibrational infrared bands are, in general, in the $\lambda \geq 3$ μm spectral range, where thermal radiation of the atmosphere is intense and absorption by the lower atmosphere takes place.

The physical aspects of molecular non-equilibrium vibrational kinetics and infrared radiation of the upper atmosphere were first briefly discussed and qualitatively analyzed in [28,29]. The first calculation of N_2 non-equilibrium vibrational temperature was carried out in [30], for altitudes 110–140 km. At the same time, the theoretical analysis of vibrational excitation of different molecules and of infrared radiation generation mechanisms in vibrational molecular bands was carried out in detail in [31]. Since then, theoretical investigations of these problems attracted increasing attention. For example, many model calculations [17,18,32–46] of the vibrational temperature of nitrogen for different altitudes, in the range 50–500 km, were carried out, both for ordinary conditions and in the presence of perturbations (auroras, electric fields). Some of these results have been discussed in a book [4] and a number of reports [47–49]. Theoretical analyses of vibrational kinetics and vibrational band emissions of minor molecular atmospheric components have been carried out in [17,18,48–73].

It should be emphasized that charge, chemical and vibrational kinetics in the ionosphere are fully interconnected. For this reason, the development of self-consistent models is the sole correct method for the theoretical modelling of such processes. Such models were developed in [17,18,38,56], using a small number (no more than 10) of input parameters characterizing the solar-geophysical conditions in the ionosphere. The altitude profiles of gas temperature $T(z)$ and of concentrations of the main components N_2, O_2, O, and of H atoms were calculated as functions of the input parameters, using empirical models. However, the kinetics of ions, electrons, minor neutral components, some metastable electronic states, and of vibrational levels of different molecules and NO^+ ions were analyzed self-consistently, together with the average electron and ion kinetic energies. To perform this analysis, a coupled system of equations of charge, chemical, electronic and vibrational kinetics was solved along with the energy balance equations for electrons and ions. This model allows one to calculate, under quasistationary conditions, the altitude profiles (for altitudes $z \simeq 80$–250 km) for:

(a) the concentration of N_2^+, O_2^+, O^+, NO^+, electrons, NO, N, $N(^2D)$, $O(^1D)$, $O_2(^1\Delta_g)$, $O^+(^2D)$, $O_2^+(a^4\Pi)$, CO_2, H_2O, O_3;
(b) the vibrational level populations, for 15 levels of N_2, 12 levels of NO, 28 levels of NO^+, 9 levels of OH, 10 levels of CO, and 3–6 levels of the triatomic molecules CO_2, N_2O, H_2O, O_3;
(c) the electron and the ion temperatures, T_e and T_i.

The model further permits the calculation of the emission intensities of infrared vibrlational–rotational bands and some bands and lines in the visible and ultraviolet spectral regions, and to explain some rocket-based experiments on infrared emissions. Possible reabsorption of some vibrational–rotational band emissions was taken into account in the "escape to space" approximation [36,74].

Results of investigations on the excitation mechanisms of vibrational infrared bands in auroras and air glows have been collected in various reports [8,49,75].

According to modern concepts, the populations of N_2 vibrational levels at ionospheric altitudes ($z \geq 60$ km) are determined by the following mechanisms:
a) excitation by the processes

$$N(^4S) + NO \rightarrow O + N_2(v \leq 12; \overline{v} \simeq 3.4) ; \tag{14.25}$$

$$N(^2D) + NO \rightarrow O + N_2(v \leq 23; \overline{v} \simeq 5.4) ; \tag{14.26}$$

14.3 Vibrational Kinetics

$$O(^1D) + N_2 \to O + N_2(v \leq 7; \bar{v} \simeq 2.2) \ ; \tag{14.27}$$

$$e_i + N_2(v = 0) \to e_i + N_2(v = 1\text{-}10) \tag{14.28}$$

(here e_i is the photoelectron or the secondary electron of ionization).
b) excitation and deactivation by collisions with plasma electrons

$$e_p + N_2(v = 0) \longleftrightarrow e_p + N_2(v = 1\text{-}10) \ ; \tag{14.29}$$

c) vibrational-vibrational exchanges (V-V processes)

$$N_2(v) + N_2(v = 0) \longleftrightarrow N_2(v-1) + N_2(v = 1) \ ; \tag{14.30}$$

d) intermolecular vibrational exchanges (V-V' processes)

$$N_2(v) + CO_2(\nu_1, \nu_2, \nu_3) \longleftrightarrow N_2(v-1) + CO_2(\nu_1, \nu_2, \nu_3 = 1) \ ; \tag{14.31}$$

$$N_2(v) + O_2(v = 0) \longleftrightarrow N_2(v-1) + O_2(v = 1) \ ; \tag{14.32}$$

$$N_2(v) + NO(v') \longleftrightarrow N_2(v-1) + NO(v'+1) \ ; \tag{14.33}$$

$$N_2(v) + NO^+(v') \longleftrightarrow N_2(v-1) + NO^+(v'+1) \ ; \tag{14.34}$$

$$N_2(v) + CO(v') \longleftrightarrow N_2(v-1) + CO(v'+1) \ ; \tag{14.35}$$

$$N_2(v = 0) + OH(v') \longleftrightarrow N_2(v = 1) + OH(v'-1) \ . \tag{14.36}$$

e) vibrational-translational energy exchanges (V-T processes)

$$N_2(v) + M \longleftrightarrow N_2(v-1) + M \quad (M = O; \ O_2; \ N_2) \ ; \tag{14.37}$$

$$N_2(v) + M^+ \longleftrightarrow N_2(v-1) + M^+ \quad (M^+ = O^+; \ O_2^+; \ NO^+) \ ; \tag{14.38}$$

f) chemical reaction

$$N_2(v \geq 12) + O \to NO + N \ . \tag{14.39}$$

The vibrational levels that can be excited, and the average number $\bar{v}$ of excited vibrational quanta are given in brackets in (14.25)–(14.27). Note that almost all the processes (14.25)–(14.39) are also typical for N_2–O_2 laboratory plasmas.

Figure 14.2 illustrates the calculated altitude profiles of nitrogen vibrational temperature $T_V^{N_2}$. The class II and III auroras correspond to vertical energy fluxes of electrons precipitating from the magnetosphere $\simeq 8$ and $\simeq 90$ erg cm^{-2}s^{-1}, respectively. The calculation was carried out for an exponential energy spectrum of these electrons $n(\epsilon) \sim \epsilon^{0.5} \exp(-\epsilon/\epsilon_0)$, with a typical value $\epsilon_0 = 2.3$ keV.

The analysis shows that reaction (14.27) makes the major contribution to $N_2(v)$ excitation during daytime, under quiet conditions (especially at altitudes $z \leq 120$ km), while for $z \geq 110$ km, the main contributors are the processes (14.25),(14.26), and (14.28). The process (14.29) also begins

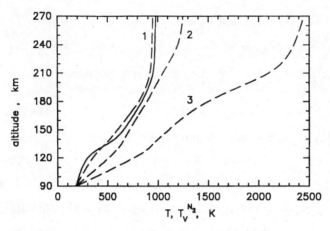

Fig. 14.2. Altitude profiles of the gas temperature T (full lines) and vibrational "temperatures" $T_V^{N_2}$ of $N_2(v=1)$ level in the quiet night ionosphere (1) and class II (2) and III auroras (3) [17]

to contribute to excitation at altitudes $z \geq 190$ km. The processes (14.25), (14.26) and (14.28) are important for $N_2(v)$ excitation in auroras. The main channels for $N_2(v)$ deactivation under any conditions are the process (14.31) of energy transfer into CO_2 vibrations (with subsequent infrared emission in the 4.3 μm band) at altitudes $z \leq 110$ km, and the process (14.37) for collisions with atomic oxygen, at $z \geq 110$ km.

It can be seen from Fig. 14.2 that the N_2 vibrational temperature in the ionosphere can be considerably different from the gas temperature T. This non-equilibrium vibrational excitation can influence different ionospheric parameters like, for example, the electron "temperature" through the processes (14.29). In turn, the change in electron temperature results in a change in electron concentration, due to the dependence of the rate coefficient for dissociative recombination of NO^+ and O_2^+ ions (the main loss processes of electrons and of these ions, especially NO^+) on the electron temperature (see Table (8.11)). For example, calculations ignoring (14.29) predict a decrease in T_e of 500 K, in the quiet ionosphere and at altitudes ≥ 150 km. This would cause a significant decrease (of several tens of percent) in electron density [45].

Vibrational excitation of N_2 molecules also influences the ion composition and electron density through reaction (14.9). The rate coefficient of this reaction depends on the degree of this excitation (see (10.22)–(10.24), Table 10.11 and Fig. 10.3). The process (14.9) and the charge exchange $O^+ + O_2 \rightarrow O_2^+ + O$ are the main O^+ loss mechanisms, at altitudes > 180 km. For example, the decrease of N_e due to the influence of N_2 vibrational excitation does not exceed 10%, at altitudes > 200 km, and at daytime, for the middle latitude ionosphere under low solar activity. However, this decrease is already equal to 30–80% under high solar activity [45,47].

The reaction (14.39) can also influence the chemical composition of the upper atmosphere. Calculations [7,38–40,42,43] show that this influence is crucial in class IV auroras or in weaker auroras, when heating of plasma electrons up to temperatures $T_e \geq 4000$ K is possible. Such heating can happen in class I, II, III auroras in the presence of electric fields parallel to the geomagnetic field, of intensity $E \geq 10^{-3}$ V/m, at altitudes ~ 120 km, or $E \geq 10^{-4}$ V/m, at $z \geq 160$ km [17,38–40]. It can also occur as a result of instabilities generated by the interaction of fast electrons precipitating from the magnetosphere with plasma electrons, due to collective effects [8,42,43].

Vibrational kinetics of the minor molecular atmospheric components at ionospheric altitudes is determined by specific processes for each component. The fact that the populations of many vibrational levels are out of equilibrium at $z \geq 60$ km is a general property, deviations from local thermodynamic equilibrium (LTE) usually growing with altitude. In this case, radiative transitions (spontaneous radiative decay and resonant absorption in vibrlational-rotational bands of infrared radiation from the sun and the lower atmosphere) play an important role with regard to the population of vibrational levels.

Let us briefly discuss the mechanisms of excitation and deactivation of NO and CO_2 vibrational levels. This is of some importance, since the infrared radiation of these molecules plays an important role in the thermal balance of ionospheric gas.

The processes

$$NO(v=0) + h\nu \Longleftrightarrow NO(v=1) , \qquad (14.40)$$

$$NO(v) + O \Longleftrightarrow NO(v+1) + O , \qquad (14.41)$$

$$N(^4S) + O_2 \rightarrow NO(v \leq 7; \bar{v} \simeq 2.7) + O , \qquad (14.42)$$

$$N(^2D) + O_2 \rightarrow NO(v \leq 12; \bar{v} \simeq 4.8) + O \qquad (14.43)$$

are the most important vibrational kinetic processes of NO molecules at ionospheric altitudes. The photo-excitation process (14.40), which is important at altitudes 70 km $\leq z \leq$ 120 km, is related to the absorption of infrared radiation from the lower atmosphere, and of solar radiation, during daytime. The process (14.41) becomes the main mechanism for excitation of level $v = 1$ at $z \geq 120$ km under quiet conditions and in weak auroras. This process, which is followed by the emission of $NO(v = 1)$ in the 5.3 μm band, plays an important role in the gas thermal balance at altitudes 120–200 km. The vibrational levels $v \geq 2$, under any solar-geophysical conditions, and the level $v = 1$ in strong auroras (III, IV classes), are mainly excited by the exothermic reactions (14.42) and (14.43). The deactivation of vibrationally excited $NO(v)$ molecules at altitudes $z \simeq 80$–110 km is essentially determined by collisions with oxygen atoms, and at higher altitudes, by spontaneous radiative decay. The rate coefficient for vibrational deactivation via process (14.41) is 6.5×10^{-11} cm^3/s [76].

The levels $00°1$, $10°1$, and $02°1$ of the CO_2 molecule are the main contributors to the ionospheric non-equilibrium radiation in the band 4.3 μm, due to the following spontaneous radiative transitions

$$00°1 \rightarrow 00°0; \quad 10°1 \rightarrow 10°0; \quad 02°1 \rightarrow 02°0 \,. \tag{14.44}$$

The absorption of infrared solar radiation in the bands 4.3 and 2.7 μm by $CO_2(00°0)$ molecules is the main mechanism for excitation of states $00°1$, $10°1$, $02°1$ during daytime. However, during nighttime, the levels $10°1$ and $02°1$ are weakly excited, and the resonant V-V' process (14.31) gives an important contribution to the excitation of state $00°1$. In this case, the electron collision pumping

$$CO_2(00°0) + e_p \rightarrow CO_2(00°1) + e_p \tag{14.45}$$

adds to this V-V' process, at $z \geq 140$ km. The deactivation of levels $00°1$, $10°1$, $02°1$ at $z \geq 80$ km is mainly caused by the spontaneous radiative transitions (14.44), and at lower altitudes, by the collision process (14.31). This is most important for levels $10°1$, $02°1$. For level $00°1$, this process provides a quasiequilibrium with N_2 vibrational levels, at $z \leq 85$ km. In general, the vibrational temperatures T_{001}, of the CO_2 asymmetric mode, and $T_V^{N_2}$ are always calculated simultaneously, due to the high rate coefficient of the V-V' process (14.31). An example of such a calculation is presented in Fig. 14.3.

Resonant absorption of infrared radiation from the lower atmosphere in the 15 μm band and collisions

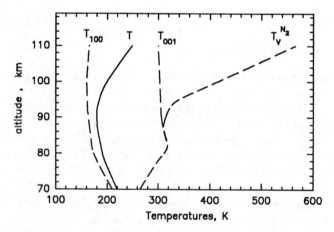

Fig. 14.3. Altitude profiles of the gas temperature T and vibrational "temperatures" T_{001}, T_{100}, and $T_V^{N_2}$, for levels $CO_2(001)$, $CO_2(100)$ and $N_2(v = 1)$, in the daytime, middle latitude ionosphere at middle solar activity [48,49]

14.3 Vibrational Kinetics

$$CO_2(00°0) + M \rightarrow CO_2(01^10) + M; \quad M = O, O_2, N_2 \quad (14.46)$$

are the most important processes for excitation of the $CO_2(01^10)$ level, under any atmospheric conditions. At altitudes $z \geq 80$ km, collisions (14.46) with oxygen atoms are the main contribution. At the same time, the deactivation of $CO_2(01^10)$ is mainly due to spontaneous radiative decay and to the reverse process of (14.46). The rate coefficient for this process, according to [59, 61–63], is $\sim 1.5 \times 10^{-11} \exp(-800/T)$ cm^3/s for M=O, while it was estimated as $\sim 5 \times 10^{-13}$ cm^3/s in [77].

Processes (14.46) followed by emission of radiation in the 15 μm band play an important role in cooling the atmosphere at altitudes $z \leq 120$ km [58–63].

The levels 10°0 and 02°0 are in quasi-equilibrium owing to the high collision probability for transitions between them (due to Fermi resonance). The deactivation of these levels to the state 01^10 is caused by the same processes as the deactivation of 01^10 to the ground state. The main mechanism for excitation of the levels 10°0 and 02°0 is pumping from 10°1 and 02°1 levels by infrared solar radiation in the 2.7 μm band, followed by the transitions 10°1 $\rightarrow$ 10°0, 02°1 $\rightarrow$ 02°0 by radiative decay or the resonant V-V$'$ process (14.31).

Figure 14.3 illustrates the dependence on altitude of the calculated vibrational "temperature" of level 10°0. It is interesting to note that the difference between "temperatures" T_{001} and T_{010}, at altitudes ≥ 80 km, is such that $T_{001}/T_{010} > 1.69$. This means that there is an inversion of population between levels 00°1 and 10°0. Calculations [48,49,78] show that the gain coefficient for the 10.6 μm radiation of transition 00°1 $\rightarrow$ 10°0 can reach $\sim 10^{-11}$ cm^{-1}, at altitudes ~ 80 km, in a quiet atmosphere exposed to sunlight. This gives a relative increase in infrared signal of $\sim 5 \times 10^{-2}$% per single pass on limb view. However, in intense auroras, the gain coefficient may even be larger. This coefficient can also be increased by the effects of powerful radio waves on this ionospheric region. The absorption of radio wave energy heats the plasma electrons, which enables them to excite N_2 vibrational levels by collisions. The population of $CO_2(00°1)$ level will then be increased due to V-V$'$ exchanges (14.31). As a result, the infrared radiation intensity in the 4.3 μm band, and the gain coefficient for the 10.3 μm radiation of the transition 00°1 $\rightarrow$ 10°0 will also increase [78].

Powerful radio waves not only cause excitation of N_2 and O_2 levels by heated electrons, but can also produce additional ionization. In fact, electrical air breakdown can occur for high radio wave electric field intensities. Such artificial, high-frequency discharges in free air can be produced at different altitudes using modern devices (for example, using radio wave beams sent from different points on the Earth's surface, which cross at a given altitude to produce the discharge). The possibility of creating and maintaining artificially ionized layers underneath the natural ionosphere (at altitudes 30–60 km) has been discussed in [79–84]. The major application of such layers would consist in their use as mirrors for metre and decimetre radio waves for communications. However, one should consider the possible ecological consequences of such artificial, high-frequency discharges in the atmosphere, including their effects on the density of nitrogen oxides and on the stratospheric ozone layer [85]. Let us mention that artificial radio frequency discharges in the lower atmosphere (troposphere) can also be used to remove from the atmosphere long-lived anthropogenic impurities like chlorofluorocarbons, which are harmful to the ozone layer [86].

To conclude, we want to emphasize that the ionosphere kinetics is a subject in continuous evolution. Numerous recent papers dealing with the kinetics of excited atomic and molecular oxygen [87-91] and triplet nitrogen states [92] in aurora emission, N_2^+ day glow emission [93], vibrational excitation of NO [94], nonequilibrium translational distribution functions of atomic nitrogen [95], and electron and ion degradation spectra [96,97] are a clear indication of the current interest in the field

References

1. Chamberlain J.W. (1961) *Physics of the Aurora and Airglow*, Academic Press, New York
2. Whitten R.C. and Poppoff I.G. (1971) *Fundamentals of Aeronomy*, John Wiley and Sons, New York
3. Ivanov-Holodnyi G.S. and Nikolsky G.M. (1969) *Sun and Ionosphere*, Nauka, Moscow, (*in Russian*)
4. Danilov A.D. and Vlasov M.N. (1973) *Photochemistry of Ionized and Excited Particles in the Low Ionosphere*, Gidrometeoizdat, Leningrad (*in Russian*)
5. Omholt A. (1971) *The Optical Aurora*, Springer, Berlin Heidelberg
6. McEwan M.J. and Phillips L.F. (1975) *Chemistry of the Atmosphere*, University of Canterbury Christchurch, New Zealand
7. Massey H.S.W. and Bates D.R. eds. (1986) *Atmospheric Physics and Chemistry. Applied Atomic Collision Physics*, Academic Press, New York
8. Mishin E.V., Ruzhin Yu.Y. and Telejin V.A. (1989) *Interaction of Electron Fluxes with Ionospheric Plasma*, Gidrometeoizdat, Leningrad (*in Russian*)
9. Steward A. (1970) J. Geoph. Res. **35**, 1755
10. Strobel D.F. (1978) J. Geoph. Res. **83**, 6225
11. Hays P. and Olivero J. (1970) Planet. Space Sci. **18**, 1729
12. Shimazaki T. and Laird A. (1970) J. Geophys. Res. **75**, 3221
13. Strobel D.F. (1971) J. Geophys. Res. **76**, 2441
14. Shimazaki T. and Laird A. (1972) Radio Sci. **7**, 23
15. Hunt B. (1973) J. Atm. Terrest. Phys. **35**, 1755
16. Jones R. and Rees M. (1973) Planet. Space Sci. **21**, 537
17. Gordiets B.F., Markov M.N. and Shelepin L.A. (1978) Proceed. Lebedev Phys. Inst. Acad. Sci. USSR (FIAN) **105**, 7 (*in Russian*)
18. Gordiets B.F., Markov M.N. and Shelepin L.A. (1978) Planet. Space Sci. **26**, 933
19. Torr D.G. and Torr M.R. (1979) J. Atm. Terr. Phys. **41**, 798
20. Kolesnik A.G. (1982) Geomagnetism and Aeronomy **22**, 601 (*in Russian*)

21. Roble R.G. and Ridley E.C. (1987) Annal. Geophys. **5A**, 369
22. Roble R.G., Ridley E.C. and Dickinson R.E. (1987) J. Geophys. Res. **92**, 8745
23. Zipf E.C., Espy P.J. and Boyle C.F. (1980) J. Geophys. Res. **85**, 687
24. Krasovsky V.I. (1949) Rep. Acad. Sci. USSR **66**, 53 (*in Russian*)
25. Shklovsky I.S. (1950) Rep. Acad. Sci. USSR **75**, 371 (*in Russian*)
26. Meinel A. (1950) Astrophys. J. **III**, 207
27. Shefov N.N. and Piterskaia N. A. (1984) in *Aurora and Night Air Glow*, Inst. Atmos. Phys. Moscow **3**, 23 (*in Russian*)
28. Dalgarno A. (1963) Planet. Space Sci. **10**, 19
29. Markov M.N. (1968) Preprint FIAN **70**, Moscow (*in Russian*)
30. Walker J.C.G. (1968) Planet. Space Sci. **16**, 321
31. Gordiets B.F., Markov M.N. and Shelepin L.A. (1970) Space Res. **8**, 437 (*in Russian*)
32. Schunk R.W. and Hays P.B. (1971) Planet. Space Sci. **19**, 1457
33. Vlasov M.N. (1972) Geomagnetism and Aeronomy **12**, 477 (*in Russian*)
34. Breig E.L., Brenner M.E. and McNeal R.J. (1973) J. Geophys. Res. **8**, 1225
35. Jamahidi E., Fisher E.R. and Kummler R.H. (1973) J. Geophys. Res. **78**, 6151
36. Kumer J.B. and James T.S. (1974) J. Geophys. Res. **79**, 638
37. Nevton G.P., Walker J.C.G. and Neijer P.H.E. (1974) J. Geophys. Res. **79**, 3807
38. Gordiets B.F., Markov M.N. and Shelepin L.A. (1976) Preprint Lebedev Phys. Inst. **N 84, N85**, Moscow (*in Russian*)
39. Gordiets B.F. (1977) Geomagnetism and Aeronomy **17**, 871 (*in Russian*)
40. Gordiets B.F. and Markov M.N. (1977) Space Res. **15**, 725 (*in Russian*)
41. Waite J.H., Nagy A.F. and Torr D.G. (1979) Planet. Space Sci. **27**, 901.
42. Vlasov M.N., Mishin E.V. and Telegin V.A. (1980) Geomagnetism and Aeronomy **20**, 57 (*in Russian*)
43. Zhdanok S.A. and Telegin V.A. (1983) Geomagnetism and Aeronomy **23**, 328 (*in Russian*)
44. Harris R.D. and Adams G.W. (1983) J. Geophys. Res. **88**, 4918
45. Richard P.G. and Torr D.G. (1986) J. Geophys. Res. **91**, 11331
46. Pavlov A.V. (1987) Geomagnetism and Aeronomy **27**, 586 (*in Russian*)
47. Pavlov A.V. and Namgaladze A.A. (1988) Geomagnetism and Aeronomy **28**, 705 (*in Russian*)
48. Gordiets B.F. (1990) Preprint Lebedev Phys. Inst. **N 117**, Moscow (*in Russian*)
49. Gordiets B.F. (1991) Proceed. Lebedev Phys. Inst. Acad.Sci. USSR (FIAN) **212**, 109 (*in Russian*)
50. Degges T.C. (1971) Appl. Opt. **10**, 1856
51. James T.C. and Kumer J.B. (1973) J. Geophys. Res. **78**, 8320
52. Kumer J.G. (1975) in *Atmospheres of Earth and the Planets*, Redel. Dordrecht, p. 347
53. Gordiets B.F. and Markov M.N. (1976) Rep. Acad. Sci. USSR **227**, 852 (*in Russian*)
54. Ogawa T. (1976) Planet. Space Sci. **24**, 749.
55. Bishop R.H., Shaw A.W., Han R. and Megill J. (1974) J. Geophys. Res. **79**, 1729
56. Gordiets B.F. (1976) Short Phys. Inf. **6**, 40 (*in Russian*)
57. Kutepov A.A. and Shved G.M. (1978) Inf.Acad.Sci.; Phys. Atm. Ocean **14**, 28 (*in Russian*)
58. Gordiets B.F., Kulikov Yu.N., Markov M.N. and Marov M.Y. (1979) Preprint Lebedev Phys. Inst. **112**, Moscow (*in Russian*)
59. Gordiets B.F. and Kulikov Yu.N. (1981) Space Res. **19**, p.249 (*in Russian*)
60. Gordiets B.F. and Kulikov Yu.N. (1981) Space Res. **19**, p.539 (*in Russian*)

61. Gordiets B.F., Kulikov Yu.N., Markov M.N. and Marov M.Ja. (1981) Adv. Space Res. **1**, 79
62. Gordiets B.F., Kulikov Yu.N., Markov M.N. and Marov M.Ja. (1982) J. Geophys. Res. **A87**, 4504
63. Gordiets B.F., Kulikov Yu.N., Markov M.N. and Marov M.Ja. (1982) Proceed. Lebedev Phys. Inst. Acad. Sci. USSR (FIAN) **130**, 3 (*in Russian*)
64. Gordiets B.F. and Kulikov Yu.N. (1982) Proceed. Lebedev Phys. Inst. Ac. Sci. USSR (FIAN) **130**, 29 (*in Russian*)
65. Kutepov A.A. and Shved G.M. (1981) Space Res. **19**, 483 (*in Russian*)
66. Kutepov A.A. and Shved G.M. (1978) Inf. Acad. Sci.; Phys. Atm. Ocean **21**, 551 (*in Russian*)
67. Sharma R.D. and Wintersteiner P.P. (1985) J. Geophys. Res. **A90**, 9789
68. Kockarts G. (1980) J. Geophys. Res. Lett. **7**, 137
69. Rawline W.T., Caledonia G.R., Gibson J.J. and Stair A.T.Jr. (1981) J. Geophys. Res. **A86**, 1313
70. Caledonia G.E. and Kennealy J.P. (1982) Planet. Space Sci. **30**, 1043
71. Stepanova G.I. and Shved G.M. (1979) Inf. Acad. Sci.; Phys. Atm. Ocean **16**, 321 (*in Russian*)
72. Manuilova R.O. and Shved G.M. (1985) J. Atm. Terr. Phys. **47**, 413
73. Manuilova R.O. and Shved G.M. (1989) in *Aurora and Airglow of Night Sky* **N 33**, 43 (*in Russian*)
74. Dickinson R.E. (1972) J. Atm. Sci. **29**, 1531
75. Gordiets B.F. (1986) Can. J. Phys. **84**, 1878
76. Fernando R.P. and Swith W.M. (1979) Chem. Phys. Lett. **66**, 218
77. Dickinson R.E. and Bongher S.W. (1986) J. Geophys. Res. **91**, 70
78. Gordiets B.F. and Stepanovich A.N. (1992) Short Phys.Inf. **7-8**, 47 (*in Russian*)
79. Ginzburg V.L. and Gurevich N.V. (1960) Adv. Phys. Sci. **70**, 201; 393 (*in Russian*)
80. Gurevich A.V. and Shvartsburg A.B. (1973) *Nonlinear Theory of Radio Wave Propagation in the Ionosphere*, Nauka, Moscow (*in Russian*)
81. Ginzburg E.I. (1967) Geomagnetism and Aeronomy **7**, 104 (*in Russian*)
82. Lombardini P.P. (1965) Radio Sci. **69 D**, 83
83. Gurevich A.V. (1980) Adv. Phys. Sci. **132**, 685 (*in Russian*)
84. Borisov N.D., Gurevich A.V. and Milih G.M. (1986) *Artificially Ionized Areas in the Atmosphere* IZMIRAN, Moscow (*in Russian*)
85. Askaryan G.A., Batanov G.M., Kossyi I.A. and Kostinsky A.Yu. (1988) Rep. Acad. Sci. USSR **302**, 566 (*in Russian*)
86. Askaryan G.A., Batanov G.M., Barkhudarov A.E., Gritsinin S.I., Korchagina E.G., Kossyi I.A., Silakov V.P. and Tarasova N.M. (1994) J. Phys. D: Appl. Phys. **27**, 1311
87. Slanger T.G. (1986) Can. J. Phys. **64**, 1657
88. Sivjee G.G., Shen D., Yee J.H., and Romick G. (1999) J. Geophys. Res. **104**, 28003
89. Onda K., Ejiri M. and Itikawa Y. (1999) J. Geophys. Res. **104**, 27991
90. Shematovich V., Gerard J.C., Bisikolo V., and Hubert B. (1999) J. Geophys. Res. **104**, 4287
91. Semeter J., Mendillo M., Banmgardner J., Holt J., Hunton D.E., and Eccles V. (1996) J. Geophys. Res. **101**, 19683
92. Morril J.S. and Benesch W.M. (1996) J. Geophys. Res. **101**, 261
93. Broadfoot A.L. and Stone T. (1999) J. Geophys. Res. **104**, 17145
94. Sharma R.D., Dothe H., von Esse F., Karchenko V.A., Sun Y., and Dalgarno A. (1996) J. Geophys. Res. **101**, 19707

95. Sharma R.D., Karchenko V.A., Sun Y., and Dalgarno A. (1996) J. Geophys. Res. **101**, 275
96. Bardway A. and Michael M. (1999) J. Geophys. Res. **104**, 24713
97. Millward G.H., Moffet R.J. and Balmforth H.F. (1999) J. Geophys. Res. **104**, 24603

15. Interaction of Space Vehicles with Atmospheric Gases

At present the problem of the interaction of space vehicles with atmospheric gases is attracting the attention of scientists due to the developments in space technology. The study of this interaction is relevant to understanding the reentry heating of the Space Shuttle and emissions near the surface of space vehicles in stationary orbits.

Gas-solid body interactions can occur in different regimes and are characterized by the relation between typical body size R and the mean free path l of the gas phase molecules and atoms. The free molecular flow regime corresponds to $R \ll l$, in which case the air molecules collide directly with the body. The hydrodynamic regime corresponds to $R \gg l$. In this case, the gas can be treated as a continuous medium.

In the Earth's atmosphere the mean free path at altitudes 80, 100, 120 and 150 km is ~ 1, ~ 10, $\sim 10^2$, $\sim 10^3$ cm, respectively. Therefore, for typical space vehicle sizes, $R \sim 100$ cm, free molecular flow occurs at altitudes $H > 140$ km, and continuous flow, at $H < 100$ km. In the first case, due to gas rarefaction, the resistance to the body motion is small and consequently this motion can persist for a long time on stationary orbits. In the second case, resistance to the motion is high, and the spacecraft interacts only briefly with the atmosphere (for example, the reentry of the Space Shuttle in the atmosphere, from 100 km down to 50 km, takes $\sim 10^3$ s).

15.1 Free Molecular Regime

Physical-chemical processes on the body surface, under the influence of direct collisions with ionospheric particles, play an important role in the gas-body interaction. The kinetic energy of heavy gas particles on the surface of the space vehicle, moving at a velocity $\sim 7.8 \times 10^5$ cm/s in the ram direction, is quite high; it equals 8.9, 10.2, and 5.1 eV, for N_2, O_2 and O, respectively. These energies often exceed the threshold energies of various processes, such as excitation of certain electronic states, exchange reactions, and dissociation. Given these favorable conditions, different processes develop on the surface. Absorption of neutral and charged ionospheric gas particles by the surface and physical-chemical processes on it, followed by desorption of some products, lead to some interesting effects. The first of these effects consists of a change in the electric characteristics of the "body-surrounding plasma" system. The electric charge of the spacecraft can change during flight, and thus a potential difference can be created between the spacecraft's surface and the surrounding ionospheric plasma. Consequently, the parameters of this plasma change. These effects were theoretically examined in [1–5].

A second effect consists in the formation of a neutral gas shell around the space vehicle, which differs in composition from the ordinary non-equilibrium ionospheric gas. The desorption of some gases (especially H_2O vapours) adsorbed on the surface before the launch of the space vehicle, and also the combustion products of steering rocket engines, can play an important role in the creation of this shell, as well as chemical reactions on the surface and consequent desorption. For example, H_2O, N_2, H_2, CO, CO_2 can be the main combustion products of rocket engines. The percentage composition of these gases in exhaust can be, respectively, 33; 31; 17; 13 and 3.6 % [6].

Another effect consists in the appearance of an airglow, in the visible, ultraviolet and infrared spectral ranges, close to the vehicle, in the ram direction. Such a glow has already been observed in 1973 on the Atmospheric Explorer satellite [7–9], and has also been observed many times during Space Shuttle flights [6,10–16].

These phenomena, observed at altitudes > 140 km under prevailing free molecular interaction conditions, are closely connected with the kinetics of physical-chemical processes on the surface of solid bodies. Some hypotheses, qualitative descriptions and semi-quantitative estimations have been made for interpreting the glow. Many references with discussions on different mechanisms for the formation of the gas shell and the glow can be found in the literature [6, 10,11]. The most credible explanation is connected with the production and emission of NO_2 molecules. Optical [10–16] and mass spectrometric [17] measurements in space experiments, as well as laboratory investigations [18–23], show that gas phase NO_2 molecules can be produced on the spacecraft surface in the ram direction in stationary orbit. Various recently developed kinetic models [24–31] include surface kinetic processes and give quantitative explanations for all the observed peculiarities of these phenomena. Agreement between model calculations [26,28] and measurements on the Atmospheric Explorer satellite [7,9] has been achieved. The kinetic surface processes considered in model [26] are the following:

$$NO + F_v \rightleftharpoons NO_f ; \qquad (15.1)$$

$$O + F_v \rightleftharpoons O_f ; \qquad (15.2)$$

$$O + (NO)_f \rightarrow NO_2^* + F_v \rightarrow NO_2 + F_v + h\nu ; \qquad (15.3)$$

$$NO + O_f \rightarrow NO_2^* + F_v \rightarrow NO_2 + F_v + h\nu ; \qquad (15.4)$$

$$M + NO_f \rightarrow M + NO + F_v ; \qquad (15.5)$$

$$M + O_f \rightarrow M + O + F_v , \qquad (15.6)$$

where O_f, NO_f stand for physisorbed O atoms and NO molecules, respectively, and F_v denote vacant physisorption sites. The direct and reverse processes (15.1), (15.2) correspond to absorption and desorption of NO and O species. The processes (15.5), (15.6) refer to the removal of physisorbed NO_f and O_f (also known as "scrubbing") by a collision with any major gas species M=N_2, O_2, O. The surface reactions (15.3), (15.4) (Eley–Rideal mechanism) produce gas phase NO_2 molecules, which can be electronically and vibrationally excited, and thus can generate a glow. This model of surface kinetic processes is simpler than that presented in Chap. 11 (see also [32,33]) and does not account for surface diffusion and Langmuir–Hinshelwood reaction mechanisms. Chemically active surface sites and reactions between physisorbed and gas phase atoms and molecules are disregarded in this model. Such reactions, however, are important if the surface density of physisorbed species is greater than the density of chemically active sites, which is why they have been included in the model presented in Chap. 11 and in [32,33]. Indeed, the surface coverage Θ_{NO} of physisorbed NO_f molecules in model [26] is appreciable, as it varies from $\sim 10^{-3}$, at altitude $Z = 200$ km, to $\sim 10^{-2}$, at $Z = 140$ km. In the framework of model [26], the total intensity I_{NO_2} of the NO_2^* glow (expressed in photons/(cm^2 s)) is given by

$$I_{NO_2} \sim [F] \left(\sigma_3 \Theta_{NO} \Phi_O + \sigma_4 \Theta_O \Phi_{NO} \right), \tag{15.7}$$

where [F] (in cm^{-2}) is the surface density of physisorbed sites, σ_3 and σ_4 are the cross-sections of reactions (15.3) and (15.4), respectively, and Φ_O and Φ_{NO} are the fluxes of O and NO to the surface. It was assumed in [26] that [F]$= 2.4 \times 10^{15}$ cm^{-2} and $\sigma_3 = \sigma_4 = 10^{-17}$ cm^2. The surface coverages Θ_{NO} and Θ_O were calculated from master equations which take into account the processes (15.1)–(15.6). The sticking coefficients for collisions of gas phase species with the surface were assumed to be 0.5 and the "scrubbing" cross-sections σ_5 and σ_6, for processes (15.5) and (15.6), were assumed equal to each other and with a magnitude in the range (0–4) $\times 10^{-16}$ cm^2. The frequencies (in s^{-1}) of desorption of NO_f and O_f were taken respectively as

$$K_{-1} \simeq 8.3 \times 10^{12} \exp\left(-\frac{E_{dNO}}{T_w}\right); \quad K_{-2} \simeq 8.3 \times 10^{12} \exp\left(-\frac{E_{dO}}{T_w}\right), \tag{15.8}$$

where E_{dM} is the activation energy for desorption of species M in Kelvin: $E_{dNO} = (8\text{--}10) \times 10^3$ K, $E_{dO} = 1500$ K, and T_w is the surface temperature, which was assumed to be 300 K. Calculations with different values of E_{dNO} and σ_5, σ_6 are given in Fig. 15.1 along with some experimental results.

The calculated results depend strongly on the activation energy E_{dNO} for desorption of NO_f molecules. This energy, in turn, strongly affects the NO_f surface coverage Θ_{NO}. Agreement between theory and experiment can be achieved by a proper selection of the value of E_{dNO}.

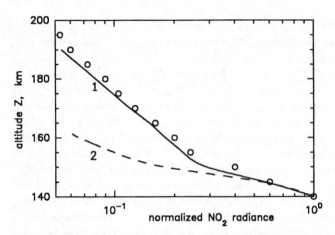

Fig. 15.1. Dependence on altitude of calculated (curves 1,2) and measured (points) relative intensities of the NO_2^* glow. The calculations and experimental data are normalized to each other at $Z = 140$ km. Curve 1 is for $E_{dNO} = 10^4$ K; $\sigma_5 = \sigma_6 = 4 \times 10^{-16}$ cm^2, and curve 2 for $E_{dNO} = 8 \times 10^3$ K; $\sigma_5 = \sigma_6 = 0$ [26]

However, it was shown in [27] that the following gas phase reactions producing NO molecules play an important role in the generation of the NO_2^* glow on the ram surface of spacecrafts, at altitudes $Z \leq 160$ km:

$$\overrightarrow{N_2} + \overleftarrow{O} \rightarrow \overrightarrow{NO} + \overrightarrow{N} ; \qquad (15.9)$$

$$\overrightarrow{O} + \overleftarrow{N_2} \rightarrow \overrightarrow{NO} + \overrightarrow{N} . \qquad (15.10)$$

Here, the arrows indicate the direction in which the atoms and molecules are moving. Reaction (15.9) involves a freestreaming $\overrightarrow{N_2}$ molecule, moving at orbital speed, colliding with an $\overleftarrow{O}$ atom that has been reflected from the surface. Reaction (15.10) involves a freestreaming $\overrightarrow{O}$ atom and a reflected $\overleftarrow{N_2}$ molecule. The relative collision energy of these reactants, which are moving at high velocities, is enough to overcome the high energy barrier E_a of the endothermic reactions (15.9), (15.10) ($E_a = 3.2$ eV). Reactions (15.9), (15.10) in fact constitute additional sources of NO molecules. They increase the NO flux to the surface, and thus also increase the intensity I_{NO_2} of the NO_2^* glow (see (15.7)).

Finally, one should also note that surface diffusion of O_f atoms and Langmuir–Hinshelwood reactions involving these atoms can be important on the surface, for currently accepted values (see [26]) of the activation energy for O_f desorption.

15.2 Hydrodynamic Regime. Vibrational and Chemical Kinetics During Space Shuttle Reentry

Shock waves are generated under the hydrodynamic interaction regime between the spacecraft and the atmosphere (at altitudes < 100 km) due to the supersonic speed of the vehicle. The gas beyond the shock front is heated to high temperatures, and different relaxation processes can develop in this gas. A thin gas boundary layer forms in front of the body, behind the shock, which provides a considerable energy flux to the surface. This flux can lead to substantial heating of the surface, and even to its melting, vaporization, and destruction (ablation). Investigation of the heating mechanisms of the body front by the gas flow is an important field of supersonic aerodynamics.

It is during the gliding path (reentry) period that most surface heating occurs during the descent of Space Shuttle vehicles. For example, this corresponds to altitudes 100–50 km and velocities 8–3 km/s (i.e., to Mach numbers $Ma = 22$–10), for the craft "Buran". Since the atmosphere is relatively rarefied at these altitudes, the typical relaxation times of the different processes (vibrational relaxation, dissociation and recombination of molecules, and other chemical reactions) occurring in the heated gas beyond the shock front are fairly large. They correspond to relaxation lengths equal to or greater than the thickness of the boundary and shock layers in front of the body surface. Therefore, a rigorous analysis of the phenomena behind the shock wave front (including the boundary layer near the surface and the trace behind the body) should be based on simultaneous solutions to the gas dynamic, vibrational, chemical kinetic, and electronic state kinetic equations.

This problem was formulated for the shock layer in front of the body and an approximate solution for the non-equilibrium vibrational energy of oxygen molecules was obtained [34]. Note that, very often, in theoretical analyses of problems of this nature only the gas dynamics along with either chemical kinetics or simple vibrational kinetics (relaxation of vibrational energy of harmonic oscillators) are considered [35–47]. For example, only a chemically reacting gas mixture was investigated for the case of a gas stream along bodies of simple form [38], or of complex form [40–47] (such as the front of the Space Shuttle). A five-component mixture of N_2, O_2, N, O, and NO was considered in [45], and a seven component mixture in [40, 42]. In these works, the ions NO^+ and the electrons e were added to the five components N_2, O_2, N, O, NO and the reaction $N+O \Longleftrightarrow NO^+ + e$ for charge production and loss was considered. Reactions involving the components C, CO, CO_2 were added in the model [47] to take into account the effects of ablation of graphite surfaces on the gas flow. In investigations of Space Shuttle frontal surface heating during its gliding reentry, greater attention was given to the effects of the surface catalytic properties (with respect to dissociation and recombination) on the amount of heat flow to this surface.

The analysis of hydrodynamics with consideration of non-equilibrium vibrational kinetics demands a greater effort, since one needs to solve a considerable number of kinetic equations for non–equilibrium populations of vibrational levels. A kinetic model for coupled vibrational relaxation and dissociation of molecules behind a shock wave was developed in [48], for harmonic oscillators, and in [49–55], for anharmonic oscillators. A discussion and comparison of different dissociation models was given in [52]. Some models have been used for illustrative calculations of non-equilibrium vibrational distribution functions and dissociation rates behind shock wave fronts. For example, this has been done for N_2, and for O_2 in O_2-Ar mixtures [50]. In the latter work, single-quantum and multi-quantum vibrational transitions, and dissociation from any vibrational levels have been included in the kinetic model. The forced harmonic oscillator (FHO) model [50,56,57] and the classical SSH theory (see [10,18,19] in Chap. 3) were used to calculate the probabilities of V-V and V-T processes, and the classical impulsive model [50,56,57] - to calculate the rate coefficient for dissociation from level v. It was found that SSH and FHO methods predict highly different vibrational distribution functions behind strong shock waves only for times shorter than the vibrational relaxation time τ_{VT}. However, such models yield small differences in the dissociation rate, since the dissociation has a time delay $\tau_{del} \sim \tau_{VT}$.

Harmonic and anharmonic oscillator models have been used in [58] to investigate a two-dimensional N_2 flow with Mach number $Ma_\infty = 6.5$ past an infinite cylinder of radius $R = 100$ cm, assuming a uniform flow in thermal equilibrium at temperature $T_\infty = 300$ K and pressure $p_\infty = 50$ Pa. The isotherms for translational T and non-equilibrium vibrational T_V temperatures behind the shock front wave have been obtained for different flow angles, from the stagnation point up to the cylinder shoulder. In particular, it was found that due to gas expansion and cooling along the cylinder wall the vibrational temperature T_V is out of equilibrium, the ratio T_V/T increasing from ~ 0.9 near the stagnation point up to ~ 1.6 at the cylinder shoulder. The kinetic analysis of the boundary layer has an additional difficulty arising from the significant role of heat conduction and diffusion to the surface. Only investigation of processes at the boundary layer will allow such an important parameter as the heat flux to the surface to be calculated.

Non-equilibrium, level-by-level vibrational kinetics in the boundary layer was investigated in detail in [59–65], using simplified hydrodynamics. A brief description of this model is given below.

To study the vibrational kinetics in the boundary layer of hypersonic reentry vehicles the relevant equations need to be solved. To emphasize the role of vibrational kinetics, the simplest flux configurations (axially symmetrical or unlimited in one direction) and the stagnation point approximation [66] can be chosen. In these conditions, the model is two-dimensional and can be described by generalized coordinates connected with ordinary cartesian coordinates by the Lees–Dorodnitsyn transformation, namely:

$$\xi = \int_0^x \rho_{ex}(x')u_{ex}(x')dx' \; ; \tag{15.11}$$

15.2 Hydrodynamic Regime. Kinetics During Space Shuttle Reentry

$$\eta = \frac{u_{ex}(x)}{\sqrt{2\xi}} \int_0^y \rho(y') dy' \, . \tag{15.12}$$

Here, x is the longitudinal cartesian coordinate, y is the transverse coordinate, ρ_{ex} is the external and ρ the local mass density and u_{ex} is the external longitudinal speed. The coordinate ξ is simply a linear transformation of x and represents the longitudinal coordinate, while η is a new transverse coordinate, which is a linear transformation of y. In these generalized coordinates, the Navier–Stokes equation is separable into different contributions. The first is the following equation for the stream function $f(\eta)$

$$\frac{\partial}{\partial \eta} \left(\frac{\rho \mu}{\rho_{ex} \mu_{ex}} \frac{\partial^2 f(\eta)}{\partial \eta^2} \right) + f(\eta) \frac{\partial^2 f(\eta)}{\partial \eta^2} = 0 \, , \tag{15.13}$$

where μ and μ_{ex} are the viscosity coefficient respectively inside and outside the boundary layer, and $f(\eta)$ is defined from the equation

$$\frac{u(\eta)}{u_{ex}} = \frac{\partial f(\eta)}{\partial \eta} \, . \tag{15.14}$$

Tabulated numerical solutions of (15.13),(15.14) can be found in the literature. These solutions are well fitted by the following quadratic expression

$$f(\eta) = \max(0; \; -0.3316 + 0.4908\eta + 0.0427\eta^2) \, . \tag{15.15}$$

The range of η values goes from 0 (at the surface) to η_{max} (at the edge of the boundary layer). The function $f(\eta)$ represents the contribution of the longitudinal flow to the transverse flow equations.

The transverse flow is described by an equation for the gas temperature,

$$\frac{\partial^2 \Theta}{\partial \eta^2} + f(\eta) Pr \frac{\partial \Theta}{\partial \eta} = -Le S_\Theta, \tag{15.16}$$

and an equation for the mass density,

$$\frac{\partial^2 C_n}{\partial \eta^2} + f(\eta) Sc \frac{\partial C_n}{\partial \eta} = -S_n \, , \tag{15.17}$$

where Pr, Sc and Le are, respectively, the Prandtl, Schmidt and Lewis numbers (in the model, these numbers can usually be considered constant and equal to 1), and Θ and C_n are, respectively, the normalized temperature and mass density, i.e.,

$$\Theta(\eta) = \frac{T(\eta)}{T_{ex}}; \quad C_n(\eta) = \frac{\rho_n(\eta)}{\rho(\eta)} \, . \tag{15.18}$$

T_{ex} denoting the temperature at the edge of the boundary layer, $\rho_n(\eta)$ the local mass density of molecules in the n-th vibrational state (or the density of atoms of type n) and $\rho(\eta)$ the local total gas density.

The second derivative terms on the left-hand side of (15.16) and (15.17) are responsible for heat conduction and diffusion to the surface. The terms on the right-hand side are the sources due to collision processes. For vibrational kinetics, these include V-V and V-T energy exchange processes, and positive and negative sources of molecules due to dissociation and recombination. For atoms, the corresponding production and loss processes need to be considered. For binary collisions (for example, V-V processes), the terms S_n and S_Θ can be written as (compare with the second term on the right-hand side of (3.66)):

$$(S_n)_{VV} = a \sum_i [(R^{i,i+1}_{n+1,n} C_i C_{n+1} - R^{i,i+1}_{n,n-1} C_i C_n) \\ - (R^{i+1,i}_{n,n+1} C_{i+1} C_n - R^{i+1,i}_{n-1,n} C_{i+1} C_{n-1})] , \tag{15.19}$$

$$(S_\Theta)_{VV} = a \sum_n \left(\frac{E_n}{c_p T_{ex}}\right) \sum_i [(R^{i,i+1}_{n+1,n} C_i C_{n+1} - R^{i,i+1}_{n,n-1} C_i C_n) \\ - (R^{i+1,i}_{n,n+1} C_{i+1} C_n - R^{i+1,i}_{n-1,n} C_{i+1} C_{n-1})] , \tag{15.20}$$

with

$$a = \frac{p\, Sc}{kT(1+J)\beta \sum_j C_j} . \tag{15.21}$$

Here, T is the local temperature, p is the pressure, J is equal to 0 for a planar flux (as in the situation analyzed in [58–62]) and equal to 1 for an axial flux, k is the Boltzmann constant, c_p is the specific heat at constant pressure (equal to $\simeq 7k/2$, taking into account only translational and rotational degrees of freedom), E_n is the energy of the vibrational level n, $\beta \equiv du_{ex}/dx$ is the velocity gradient along the surface (which is related to the characteristic decreasing time of the parallel flux) and $R^{i,i+1}_{n+1,n}$ is the mass rate coefficient, which is related to the classical rate coefficients $Q^{i,i+1}_{n+1,n}$ by

$$R^{i,i+1}_{n+1,n} = Q^{i,i+1}_{n+1,n} \frac{\rho k T}{pm} , \tag{15.22}$$

m denoting the mass of the molecule.

To solve (15.16),(15.17) one has to consider boundary conditions taking into account the physical quantities outside the boundary layer and heat exchange and catalytic processes on the space vehicle surface. Outside the boundary layer (for $\eta = \eta_{max}$), a Boltzmann vibrational distribution and an equilibrium composition of atoms and molecules were assumed to hold at $T = T_{ex}$ [59–64]. Further, the surface ($\eta = 0$) was assumed to be at constant temperature and both atom recombination and vibrational relaxation were assumed ineffective (non-catalytic surface). In this case, the appropriate boundary conditions are the following:

$$T(\eta = 0) = T_w; \qquad T(\eta_{max}) = T_{ex} ;$$

15.2 Hydrodynamic Regime. Kinetics During Space Shuttle Reentry

$$\frac{dC_n}{d\eta}\bigg|_{\eta=0} = 0; \qquad C_n(\eta_{\max}) = C_0 e^{-E_n/kT_{\mathrm{ex}}} . \tag{15.23}$$

The boundary conditions at $\eta = 0$ would be different in the case of a catalytic surface. The derivatives $\frac{dC_n}{d\eta}\big|_{\eta=0}$ would not vanish, but would be proportional to the probabilities of wall losses (vibrational deactivation or atom recombination) (compare with (11.7); see [58] for details).

Therefore, the above model of gas dynamic and kinetic processes at the boundary layer of a body flying at supersonic velocity reduces to a system of non-linear, coupled, parabolic differential equations with two boundary conditions. This was used in [59–63,65] for the investigation of different effects in pure N_2 with non-equilibrium vibrational kinetic and dissociation-recombination processes. The conditions were typical for reentry of Space Shuttles (altitudes 86–50 km, speeds 7.5–3.8 km/s, incident gas flow density $\rho_\infty = 6.4 \times 10^{-6} - 6.8 \times 10^{-4}$ kg/m^3. It was shown [59] that allowing for a finite relaxation time of vibrationally excited N_2 molecules leads to a reduction of up to $\sim 30\%$ in the calculated heat flux to the surface. For a catalytic surface, the fast wall recombination of N atoms can result in a heat flux on the surface up to ~ 2 times as large as that in the absence of such recombination [60]. Detailed investigations of non-equilibrium vibrational distribution functions of N_2 were carried out in [60–65]. It was shown, for example, that V-T processes in N_2–N collisions have a marked effect on the vibrational distribution. The probabilities of these multi-quantum processes used in the computations were taken from [64] (see also (7.13),(7.14)) In spite of the large values used for these probabilities, the vibrational distributions of N_2 molecules in the boundary layer are far from equilibrium, as can be seen from Fig. 15.2.

The calculations shown in Fig. 15.2 were carried out using the "ladder climbing" model (vibrational dissociation and recombination of N_2 molecules taking place only through the upper boundary vibrational level). The limiting values $\eta = 4$ and $\eta = 0$ correspond to the boundary layer edge and to the body surface, respectively. We can see that the vibrational distribution has a strong non-equilibrium character near the surface ($\eta = 0$). This behaviour is due to recombination (strong positive source of vibrationally excited molecules in the boundary level $v = 45$). Note that such a strong non-equilibrium vibrational distribution leads to a strong effective dissociation rate, which differs from the usual Arrhenius behaviour as a function of the gas temperature T [63,65].

The above model was also used in [64] to investigate the electron energy distribution function (EEDF) and the influence of free electrons in the boundary layer on gas dynamic parameters and vibrational kinetics. The degree of ionization was taken as an independent input parameter. It was found that the overall effect consists in an energy flux from translational to vibrational degrees of freedom, and then to electrons. In this way, the heat flux to the vehicle surface decreases by up to $\sim 15\%$, whereas the degree of ionization increases up to 10^{-3}.

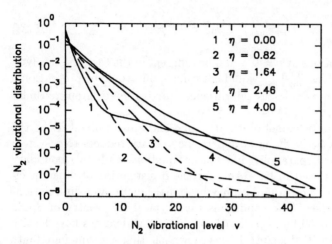

Fig. 15.2. Normalized vibrational distributions of N_2 molecules versus vibrational quantum number, for different values of η, T_w=1000 K, T_{ex}=7000 K, p_{ex}=10^5 N/m^2, and β= 10^5 s^{-1} [63]

The above model was also used for analyzing the boundary layer gas dynamics, the level-by-level vibrational kinetics of N_2 and O_2, the dissociation-recombination processes of N_2 and O_2, and exchange reactions involving N_2, O_2, NO molecules and N and O atoms. Preliminary results were reported in [63,65]. The calculations exhibit a strong non-Arrhenius temperature dependence for the rate coefficient of NO production via the reaction $N_2(v)+O \rightarrow NO+N$.

The analysis of shock wave hydrodynamics in front of a moving body (including the boundary layer or not) for N_2–O_2 air mixtures has usually been carried out taking into account chemical kinetics but using only a simple vibrational energy relaxation model (without detailed calculations of the vibrational distribution function) [68–79]. Such an approach was used, for example, in a model [71–75] for computing the heat flux on the bow of the Russian Space Shuttle "Buran" during reentry, at altitudes 100–52 km. Processes involving charged particles and electronic excited states were not considered, but five air components (N_2, O_2, N, O, NO) were taken into account. It was assumed that these components participate in the following exchange, dissociation and recombination reactions:

$$O_2 + M \Longleftrightarrow 2O + M ; \tag{15.24}$$

$$N_2 + M \Longleftrightarrow 2N + M ; \tag{15.25}$$

$$NO + M \Longleftrightarrow N + O + M ; \tag{15.26}$$

$$NO + O \Longleftrightarrow N + O_2 ; \tag{15.27}$$

$$N_2 + O \Longleftrightarrow N + NO ; \tag{15.28}$$

$$N_2 + O_2 \Longleftrightarrow 2\,NO\,, \tag{15.29}$$

where M is any of the particles in the mixture. The rate coefficients of these processes were given in Tables 10.1–10.3. Similar values were used in [74–78]. The exponential factors in the rate coefficients of dissociation in these tables depend upon both translational and vibrational temperatures. The anharmonic oscillator model was therefore used (see (3.126)–(3.129)). However, vibration-translation exchanges were described by equation (3.35) for the harmonic model, using an approximate formula taken from [80] for the relaxation times τ_{VT}. It was also assumed in [74–77] that the vibrational temperatures of all molecular air components are nearly the same, therefore the same average vibrational temperature T_V was assumed for all molecules. However, calculations using different vibrational temperatures for N_2, O_2 and NO were carried out in [78].

A model of a thin, viscous shock layer (including the boundary layer) was developed in [74–78]. This model is based on the asymptotic form of the Navier–Stokes equations for large Mach and Reynolds numbers, and a large density ratio behind and ahead of the shock wave front. These are typical conditions for the main heating regions during gliding path reentry. At altitudes 50–90 km, this model yields similar results to those from the parabolic Navier–Stokes equations. The boundary conditions for the gas component concentrations on the body surface can correspond either to ideal catalytic or non-catalytic heterogeneous recombination reactions. The boundary condition for the translational temperature was determined by a thermal balance, that is, by equating the heat flux to the surface to the radiated outward energy flux. Concerning the vibrational temperature T_V, either equilibrium $T_V = T_w$ (T_w is the wall temperature) or the non-catalyticity of the wall relative to internal degrees of freedom (that is, $\frac{\partial T_V}{\partial y}|_{y=0} = 0$, where y is the distance from the body) was assumed. The generalized Rankine-Hugoniot conditions [81,82] were used as boundary conditions at the shock wave front.

The results obtained with this model show that the translational temperature exceeds the vibrational one behind the shock wave on the stagnation line up to the boundary layer, which significantly affects the heat flux to the surface. This situation arises along all the heating path of reentry into the Earth's atmosphere. Figures 15.3 and 15.4 illustrate this fact. The increase in heat flux under non-equilibrium flow relative to the equilibrium case can be explained by the smaller dissociation, due to vibrational temperature being lower than translational one. The decrease in vibrational temperature is explained by the losses of vibrational energy due to dissociation. The decrease in vibrational temperature and the slowing down of dissociation result in a higher translational temperature (see Fig. 15.3). This effect increases the dissociation rate, but this growth is not enough to overcome the decrease caused by the drop in vibrational temperature. Dissociation due to non-equilibrium vibrational energy takes place along almost all the gliding reentry path.

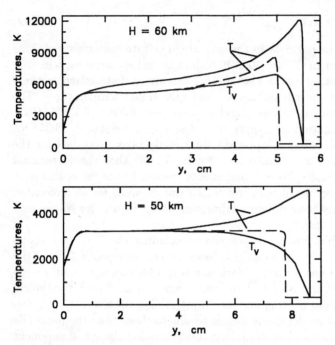

Fig. 15.3. Profiles of vibrational (T_V) and translational (T) temperatures in the thin shock layer in front of the body at different gliding reentry altitudes. The gas flow is directed from the right to the left. The point $y = 0$ corresponds to the body surface. The full curves were calculated considering vibrational and chemical non-equilibrium. The dashed curves were calculated assuming vibrational-translational equilibrium [78]

Besides investigations of the shock layer in front of supersonic bodies, analyses of physical-chemical processes in the track are also of great interest. The air composition and radiation remain out of equilibrium for a long time in this region, at altitudes 60–100 km. However, this problem has not yet been studied in detail. A simultaneous analysis of non-equilibrium chemical kinetics and vibrational relaxation in a spherical body's track was carried out in [83], using the reactions (15.24)–(15.29) and the harmonic oscillator model. However, the gas dynamic problem was not solved in this work. Using the analogy of a cylindrical explosion [84–86] and analytical approximations for the numerical solutions of equilibrium gas dynamic equations [84,85], the translational temperature and gas pressure behind the shock front were determined into the body's track. Such an approach makes this problem somewhat simpler and allows the concentrations of N_2, O_2, N, O, NO, the vibrational temperatures of N_2, O_2, NO and the volume intensity of NO vibrational band infrared radiation to be calculated. In this situation, the effects of the chemical and vibrational kinetics on the gas temperature behind the shock wave front can be estimated by varying the equilibrium specific heat

15.2 Hydrodynamic Regime. Kinetics During Space Shuttle Reentry

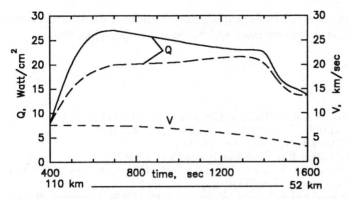

Fig. 15.4. Velocity of the spacecraft "Buran" and heat flux on its front as a function of time of reentry. The full curve is for non-equilibrium gas flow, with boundary condition $T_V = T_w$. The dashed line is for equilibrium, $T_V = T$. The surface is assumed non-catalytic for chemical processes [76]

ratio γ, which depends on the chemical gas composition and the vibrational temperature.

The behavior of gas temperature and chemical composition in the gas flow under the bottom of a body have also recently been investigated [87]. A numerical algorithm for solving the Navier–Stokes equations in this region was developed, taking into account a chemical non-equilibrium air model. The effects of catalytic surface properties, with respect to surface recombination of N and O atoms, on the profiles of gas temperature, velocity and concentrations of O and NO behind the body were analyzed. However, the vibrational kinetics was not investigated in this model.

The works [83,87] constitute merely the first steps towards solving the problem. Further research should concentrate on self-consistent analyses of gas dynamics and kinetic processes (including non-equilibrium vibrational kinetics). Considerable improvement in kinetic modelling would, for example, result from introducing an anharmonic vibration model, extending the list of reactions, using more accurate recent data on rate coefficients, and also including charge kinetics in the analysis.

Before ending this chapter we wish to stress that state-to-state vibrational kinetics has been analysed in a one-dimensional nozzle in air [88,89] and in nitrogen [90–92] and oxygen [93] systems. The results in [54,88–92] confirm those obtained for the boundary layer of reentring bodies [59–65]. In particular, nonequilibrium vibrational distribution and non-Arrhenius behavior of the different rates have been calculated along the nozzle axis. These effects are mainly due to preferential pumping of the high-lying vibrational levels by the recombination process. On the other hand, highly non-Boltzmann vibrational distributions were found in the stagnation region of a blunt body [94]. Further, it should be noted that also electrons and ions can play an important role in reentry problems [64,65,91,95–98]. Finally, the effects on the transport properties of the nonequilibrium vibrational distribution have been analysed under boundary layer and nozzle conditions [99,101].

References

1. Bread D.B., Johnson F.S. (1961) J. Geophys. Res. **66**, 4113
2. Parker L.W., Murphy B.L. (1967) J. Geophys. Res. **72**, 1631
3. Linson L.M. (1969) J. Geophys. Res. **74**, 2368
4. Winckler G.R. (1980) Rev. Geophys. Space Phys. **18**, 659
5. Alpert Ya.L., Gurevich A.U. and Pitaevsky L.P. (1964) *Sattelites in Rarefied Plasmas*, Nauka, Moscow (*in Russian*)
6. Green B.D., Caledonia G.E. and Wilkerson T.D. (1985) J. Spacecraft and Rockets **22**, 500
7. Hays P.B., Carignan G. and Kennedy B.C. (1973) Radio Sci. **8**, 369
8. Torr M.R., Hays P.P., Kennedy B.C. and Walker J.C.G. (1977) Planet. Space. Sci. **25**, 173
9. Yee J.H. and Abreu V.J. (1983) Geophys. Res. Lett. **10**, 126
10. Garret H.B., Chatdjan A. and Gabrial C.B. (1988) J. Spacecraft and Rockets **25**, 321
11. Mende S.B., Swenson G.R. and Llewellyn E.J. (1988) Adv. Space Res. **8**, 229
12. Torr M.R. (1983) Geophys. Res. Lett. **10**, 114
13. Ahmadjian M. and Jennings D.E. (1995) J. Spacecraft and Rockets **32**, 507
14. Swenson G.R., Mende S.B. and Clifton K.S. (1985) Geophys. Res. Lett. **12**, 97
15. Swenson G.R., Rairden R.L., Jennings D.E. and Ahmadjian M. (1996) J. Spacecraft and Rockets **33**, 240
16. Kofsky I.L. and Barrett J.I. (1986) Planet. Space Sci. **34**, 665
17. Von Zanh U. and Murad E. (1986) Nature **321**, 147
18. Arnold G.S. and Coleman D.J. (1988) J. Chem. Phys. **88**, 7147
19. Arnold G.S. and Coleman D.J. (1991) Chem. Phys. Lett. **177**, 279
20. Caledonia G.E., Holtzclaw K.W., Green B.D., Krech R.H., Leone A. and Swenson G.R. (1990) Geophys. Res. Lett. **17**, 1881
21. Swenson G.R., Leone A., Holtzclaw K.W. and Caledonia G.E. (1991) J. Geophys. Res. **96**, 7603
22. Orient O.J., Martus K.E., Chutjian A. and Murad E. (1992) Phys. Rev. A **41**, 4106
23. Caledonia G.E., Holtzclaw K.W., Krech R.H., Sonnenfroh D.M., Leone A. and Blumberg W.A.M. (1993) J. Geophys. Res. **98**, 3725
24. Collins R.J., Dogra V.K. and Levin D.A. (1997) AIAA Paper **97-0987**
25. Dogra V.K., Collins R.J. and Levin D.A. (1998) AIAA Paper **98-0834**
26. Collins R.J., Levin D.A. and Dogra V.K. (1998) in Proceed. 21st International Symposium on Rarefied Gas Dynamics, Marseille
27. Karipides D.P., Boyd I.D. and Caledonia G.E. (1998) J. Thermophys. and Heat Transfer **12**, 30
28. Karipides D.P., Boyd I.D. and Caledonia G.E. (1998) *AIAA Paper*, **98-2848**
29. Karipides D.P., Boyd I.D. and Caledonia G.E. (1998) in Proceed. 21st International Symposium on Rarefied Gas Dynamics, Marseille
30. Kortsenstein N.M., Plastinin Yu.A. and Samuilow E.W. (1998) in Proceed. 21st International Symposium on Rarefied Gas Dynamics, Marseille

31. Friedlander O.G., Perminov V.D., Karabadjak G.F.and Plastinin Yu.A. (1998) in Proceed. 21st International Symposium on Rarefied Gas Dynamics, Marseille
32. Gordiets B.F. and Ferreira C.M. (1997) AIAA Paper **97-2504**
33. Gordiets B.F. and Ferreira C.M. (1998) AIAA Journal **36**, 1643
34. Koleshko C.B., Lukashin Yu.P. and Popov F.D. (1966) *Aerophysical Investigations of Supersonic Flows*, Nauka, Moscow (*in Russian*)
35. Hall J.G., Eschenraeder A.Q. and Marrone P.V. (1962) JAS **29**, 1038
36. Belotserkovsky O.M. and Dumin V.K. (1962) J. Cal. Math.and Math. Phys. **4**, 61 (*in Russian*)
37. Lukashin Yu.P. and Popov F.D. (1984) J. Cal. Math. and Math. Phys. **4**, 896 (*in Russian*)
38. Young L.A. (1968) JQSRT **8**, 105
39. Davis R.T. (1970) AIAA Paper **N 70-805**
40. Minner E.W. and Lewis C.H. (1975) NASA Rep. **CR**-2550
41. Rakich J.V. and Lanfranco M.J. (1975) J. Spacecraft and Rockets **14**, 265
42. Swaminathan S., Kim M.D. and Lewis C.H. (1983) J. Spacecraft and Rockets **20**, 331
43. Swaminathan S., Kim M.D. and Lewis C.H. (1984) AIAA Journal **22**, 754
44. Kim M.D., Swaminathan S. and Lewis C.H. (1984) J. Spacecraft and Rockets **21**, 29
45. Gershbein E.A., Schelin V.S. and Unitski S.A. (1985) Space Res. **23**, 416 (*in Russian*)
46. Scott C.D. (1985) J. Spacecraft and Rockets **22**, 489
47. Song D.J. and Lewis C.H. (1986) J. Spacecraft and Rockets **23**, 47
48. Landrum D.B. and Cander G.V. (1992) AIAA Paper **92-2853**
49. Sharma S.P., Huo W.H. and Park C. (1992) J. Thermophysics **6**, 9
50. Adamovich I.V., Macheret S.O., Rich J.W. and Treanor C.E. (1995) AIAA Journal **33**, 1064; ibid **33**, 1070.
51. Varghese P.L. and Gonzales D.A. (1995) in *Molecular Physics and Hypersonic Flows*, ed. M.Capitelli, Kluwer, Dordrecht, NATO ASI Series, Vol. 482, p. 105
52. Losev S.A., Kovach E.A., Makarov V.N., Pogosbekjan M.Ju. and Sergievskaia A.L. (1995) in *Molecular Physics and Hypersonic Flows*, ed. M.Capitelli, Kluwer, Dordrecht, NATO ASI Series, Vol. 482, p. 597
53. Zhluctov S.V. and Smechov G.D. (1995) in *Molecular Physics and Hypersonic Flows*, ed. M.Capitelli, Kluwer, Dordrecht, NATO ASI Series, Vol. 482, p. 615
54. Park C. (1995) in *Molecular Physics and Hypersonic Flows*, ed. M.Capitelli, Kluwer, Dordrecht, NATO ASI Series, Vol. 482, p. 665
55. Gordiets B.F. and Sergievskaia A.L. (1997) Chemical Physics **16**, 11 (*in Russian*)
56. Macheret S.O. and Rich J.W. (1993) Chemical Physics **174**, 25
57. Macheret S.O., Fridman A.A., Adamovich I.V., Rich J.W. and Treanor C.E (1994) AIAA Paper **94-1984**
58. Giordano D.,Bellucci V., Colonna G., Capitelli M., Armenise I. and Bruno C. (1997) J. Thermophys. and Heat Transfer **11**, 27
59. Doroshenko V.M., Kudryavtsev N.N., Novikov S.S. and Smetanin V.V. (1990) High Temp. **28**, 82
60. Armenise I., Capitelli M., Colonna G., Koudriavtsev N. and Smetanin V. (1995) Plasma Chem. Plasma Process. **15**, 501
61. Armenise I., Capitelli M., Celiberto R., Colonna G., Gorse C. and Lagana A. (1994) Chem. Phys. Lett. **227**, 157
62. Armenise I., Capitelli M., Celiberto R., Colonna G. and Gorse C. (1994) AIAA Paper **94-1987**

63. Armenise I., Capitelli M. and Gorse C. (1995) Proceed. Second European Symposium on Aerothermodynamics for Space Vehicles, **ESA SP-367**; Armenise I., Capitelli M. and Gorse C. (1998) J. Thermophys. and Heat Transfer **12**, 45
64. Colonna G. and Capitelli M. (1995) AIAA Paper **95-2071**; Colonna G. and Capitelli M. (1996) J. Thermophys. and Heat Transfer **10**, 406
65. Armenise I., Capitelli M. and Gorse C. (1995) in *Molecular Physics and Hypersonic Flows*, ed. M.Capitelli, Kluwer, Dordrecht, NATO ASI Series, Vol. 482, p. 703; Armenise I., Capitelli M. and Gorse C. (1996) J. Thermophys. and Heat Transfer **10**, 397; Armenise I., Capitelli M. and Gorse C. (1997) J. Thermophys. and Heat Transfer **11**, 570
66. Agafonov V.P., Vertuchkin V.K., Gladkov A.A. and Poliansky O.Yu. (1972) *Non-equilibrium Physical and Chemical Processes in Aerodynamics*, Machinostroenie, Moscow (*in Russian*)
67. Lagana A. and Garcia E. (1994) J. Phys. Chem. **98**, 502
68. Tong H. (1966) AIAA Journal **4**, 14.
69. Ladnova L.A. (1966) Inf. Leningrad State Univ. **N 13**, 105 (*in Russian*)
70. Zalogin G.N. (1976) Inf. Acad, Sci. USSR **6**, 105 (*in Russian*)
71. Park C. (1984) AIAA Paper **84-0306**
72. Park C. (1984) AIAA Paper **84-1730**
73. Park C. (1990) *Non-equilibrium Hypersonic Aerodynamics*, John Wiley and Sons. New York.
74. Tirskiy A.G. and Scherbak V.G. (1990) J. Appl. Mech. Techn. Phys. **6**, 55 (*in Russian*)
75. Tirskiy A.G. and Scherbak V.G. (1990) MJG **1**, 151 (*in Russian*)
76. Tirskiy A.G., Schelin V.S. and Scherbak V.G. (1990) Mathematical Modelling **2**, 28 (*in Russian*)
77. Tirskiy A.G. and Scherbak V.G. (1991) Thermophys. High Temp. **29**, 317 (*in Russian*)
78. Shcherbak V.G. (1993) J. Appl. Mech. Techn. Phys. **N 1** (*in Russian*)
79. Knab O., Fruhauf H.-H. and Jonas S. (1992) *AIAA Paper* **92-2947**
80. Millikan R.C. and White D.R. (1963) J. Chem. Phys **39**, 3209
81. Zeldovich Y.B. and Raizer Yu.P. (1966) *Physics of Shock Waves and High Temperature Hydrodynamic Flows*, Nauka, Moscow (*in Russian*)
82. Stupochenko E.A., Losev S.A. and Osipov A.I. (1967) *Relaxation Processes in Shock Waves*, Springer, Berlin Heidelberg
83. Birukov A.S., Gordiets B.F., Markov M.N. and Shelepin L.A. (1971) Preprint Lebedev Phys. Inst. (FIAN) **N 59** (*in Russian*)
84. Feldman S. (1960) ARS Journal **30**, 463
85. Feldman S. (1961) J. Aerospace Sci. **28**, 433
86. Tsikulin M.A. (1969) *Shock Waves under the Movement of Large Meteorites Through the Atmosphere*, Nauka, Moscow (*in Russian*)
87. Ivanov D.V., Yegorov I.V. and Yegorova M.V. (1997) IAAA Paper **97-2583**
88. Colonna G., Tuttafesta M., Capitelli M., and Giordano D. (1999) J. Thermophys. and Heat Transfer **13**, 372
89. Colonna G., Tuttafesta M., Capitelli M., and Giordano D. (1999) AIAA Paper **99-3685** (2000) J. Thermophys. Heat Transf. **14** (in press)
90. Colonna G., Tuttafesta M., Capitelli M., and Giordano D. (1999) in *Rarefied Gas Dynamics*, eds. R. Brun, R. Campargue, R. Gatignol, and J. C. Lengrand, Cépadues Editions, Toulouse, Vol. 2, p. 281
91. Capitelli M., Colonna G., Gorse C., and Esposito F. (1999) AIAA Paper **99-3568**

92. Buzykin O. G., Makashev N. K. and Nosik V. I. (1999) in *Rarefied Gas Dynamics*, eds. R. Brun , R. Campargue, R. Gatignol , and J. C. Lengrand, Cépadues Editions, Toulouse, Vol. 2, p. 305
93. Shizgal B. and Lordet F. (1996) J. Chem. Phys. **104**, 3579
94. Candler G. V., Olejniczak J. and Harrold B. (1997) Phys. Fluids **9**, 2108
95. Capitelli M., Gorse C., Longo S., Dyatko N., and Hassouni K. (1998) J. Thermophys. and Heat Transfer **12**, 478
96. Laux C.O., Yu L., Packhan D.M., Gessman R.J., Pierrot L., and Kruger C.H. (1999) AIAA Paper **99-3476**
97. Mertens J. D. (1999) J. Thermophys. and Heat Transfer **13**, 204
98. Colonna G., Armenise I., Catella M., and Capitelli M. (2000) J. Techn. Phys. **41**, Special Issue, 203
99. Capitelli M., Bruno D., Colonna G., Giordano D., Kustova E., and Nagnibeda E. (1999) Mathematical Modelling **11**, 45
100. Armenise I., Capitelli M., Kustova E., and Nagnibeda E. (1999) J. Thermophys. and Heat Transfer **13**, 210
101. Bruno D., Capitelli M., Kustova E., and Nagnibeda E. (1999) Chem. Phys. Lett. **308**, 463

16. Acoustic and Shock Waves in Non-equilibrium Gases

As mentioned above, non-equilibrium molecular gases and low-temperature plasmas are unusual states of matter. The peculiarities of these states, together with the non-equilibrium magnitude of the internal energy of gas particles, can strongly affect the macroscopic characteristics and the behaviour of the gas, i.e. its chemical activity, optical parameters, transport coefficients and hydrodynamics.

In this chapter we will discuss the properties of a non-equilibrium vibrationally excited molecular gas when acoustic and shock waves propagate through it. In such a gas, the transformation of internal energy into energy of hydrodynamic motion can occur, under the influence of hydrodynamic perturbations. In this way, hydrodynamic perturbations can increase. This leads to new effects, which will now be briefly discussed.

Let us note that the theory presented here is valid whatever the mechanisms leading to vibrational non-equilibrium and the nature of the molecular gas. However, the effects to be discussed have been studied, both theoretically and experimentally, only for N_2, N_2-O_2, and N_2-Ar. Usually, vibrational non-equilibrium has been produced by a gas discharge.

16.1 Propagation of Small Perturbations in Non-equilibrium Gases

Two types of waves can develop as a result of small hydrodynamic perturbations in a homogeneous gas flow of structureless particles: acoustic waves, propagating with velocity equal to the sum of the flow and sound velocities, and entropic waves propagating at the flow velocity [1]. Only the density and the temperature change in the case of entropic waves, the pressure and velocity remain constant. Mathematically, the existence of acoustic and entropic waves is related to the properties of the solutions to linearized hydrodynamic equations. In the simple case of one-dimensional motion, the system of such equations consists of three first-order equations (the mass, momentum, and energy conservation equations) involving three independent variables, for example, perturbations of pressure, mass density, and velocity, p', ρ' and v', respectively. The solution of this system consists of three eigenfunctions

288   16. Acoustic and Shock Waves in Non-equilibrium Gases

and the corresponding eigenvalues. Two of these eigenfunctions correspond to acoustic waves propagating in opposite directions, and the third to the entropic wave.

The analysis is more complicated for the propagation of perturbations in a molecular gas in which the internal and the translational degrees of freedom are not in equilibrium with each other. The non-equilibrium energy distribution changes the adiabatic compression which determines the sound velocity. This change leads to anomalous dispersion of sound waves. This effect has been experimentally investigated in detail. Herzfeld and Rice [2] and Knezer [3] pioneered the development of the corresponding theory. A general thermodynamic theory of relaxation processes in sound waves was first developed by Mandelstham and Leontovich [4].

New effects can appear in connection with the propagation of hydrodynamic perturbations in a non-equilibrium gas, such as sound wave amplification. The physical mechanisms leading to this amplification can be explained as follows. The energy exchanges between internal and translational degrees of freedom, during different half periods of the sound wave, proceed in different ways. The energy converted into translational motion during a half period, corresponding to an increase in temperature, can exceed that converted back by the reverse process, when the temperature is smaller than the equilibrium temperature. The gas temperature is the main factor determining the energy exchange rate, the latter increasing with temperature.

To understand the main peculiarities of sound wave propagation, let us consider the simple case of a plane sound wave in a non-equilibrium gas. It will be assumed that the stationary, non-equilibrium state of the gas is sustained by some energy pumping source (with power I per unit mass) into the vibrational degrees of freedom, while energy losses will be assumed to take place via the translational degrees of freedom, by thermal conductivity (with power Q per unit mass) [5,6]. The conservation equations take the following form:

$$\frac{\partial \rho}{\partial t} + \text{div}(\rho \boldsymbol{v}) = 0 ; \tag{16.1}$$

$$\rho \frac{\partial \boldsymbol{v}}{\partial t} = -\nabla p ; \tag{16.2}$$

$$\frac{dH}{dt} - \frac{1}{\rho}\frac{dp}{dt} = I - Q , \tag{16.3}$$

where: $H = \gamma p/[(\gamma - 1)\rho] + \epsilon_{\text{vib}}$; γ is the adiabatic coefficient (without consideration of the vibrational degrees of freedom); ρ, p, $\boldsymbol{v}$ are, respectively, the mass density, pressure and gas velocity; and ϵ_{vib} is the vibrational energy per unit mass. The relaxation equation for ϵ can be written as

$$\frac{d\epsilon_{\text{vib}}}{dt} = \frac{\epsilon_{\text{vib}}^0(T) - \epsilon_{\text{vib}}}{\tau_{\text{VT}}} + I , \tag{16.4}$$

16.1 Propagation of Small Perturbations in Non-equilibrium Gases

where $\epsilon_{\text{vib}}^0(T)$ is the equilibrium vibrational energy at translational temperature T, and $\tau_{\text{VT}}(\rho, T)$ is the vibrational relaxation time. The system (16.1)–(16.4) together with the equation of state forms a closed system of equations. Small perturbations can be described by linearized equations, obtained by inserting into (16.1)–(16.4) plane wave solutions of the type

$$A + A' \exp(\mathrm{i} k' x - \mathrm{i}\omega t) , \tag{16.5}$$

where $A \equiv \rho, p, T, v, \epsilon_{\text{vib}}$ is an unperturbed medium parameter (in the absence of the perturbation), and A' is the amplitude of the perturbation. By this procedure one obtains the dispersion relation, that is, the relationship between the frequency ω and the wavenumber k'. The specific relationship is determined by the boundary conditions. The wavenumber is generally a complex quantity of the form $k'(\omega) = \mathrm{Re}\, k'(\omega) + i\, \mathrm{Im}\, k'(\omega)$, whose real and imaginary parts determine the phase velocity and the absorption (or amplification) coefficient, respectively. The condition for amplification at high frequencies ($\omega \tau_{\text{VT}} \gg 1$) is

$$\mathrm{Im}\, k' = \frac{m(\gamma-1)^2}{2k(\gamma p/\rho)^{1/2} \gamma \tau_{\text{VT}}}$$

$$\left[\left(\frac{2}{\gamma-1} - \frac{\partial \ln \tau_{\text{VT}}}{\partial \ln T} \right) \frac{\epsilon_{\text{vib}} - \epsilon_{\text{vib}}^0}{T} - \frac{\partial Q}{\partial T} \tau_{\text{VT}} - c_{\text{vib}} \right] > 0 , \tag{16.6}$$

where m is the mass of the molecule, k is the Boltzmann constant, and $c_{\text{vib}} = \partial \epsilon_{\text{vib}}^0/\partial T$ is the thermal capacity of the vibrational degrees of freedom. The amplification is larger if $\partial \ln \tau_{\text{VT}}/\partial \ln T < 0$ and the degree of vibrational non-equilibrium $(m(\epsilon_{\text{vib}} - \epsilon_{\text{vib}}^0)/kT)$ increases. The last term in (16.6) corresponds to the ordinary dissipation of sound in a gas with internal degrees of freedom. The amplification coefficient $\mathrm{Im}\, k$ also increases as the relaxation time τ_{VT} decreases. In this case, the pumping power I must be increased to provide the same non-equilibrium, since $\epsilon_{\text{vib}} - \epsilon_{\text{vib}}^0 = I\tau_{\text{VT}}$ (see (16.4)). For example, for CO at $p = 100$ Torr, $T = 500$ K, and vibrational temperature $T_V = 700$ K, the amplification coefficient is $\mathrm{Im}\, k' \simeq 0.1$ cm^{-1}. For comparison, the amplification coefficient for infrared radiation of the CO laser is ~ 0.1 m^{-1}.

For frequencies such that $\omega \tau_{\text{VT}} \leq 1$, the function $\mathrm{Im}\, k'(\omega)$ was analyzed in [5]. This function has a complicated form and depends primarily on the energy pumping and loss characteristics.

The possibility of amplification of sound waves was first considered in the case of low-temperature non-equilibrium plasmas of ordinary gas discharges, with electron temperature $T_e \gg T$ [7,8] (see also [9–23]). Note that all typical problems of acoustics (for example, standing sound waves, sound reflection, etc.) also arise in the case of sound propagation in non-equilibrium gases. These problems have yet to be studied in detail. However, in several works [24–26], the results obtained from analysis of light wave propagation in active media were used to solve these problems. This type of analogy has been in use for some time. For example, the possibility of sound wave amplification was considered in [10], having in mind the creation of an acoustic analogue of the laser. Supersaturated vapors have been proposed [27] as non-equilibrium media for such a system. However, the results of optics should be used with care in acoustics. Such an analogy is correct for stationary medium vibrations [24], but it can be erroneous for sound reflection problems. For example, vibrations of the boundary, which can significantly affect the calculations, do appear when sound is reflected from the boundary between equilibrium and non-equilibrium gases.

The imaginary part of the wavenumber k' at low frequencies ($\omega \tau_{VT} \ll 1$) is proportional to ω^2, the same dependence as that of viscous effects. There is therefore an analogy between relaxation absorption and the action of viscous forces [28,29]. For $\omega \tau_{VT} \ll 1$, one can consider the relaxation process in the zero-order approximation only (with respect to $\omega \tau_{VT}$). The correction of first order in $\omega \tau_{VT}$ appears in the equation of motion as an additional term corresponding to the second viscosity. Note that, in such an approach, there are no restrictions to the magnitude of the perturbation. The quantity $\omega \tau_{VT}$ is the small parameter of the development (in this case, ω^{-1} is to be considered as some characteristic time scale of the problem). This approach was developed in detail for equilibrium in [30], and for non-equilibrium in [31].

There exist few experimental data concerning sound wave amplification. The study of acoustic instabilities in discharges started thirty years ago [32–36] and is now undergoing intensive development [37–43]. The amplification coefficient for molecular gases N_2, N_2-O_2, N_2-Ar was measured in [43]. It is worth pointing out some experimental investigations on sound wave propagation in a reacting gas mixture with the exothermic reaction $H_2+Cl_2 \Longleftrightarrow 2HCl$ [44–46]

16.2 Nonlinear Hydrodynamic Waves in Non-equilibrium Gases

The nonlinear equations of motion (Euler's equations), not assuming small nonlinearities, were first integrated by Riemann (see [1]), who solved the system of one-dimensional hydrodynamic equations and the equation of state $p = p(\rho)$. The method used by Riemann consists in linearizing the system of equations by substitution of variables, which can be done in the one-dimensional case. Analysis of the solution shows that the regions of compression move faster than the low pressure regions, which distorts the wave profile as the wave propagates. The wave front becomes steeper and steeper until the compression wave is transformed into a shock wave. However, Riemann's solution is not correct in the vicinity of this transformation, because it fails to take into account viscosity and thermal conductivity, the role of which become greater as the gradients increase.

16.2 Nonlinear Hydrodynamic Waves in Non-equilibrium Gases

Although Riemann's solution does not assume small nonlinearity, its validity is also limited, from the physical point of view, to the range where the adiabatic equation of state $p = p(\rho)$ applies. The adiabatic equation of state describes isentropic processes, while the change of entropy for acoustic waves is of order three in smallness with respect to the Mach number, $Ma = v/c_s$ (c_s is the local sound speed). Therefore, the Riemann solution describes the behaviour of intense sound waves only as a second-order approximation.

Nonlinear and dissipative effects can be considered simultaneously using Burgers equation [47]. This equation is not exact, since it results from some simplifications of the hydrodynamic equations for a viscous gas. But, at the same time, this equation is reasonably accurate for describing nonlinear waves in dissipative media, since it takes into account all basic small terms to the second order.

Using a prime to designate the perturbations in the hydrodynamic parameters, Burgers equation for p' can be written as

$$\frac{\partial p'}{\partial x} - bp'\frac{\partial p'}{\partial \theta} - \delta \frac{\partial^2 p'}{\partial \theta^2} = 0, \qquad (16.7)$$

where $\theta = t - (x/c_s)$, δ is the dissipation coefficient, and $b = (\gamma+1)/2c_s^3$ is a nonlinear parameter.

For small nonlinearity, Burgers equation transforms into a well-known linear equation for dissipative media, and, in the absence of dissipation, into the Riemann's equation for ordinary waves [47].

The evolution of a sound wave with finite amplitude as described by Burgers equation can be qualitatively explained as follows. During the first stage of propagation, the wave front profile is progressively distorted until a saw-tooth signal forms. In the second stage, the front form stabilizes. This equilibrium is reached through competitive nonlinear and dissipative effects. However, the oscillating wave amplitude progressively decreases, because the dissipative effects are not compensated. In the third stage the wave amplitude no longer depends on the initial amplitude. The wave becomes harmonic again and attenuates according to the laws of linear acoustics. A solution of Burgers' equation describing the evolution of a harmonic initial perturbation was obtained in [47].

Energy exchanges between vibrational degrees of freedom and hydrodynamic motion are a new problem to be considered with respect to nonlinear hydrodynamic wave propagation in non-equilibrium gases, with a surplus of vibrational energy. This additional energy source for wave processes can lead to two new effects. First of all, wave breaking is accelerated. Wave amplification in the vibrationally excited gas increases the role of the nonlinear factors, and, consequently, nonlinear effects show up earlier. Secondly, the energy pumping tends to compete with dissipative processes. After the wave breaks, this can lead to the establishment of a stationary regime in which dissipation losses are compensated by an energy input from the internal degrees of freedom.

A quantitative description of nonlinear hydrodynamic perturbations in non-equilibrium vibrationally excited gases was presented in [28,29]. The main equation is a generalized Burgers equation, for $\omega\tau_{VT} \gg 1$:

$$\frac{\partial p'}{\partial x} - bp'\frac{\partial p'}{\partial \theta} - \delta\frac{\partial^2 p'}{\partial \theta^2} - \beta p' = 0 \qquad (16.8)$$

which differs from (16.7) by the term $\beta p'$, which takes into account wave amplification due to energy exchanges between internal degrees of freedom and hydrodynamic modes. In (16.8), the parameter β is given by (compare with (16.6))

$$\beta = -\frac{(\gamma-1)\rho}{2(\gamma p/\rho)^{1/2}\tau_{VT}}\left[\frac{T(\gamma-1)c_{\text{vib}}}{\gamma p} + I\frac{\partial \tau_{VT}}{\partial p} - \frac{\partial Q}{\partial p}\tau_{VT}\right]. \qquad (16.9)$$

The solution of equation (16.8) shows, for example, that an initial sinusoidal signal, propagating in a vibrationally excited gas, transforms into a stationary saw-tooth signal, the shape of which does not subsequently change. It should be emphasized that this solution does not take into account changes in the gas temperature and that it describes the creation of "gaps" moving with Mach number ~ 1.

Equations of the same type as (16.8) are known in plasma [48] and chemically reacting gas [49] theories. The solution of equation (16.8) was investigated in [50]. An equation similar to (16.8) was also obtained in [51] for a medium with a heat source. The results of [28] were generalized in [52] for the case of cylindrical and spherical waves. The propagation of nonlinear perturbations in a vibrationally non-equilibrium gas, when $\omega\tau_{VT} \gg 1$, was analytically studied in [53]. The propagation of sound waves in a vibrationally non-equilibrium gas was also considered in [54], for a nonlinear regime, when the relaxation time τ_{VT} depends strongly on the gas temperature.

For calculations of finite nonlinear perturbations in a non-equilibrium flowing gas mixture, with changing parameters, a more complicated (as compared with (16.4)) system of gas kinetic equations has to be considered. Such calculations were carried out in [55] for the CO_2–N_2–H_2O mixture used in hydrodynamic CO_2 lasers. The propagation of sound and entropic waves, and the influence of different parameters on the creation of shock waves were studied in [55].

At the present time, there are no direct experiments on the propagation of nonlinear hydrodynamic perturbations in a non-equilibrium gas. Only the interaction between sound waves and a discharge was investigated in the experimental works [56–58].

16.3 Propagation of Shock Waves in Non-equilibrium Gases

The propagation of a shock wave is accompanied by heating of the gas. The heating process in diatomic and polyatomic gases has two stages. At the

16.3 Propagation of Shock Waves in Non-equilibrium Gases

beginning, the translational temperature T increases in the viscous compression jump behind the shock wave front. The thickness of this jump layer, for relatively strong shock waves, is of the same order as the mean free path. Changes in the energy of internal degrees of freedom then occur in the relaxation zone. The rotational degrees of freedom have some peculiarities. For most molecules, the rotational relaxation time τ_{RT} is not significantly different from the mean free time τ_0 (see Chap. 2), and thus the translational and rotational degrees of freedom are generally considered simultaneously in the investigation of kinetic processes.

Molecular vibrations will be excited behind the wave front by V-T processes, if the initial (before the shock wave front) vibrational energy of the molecules corresponds to the equilibrium value. In this case, due to the energy exchanges, the translational temperature T will decrease and the vibrational temperature T_V will increase.

Let us consider the propagation of shock waves in a non-equilibrium vibrationally excited molecular gas. Assume that the shock wave has a velocity such that the translational temperature T behind the wave front is smaller than the vibrational temperature T_V before it. In this case, V-T energy exchanges occur but yield an opposite result: the vibrational degrees of freedom cool down, while the translational degrees of freedom are heated up. This is a peculiar result.

In the case of shock wave propagation in an equilibrium gas, the distribution of parameters behind the shock wave is fully determined by the shock wave velocity or by the Mach number. For example, the gas temperatures before the front (T_1) and behind it (T_2), for an one-dimensional ideal gas flow, are connected by the relation [1]

$$\frac{T_2}{T_1} = \frac{[2\gamma Ma^2 - (\gamma - 1)][(\gamma - 1)Ma^2 + 2]}{(\gamma + 1)^2 Ma^2}, \qquad (16.10)$$

where $Ma = v_1/c_s$ is the Mach number, v_1 is the velocity of the shock wave and c_s is the local sound velocity in the unperturbed gas. The situation changes markedly in a non-equilibrium gas. In this case, the gas temperature T_2 depends not only on the Mach number, but also on the initial non-equilibrium conditions. Under some conditions, the velocity of a shock wave in a non-equilibrium gas changes until all the parameters of the gas and the wave adjust to self-consistent values. This is only possible for velocities greater than or equal to some minimum magnitude, which is determined by the initial gas and vibrational temperatures, and by the type of gas.

The calculations [59,60] show that stationary shock waves can exist provided that

$$\left(\frac{Ma^2 - 1}{Ma}\right)^2 \geq \frac{9.6m(\epsilon_{vib}^{(1)} - \epsilon_{vib}^{(2)})}{7kT_1}, \qquad (16.11)$$

where $\epsilon_{vib}^{(1)}$ and $\epsilon_{vib}^{(2)}$ are the average vibrational energies per unit mass before the shock wave front and at the end of the relaxation region, respectively.

In the case of shock wave propagation in an equilibrium gas, $\epsilon_{\text{vib}}^{(2)} > \epsilon_{\text{vib}}^{(1)}$, so that inequality (16.11) is satisfied for any Mach number. However, for shock waves in a vibrationally excited gas, (11.6) can be satisfied only if the Mach number exceeds some minimal value, when the conditions are such that $\epsilon_{\text{vib}}^{(2)} < \epsilon_{\text{vib}}^{(1)}$. For example, this minimal value is $Ma = 1.8$ in HF, for the initial conditions $T_1 = 300$ K, $T_V^{(1)} = 2000$ K and $p = 5$ Torr. For the same values of T_1 and p as above, but for $T_V^{(1)} = 3000$ K, the minimal value is $Ma = 2.6$.

Numerical modelling of shock wave development and unsteady propagation in a vibrationally non-equilibrium gas was carried out in [61-64]. Numerical calculations for nitrogen, at fixed translational and vibrational temperatures of 300 K and 2000 K, respectively, have shown that the time needed to establish a stationary regime is ~ 7 s [62]. This time corresponds to a length of several kilometers in the relaxation zone. However, this time is only $\simeq 20$ msec for a vibrational temperature of 4000 K. This considerable decrease in relaxation time is due to anharmonic effects (see Chap. 3).

Experimental investigations of shock waves in vibrationally non-equilibrium gases have also been carried out. For example, a marked increase in wave velocity and a considerable decrease in its amplitude were experimentally found [65-68] in the case of shock wave propagation through a glow discharge plasma. The decrease in amplitude and the broadening of the wave front were measured in impulsive gas dynamic perturbation experiments, in high-frequency post-discharges in N_2-O_2 air mixtures. These authors have connected their observations with relaxation phenomena effects. Agreement between calculations and measurements was obtained in [70,71] concerning wave acceleration by a temperature gradient in N_2-O_2 glow discharges. The works [72,73], where the interaction between a shock wave and a laser spark plasma in air was investigated, are also worthy of mention. A detailed review of theoretical and experimental works in this field is given in [74].

In conclusion, the study of the propagation mechanisms of shock waves and other hydrodynamic perturbations in non-equilibrium gases is a new trend in hydrodynamics. From a physical point of view, two groups of problems leading to new effects can be selected. First are problems associated with the stability of motion for a non-equilibrium gas. The presence of non-equilibrium can either damp or amplify any small perturbations appearing in gas dynamic flows. Second are problems connected with the control of shock wave intensity. By changing the properties of the medium, it is possible to either attenuate or amplify the wave. Finally, it should be emphasized that an accurate quantitative description of hydrodynamic effects in non-equilibrium molecular gases calls for detailed numerical models accounting for the different kinetic processes that determine the properties of the medium.

References

1. Landau L.D. and Lifshits E.M. (1986) *Hydrodynamics*, Nauka, Moscow (*in Russian*)
2. Rice F.O. (1928) Phys. Rev. **31**, 691
3. Mihailov I.G., Soloviev V.A., Syrnikov Y.A. (1964) *The Basis of Molecular Acoustics*, Nauka, Moscow (*in Russian*)
4. Mandelshtam L.I. and Leontovich M.A. (1937) J. Exper. Theor. Phys. **7**, 438 (*in Russian*)
5. Osipov A.I. and Uvarov A.V. (1984) Proceed. Moscow State University **25**, 74 (*in Russian*)
6. Osipov A.I. and Uvarov A.V. (1988) Engineer. Phys. J. **55**, 149 (*in Russian*)
7. Tsendin L.D. (1965) J. Tech. Phys. **35**, 1972 (*in Russian*)
8. Ingard U. (1966) Phys. Rev. **145**, 41
9. Schulz M. (1968) Phys. Fluids **11**, 676
10. Bauer H.J. and Bass H.E. (1973) Phys. Fluids **16**, 988
11. Haas R.A. (1973) Phys. Rev. A **8**, 1017
12. Brinivasan J. and Vincenti W.G. (1975) Phys. Fluids **18**, 1670
13. Ingard U. and Gentle K.W. (1965) Phys. Fluids **8**, 1396
14. Ingard U. and Schulz M. (1967) Phys. Rev. **158**, 106
15. Ellis R.J. and Gilbert R.G. (1977) J. Acoust. Soc. Am. **62**, 245
16. Kogan E.Ya. and Malnev V.N (1977) J. Tech. Phys. **47**, 653 (*in Russian*)
17. Napartovich A.P. and Starostin A.N. (1979) in *Plasma Chemistry*, ed. B.M. Smirnov, Atomizdat, Moscow, Vol. 22, p. 153 (*in Russian*)
18. Korshunov S.E. (1982) Proceed. Acad. Sci. USSR, MJG **5**, 176 (*in Russian*)
19. Kogan E.Ya., Moiseev S.S., Molevich N.E. and Tur A.V. (1985) J. Tech. Phys. **55**, 2036 (*in Russian*)
20. Kogan E.Ya. and Molevich N.E. (1985) J. Tech. Phys. **55**, 754 (*in Russian*)
21. Kogan E.Ya. and Molevich N.E. (1986) Proceed. Univers. Phys. **7**, 53 (*in Russian*)
22. Demidov V.I., Rytenkov S.K. and Skrebov V.N (1988) J. Tech. Phys. **58**, 1413 (*in Russian*)
23. Blinov N.A., Lezin A.Yu., Zolotkov V.N. and Cheburkin N.V. (1989) J. Tech. Phys. **59**, 79 (*in Russian*)
24. Kogan E.Ya. and Molevich N.E. (1988) Acoust. J. **34**, 690 (*in Russian*)
25. Kogan E.Ya. and Molevich N.E. (1987) Acoust. J. **33**, 252 (*in Russian*)
26. Molevich N.E. and Oraevsky A.N. (1988) Acoust. J. **34**, 547 (*in Russian*)
27. Kotusov A.N. and Nemtsov B.E. (1991) Acoust. J. **37**, 123 (*in Russian*)
28. Osipov A.I. and Uvarov A.V. (1987) Chemical Physics **6**, 385 (*in Russian*)
29. Osipov A.I. and Uvarov A.V. (1988) Chem. Phys. Lett. **145**, 247
30. Stupochenko E.V., Losev S.A. and Osipov A.I. (1967) *Relaxation Processes in Shock Waves*, Springer, Berlin Heidelberg
31. Osipov A.I. and Uvarov A.V. (1987) Proceed. Moscow State University **28**, 52 (*in Russian*)
32. Strickler S.D. and Stewart A.B. (1963) Phys. Rev. Lett. **11**, 527
33. Derlande J., Goldan P.D. and Goldstein L. (1964) Appl. Phys. Lett. **5**, 51
34. Alexeff I. and Neidigh R.V. (1963) Phys. Rev. **129**, 516
35. Wojaczek K. (1960/61) Beitr. Plasmaphys. **1**, 127
36. Gentle K.W. and Ingard U. (1964) Appl. Phys. Lett. **5**, 105
37. Mkrtchyan A.R., Galechyan G.A.and Divanyan E.G. (1987) Proceed. Acad. Sci. Arm.SSR. Physics **22**, 231 (*in Russian*)
38. Mkrtchyan A.R., Hatsagortsian K.Z. and Divanian E.G. (1989) Acoustica **69**, 124

39. Antinian M.A., Galechian G.A. and Tavakalian L.B. (1991) Thermophys. High Temp. **29**, 659 (*in Russian*)
40. Yatsui K., Kobayashi T. and Inuishi Y. (1968) J. Phys. Soc. Jap. **24**, 1186
41. Fitaire M. and Mantei T.D. (1972) Phys. Fluids. **15**, 464
42. Hasegawa M. (1974) J. Phys. Soc. Jap. **37**, 193
43. Aleksadrov N.L, Napartovich A.P., Pal A.F. et al. (1990) Plasma Phys. **16**, 862 (*in Russian*)
44. Patureau J.P., Toong T.Y. and Garris C.A. (1977) Proceed. 16th Int. Symp. on Combustion, The Combustion Institute
45. Abouseif G.E. , Toong T.Y. and Converti J. (1978) Proceed. 17th Int. Symp. on Combustion, University of Leeds, UK
46. Detsch R.M. and Bass H.E. (1985) J. Acoust. Soc. Am. **77**, 512
47. Rudenko O.V. and Soluyan S.I. (1975) *Theoretical Basis of Nonlinear Acoustics*, Nauka, Moscow (*in Russian*)
48. Ott E., Manheimer W.M, Book D.L. and Boris J.P. (1973) Phys. Fluids **16**, 855
49. Buevich Y.A. and Fedotov S.P. (1985) Phys. Combust. Explos. **5**, 64 (*in Russian*)
50. Pelinovsky E.N. and Fridman V.E. (1987) Acoust. J. **33**, 365 (*in Russian*)
51. Krasnobaev K.V. and Tarev V.Yu.(1987) Astron. J. **64**, 1210 (*in Russian*)
52. Krasnobaev K.V. and Tarev V.Yu. (1990) Proceed. Acad. Sci. USSR., MJG **2**, 128 (*in Russian*)
53. Bogdanov A.N. and Kulikovsky V.A. (1990) J. Appl. Mech. Techn. Phys. **5**, 26 (*in Russian*)
54. Eletsky A.V. and Stepanov E.V. (1988) Preprint Inst. Atom. Energy **4638/12** (*in Russian*)
55. Kirmusov I.P., Levin V.A. and Starik A.M. (1990) Proceed. Acad. Sci. USSR., MJG **2**, 128 (*in Russian*)
56. Galechyan G.A., Aramian A.R. and Mkrtchyan A.R. (1990) J. Tech. Phys. **60**, 207 (*in Russian*)
57. Aramian A.R., Galechyan G.A. and Mkrtchyan A.R. (1990) Plasma Phys. **16**, 383 (*in Russian*)
58. Galechyan G.A., Divanian E.G. and Mkrtchyan A.R. (1990) Preprint IPPF Acad. Sci. Arm.SSR, Erevan (*in Russian*)
59. Buyanova E.A., Lovetsky E.E., Silakov V.P. and Fetisov V.S. (1982) Chemical Physics **1**, 1701 (*in Russian*)
60. Silakov V.P. and Fetisov V.S. (1983) Chemical Physics **2**, 96 (*in Russian*)
61. Andreeva T.E., Gritsinin S.I., Kossyi I.A. and Chebotarev A.V. (1983) Short Inform. Phys. **7**, 3 (*in Russian*)
62. Ruhadze A.A., Silakov V.P. and Chebotarev A.V. (1983) Short Inform. Phys. **6**, 18 (*in Russian*)
63. Zhdanok S.A. and Porshev P.I. (1985) in *Thermal Exchange and Physical-Chemical Processes in Energy Devices*, Minsk, p. 90 (*in Russian*)
64. Silakov V.P. and Chebotarev A.V. (1986) Preprint MIFI **055-86**, Moscow (*in Russian*)
65. Klimov A.I., Koblov A.N., Mishin G.I. et al. (1982) Letters J. Tech. Phys. **8**, 439 (*in Russian*)
66. Klimov A.I., Koblov A.N., Mishin G.I. et al. (1982) Letters J. Tech. Phys. **8**, 551 (*in Russian*)
67. Basargin I.V. and Mishin G.I. (1985) Letters J. Tech. Phys. **11**, 209 (*in Russian*)
68. Basargin I.V. and Mishin G.I. (1985) Letters J. Tech. Phys. **11**, 1297 (*in Russian*)

69. Grachev L.P., Esakov I.I., Mishin G.I. et al. (1985) J. Tech. Phys. **55**, 972 (*in Russian*)
70. Evtuhin N.B., Margolin A.D. and Shmelev V.M. (1984) Chemical Physics **3**, 1322 (*in Russian*)
71. Voinovich P.A., Ershov A.P., Ponomareva S.E. and Shibkov B.M. (1991) Thermophys. High Temp. **29**, 582 (*in Russian*)
72. Barhudarov E.M., Berezovsky V.P., Mdivnishvili M.O. et al. (1984) Letters J. Tech. Phys. **10**, 1178 (*in Russian*)
73. Kondrashov V.N., Rodionov N.V., Sitnikov S.F. and Sokolov V.I. (1986) J. Tech. Phys. **56**, 89 (*in Russian*)
74. Osipov A.I. and Uvarov A.V. (1992) Adv. Phys. Sci. **162**, 1 (*in Russian*)

Index

Absorption   14, 16, 54, 100, 105, 143, 252, 257, 258, 261–263, 269, 271, 289, 290
Adiabacity function   114
Adiabatic factor   37
Ambipolar diffusion coefficient   98
Anharmonicity   29, 31, 36, 37, 39, 48, 53, 109, 214
Arrhenius law   52
Associative ionization   52, 111, 168, 212, 240
Attachment   64, 120, 134, 135, 139, 140, 142, 148, 206, 215, 216, 229, 239, 242, 246
Aurora   251, 254, 258–261, 263–266

Blanc's law   98, 233
Boltzmann equation   7–10, 59, 62, 64–66, 68, 71, 74, 78, 123, 124, 131, 150, 231, 234
Boltzmann H-theorem   39
Boundary layer   52, 273–279, 281, 282
Breit–Wigner formula   76, 120
Burgers equation   291, 292

Canonical invariance   9, 18, 27
Catalytic surface   276, 277
Cauchy problem   8
Chapman function   252
Characteristic energy   98, 134, 137, 217, 231
Chemically active sites   197, 199, 206, 238–240, 271
Collision integral   17, 99, 231
Continuous approximation   74, 138
Convective Scheme   73
Coulomb logarithm   62, 145

Debye length   62
Degeneracy   15, 35, 170
Degradation spectrum   74, 76–78, 242
Desorption   199–203, 269–272

Detailed balance   16, 26, 27, 29, 32, 61, 162
Differential ionization frequency   63
Diffusion approximation   8–10, 16, 17, 19
Diffusion coefficient   17–19, 68, 75, 95–99, 123, 134, 139, 200, 209, 217, 231, 233
Dissociation energy   85, 86, 91, 171
Dissociative recombination   140, 141, 215, 246, 260
Drift velocity   68, 69, 78, 124, 127, 134, 137, 139, 178, 231, 234
Druyvesteyn distribution   67

EEDF   60, 62, 64–74, 76–78, 119, 123, 125, 126, 131, 133, 143, 145, 146, 209, 210, 212, 215, 222, 224, 229, 231, 246, 277
Effective field approximation   66, 131, 134
Effective probability   49
Elastic collisions   7, 11, 60, 62, 65, 70–72, 77, 78, 105, 124, 129, 132, 143, 147, 214, 229, 242
Electron affinity   85
Electron detachment   182, 183
Electron power balance   124, 126, 132, 137, 138, 145, 209
Electron–electron collisions   62, 64, 72, 125, 131, 231
Electronegative gases   215, 216, 220
Electronegativity   215
Eley–Rideal mechanism   199, 205, 271
Energetic cost   146, 229, 242
Energy cost   78, 146, 148, 248
Energy dissipation function   149, 242
Enrichment coefficient   246–248

Fokker–Planck equation   9, 16, 38
Fourier-component   25
Fractional power losses   146, 148

Fractional power transfer   129
Franck–Condon factor   88, 91, 92, 155

Gibbs canonical distribution   39
Gotlieb polynomials   28

Impact parameter   7, 178
Impulsive collisions   14, 17
Inelastic collisions   61
Inelastic scattering of electrons   119, 121, 122
Instability   214, 215
Interaction potential   25
Ion conversion   182, 183
Ionization potential   85
Isotope separation   52, 53, 176, 246, 247
Isotopic frequency shift   247

Laguerre polynomials   18
Landau–Teller equation   29
Landau–Teller formula   26, 45
Landau–Zener formula   158
Langevin theory   178
Langmuir–Hinshelwood mechanism   199, 205
Le Roy formula   174
Lees–Dorodnitsyn transformation   274
Lorentz approximation   59
Lorentz gas   6, 10
Losev formula   38

Massey parameter   23, 25, 101
Mean free path   95, 194, 233, 269, 293
Mean free time   5, 6, 95, 101, 293
Mobility   60, 75, 98, 103, 123, 124, 134, 209, 216, 220, 231
Morse oscillator   39, 49, 91, 107
Morse potential   88, 91

Navier–Stokes equation   275, 279, 281
Non–equilibrium factor   49, 50

Optical transitions   92, 93, 155–158
Ozone layer   241, 245, 264

Partition function   170
PIC/MCC model   73, 214
Polar molecules   19, 102
Polarizability   98, 178

Polarization interaction   97
Predissociation   53, 157, 158

Radial density distribution   219, 221
Radiative recombination   140
Rayleigh gas   6, 9, 10
Reentry   269, 273, 274, 277–281
Removal efficiency   243–245
Repulsive potential   109
Resonant scattering   120
Rotational relaxation time   99–101
Roughness factor   204

Scaling function   107, 114, 115
Secondary electrons   63, 65, 74, 76, 229, 242, 251
Shock wave   273, 274, 278–280, 284, 287, 290, 292–295
SSH theory   35, 37, 106, 107, 110, 112–115, 274
Stagnation line   279
Stagnation point   274
Statistical weight   15, 61, 75, 162
Stream function   275
Strong collision approximation   8, 16
Superelastic collisions   65
Surface coverage   271
Surface diffusion   199, 200, 238, 239, 271, 272

Three–body recombination   140, 160
Titration method   230
Treanor distribution   38–40, 42, 45, 47, 50–52, 126
Truncated anharmonic oscillator   50
Two–quantum exchange   37, 41, 44

Upper atmosphere   252, 255, 257, 261

V-T exchange   23, 26, 29, 30, 35, 36, 39, 210
V-V exchange   29–35, 38–42, 44–46, 211
V-V' exchange   32–36, 114
Vibrational relaxation time   100, 105, 274, 289

Wall loss probability   194, 196, 203, 205, 206
Wavenumber   289, 290

Springer Series on
Atoms+Plasmas

Editors: G. Ecker P. Lambropoulos I. I. Sobel'man H. Walther
Founding Editor: H. K. V. Lotsch

1 **Polarized Electrons** 2nd Edition
By J. Kessler

2 **Multiphoton Processes**
Editors: P. Lambropoulos and S. J. Smith

3 **Atomic Many-Body Theory**
2nd Edition
By I. Lindgren and J. Morrison

4 **Elementary Processes
in Hydrogen-Helium Plasmas**
Cross Sections
and Reaction Rate Coefficients
By R. K. Janev, W. D. Langer, K. Evans Jr.,
and D. E. Post Jr.

5 **Pulsed Electrical Discharge
in Vacuum**
By G. A. Mesyats and D. I. Proskurovsky

6 **Atomic and Molecular Spectroscopy**
2nd Edition
Basic Aspects and Practical Applications
By S. Svanberg

7 **Interference of Atomic States**
By E. B. Alexandrov, M. P. Chaika
and G. I. Khvostenko

8 **Plasma Physics** 3rd Edition
Basic Theory
with Fusion Applications
By K. Nishikawa and M. Wakatani

9 **Plasma Spectroscopy**
The Influence of Microwave
and Laser Fields
By E. Oks

10 **Film Deposition
by Plasma Techniques**
By M. Konuma

11 **Resonance Phenomena
in Electron-Atom Collisions**
By V. I. Lengyel, V. T. Navrotsky
and E. P. Sabad

12 **Atomic Spectra
and Radiative Transitions** 2nd Edition
By I. I. Sobel'man

13 **Multiphoton Processes
in Atoms** 2nd Edition
By N. B. Delone and V. P. Krainov

14 **Atoms in Plasmas**
By V. S. Lisitsa

15 **Excitation of Atoms
and Broadening of Spectral Lines**
2nd Edition
By I. I. Sobel'man, L. Vainshtein,
and E. Yukov

16 **Reference Data
on Multicharged Ions**
By V. G. Pal'chikov and V. P. Shevelko

17 **Lectures
on Non-linear Plasma Kinetics**
By V. N. Tsytovich

18 **Atoms and Their
Spectroscopic Properties**
By V. P. Shevelko

19 **X-Ray Radiation
of Highly Charged Ions**
By H. F. Beyer, H.-J. Kluge,
and V. P. Shevelko

20 **Electron Emission
in Heavy Ion–Atom Collision**
By N. Stolterfoht, R. D. DuBois,
and R. D. Rivarola

21 **Molecules and Their
Spectroscopic Properties**
By S. V. Khristenko, A. I. Maslov,
and V. P. Shevelko

22 **Physics of Highly Excited Atoms
and Ions**
By V. S. Lebedev and I. L. Beigman

23 **Atomic Multielectron Processes**
By V. P. Shevelko and H. Tawara

24 **Guided-Wave-Produced Plasmas**
By Yu. M. Aliev, H. Schlüter, and
A. Shivarova

25 **Quantum Statistics of Strongly
Coupled Plasmas**
By D. Kremp, W. Kraeft, and
M. Schlanges

26 **Atomic Physics with Heavy Ions**
By H. F. Beyer and V. P. Shevelko

Springer Series on
ATOMIC, OPTICAL, AND PLASMA PHYSICS

Editors-in-Chief:

Professor G.F. Drake
Department of Physics, University of Windsor
401 Sunset, Windsor, Ontario N9B 3P4, Canada

Professor Dr. G. Ecker
Ruhr-Universität Bochum, Fakultät für Physik und Astronomie
Lehrstuhl Theoretische Physik I
Universitätsstrasse 150, 44801 Bochum, Germany

Editorial Board:

Professor W.E. Baylis
Department of Physics, University of Windsor
401 Sunset, Windsor, Ontario N9B 3P4, Canada

Professor R.N. Compton
Oak Ridge National Laboratory
Building 4500S MS6125, Oak Ridge, TN 37831, USA

Professor M.R. Flannery
School of Physics, Georgia Institute of Technology
Atlanta, GA 30332-0430, USA

Professor B.R. Judd
Department of Physics, The Johns Hopkins University
Baltimore, MD 21218, USA

Professor K.P. Kirby
Harvard-Smithsonian Center for Astrophysics
60 Garden Street, Cambridge, MA 02138, USA

Professor P. Lambropoulos, Ph.D.
Max-Planck-Institut für Quantenoptik, 85748 Garching, Germany, and
Foundation for Research and Technology – Hellas (F.O.R.T.H.),
Institute of Electronic Structure & Laser (IESL),
University of Crete, PO Box 1527, Heraklion, Crete 71110, Greece

Professor P. Meystre
Optical Sciences Center, The University of Arizona
Tucson, AZ 85721, USA

Professor J. Mlynek
Universität Konstanz
Universitätsstrasse 10, 78434 Konstanz, Germany

Professor Dr. H. Walther
Sektion Physik der Universität München
Am Coulombwall 1, 85748 Garching/München, Germany